T0232825

Lecture Notes in Mathematics

Edited by J.-M. Morel, F. Takens and B. Teissier

Editorial Policy
for the publication of monographs

1. Lecture Notes aim to report new developments in all areas of mathematics – quickly, informally and at a high level. Monograph manuscripts should be reasonably self-contained and rounded off. Thus they may, and often will, present not only results of the author but also related work by other people. They may be based on specialized lecture courses. Furthermore, the manuscripts should provide sufficient motivation, examples and applications. This clearly distinguishes Lecture Notes from journal articles or technical reports which normally are very concise. Articles intended for a journal but too long to be accepted by most journals, usually do not have this "lecture notes" character. For similar reasons it is unusual for doctoral theses to be accepted for the Lecture Notes series.

2. Manuscripts should be submitted (preferably in duplicate) either to one of the series editors or to Springer-Verlag, Heidelberg. In general, manuscripts will be sent out to 2 external referees for evaluation. If a decision cannot yet be reached on the basis of the first 2 reports, further referees may be contacted: the author will be informed of this. A final decision to publish can be made only on the basis of the complete manuscript, however a refereeing process leading to a preliminary decision can be based on a pre-final or incomplete manuscript. The strict minimum amount of material that will be considered should include a detailed outline describing the planned contents of each chapter, a bibliography and several sample chapters.
Authors should be aware that incomplete or insufficiently close to final manuscripts almost always result in longer refereeing times and nevertheless unclear referees' recommendations, making further refereeing of a final draft necessary.
Authors should also be aware that parallel submission of their manuscript to another publisher while under consideration for LNM will in general lead to immediate rejection.

3. Manuscripts should in general be submitted in English.
Final manuscripts should contain at least 100 pages of mathematical text and should include
– a table of contents;
– an informative introduction, with adequate motivation and perhaps some historical remarks: it should be accessible to a reader not intimately familiar with the topic treated;
– a subject index: as a rule this is genuinely helpful for the reader.

Continued on inside back-cover

Lecture Notes in Mathematics 1762

Editors:
J.-M. Morel, Cachan
F. Takens, Groningen
B. Teissier, Paris

Springer
Berlin
Heidelberg
New York
Barcelona
Hong Kong
London
Milan
Paris
Singapore
Tokyo

Sandra Cerrai

Second Order PDE's in Finite and Infinite Dimension

A Probabilistic Approach

 Springer

Author

Sandra Cerrai
Dipartimento di Matematica per le Decisioni
Università di Firenze
Via Cesare Lombroso 6/17
50134 Firenze, Italy

E-mail: cerrai@cce.unifit.it

Cataloging-in-Publication Data applied for

Die Deutsche Bibliothek - CIP-Einheitsaufnahme

Cerrai, Sandra:
Second order PDE's in infinite dimension : a probabilistic approach /
Sandra Cerrai. - Berlin ; Heidelberg ; New York ; Barcelona ; Hong
Kong ; London ; Milan ; Paris ; Singapore ; Tokyo : Springer, 2001
  (Lecture notes in mathematics ; 1762)
  ISBN 3-540-42136-X

Mathematics Subject Classification (2000): 35K15, 35J, 35R15, 47A35, 49L20,
60H10, 60H15, 60J35, 93C20

ISSN 0075-8434
ISBN 3-540-42136-X Springer-Verlag Berlin Heidelberg New York

This work is subject to copyright. All rights are reserved, whether the whole or part
of the material is concerned, specifically the rights of translation, reprinting, re-use
of illustrations, recitation, broadcasting, reproduction on microfilms or in any other
way, and storage in data banks. Duplication of this publication or parts thereof is
permitted only under the provisions of the German Copyright Law of September 9,
1965, in its current version, and permission for use must always be obtained from
Springer-Verlag. Violations are liable for prosecution under the German Copyright
Law.

Springer-Verlag Berlin Heidelberg New York
a member of BertelsmannSpringer Science+Business Media GmbH

http://www.springer.de

© Springer-Verlag Berlin Heidelberg 2001
Printed in Germany

Typesetting: Camera-ready TeX output by the authors
SPIN: 10839621   41/3142-543210/du - Printed on acid-free paper

Preface

The main objective of this monograph is the study of a class of stochastic differential systems having unbounded coefficients, both in finite and in infinite dimension. We focus our attention on the regularity properties of the solutions and hence on the smoothing effect of the corresponding transition semigroups in the space of bounded and uniformly continuous functions. As an application of these results, we study the associated Kolmogorov equations, the large-time behaviour of the solutions and some stochastic optimal control problems together with the corresponding Hamilton-Jacobi-Bellman equations.

In the literature there exists a large number of works (mostly in finite dimension) dealing with these arguments in the case of bounded Lipschitz-continuous coefficients and some of them concern the case of coefficients having linear growth. Few papers concern the case of non-Lipschitz coefficients, but they are mainly related to the study of the existence and the uniqueness of solutions for the stochastic system. Actually, the study of any further properties of those systems, such as their regularizing properties or their ergodicity, seems not to be developed widely enough. With these notes we try to cover this gap.

This work is structured in two parts, which are fairly independent of each other and which may be read separately. The first one is devoted to the study of stochastic ordinary differential equations. The second one is devoted to the study of stochastic partial differential systems of reaction-diffusion type. In both cases we start from the stochastic problem, we show that it admits a unique solution and then introduce the corresponding transition semigroup. We analyze the smoothing properties of such a semigroup and this allows us to proceed with the applications to the study of the optimal regularity of the corresponding PDE's, of the ergodicity of the system and of some stochastic optimal control problems.

It is important to stress that the study of the infinite-dimensional case is not a straightforward generalization of the study in finite dimension. In fact, many infinite-dimensional diffusion semigroups that arise e.g. in mathematical physics lack smoothing properties and, in contrast to the finite-dimensional case, there is no general analytic approach to the regularity properties in the infinite-dimensional case. In the present work we show how in some special cases, which are of interest in the applications, a theory with certain similarities to the finite-dimensional case can be developed, by using mainly stochastic methods.

This book started as a PhD thesis at the Scuola Normale Superiore of Pisa. In the years since I began work on this subject, I found at the Scuola Normale a stimulating environment for my research and had the opportunity there to meet many people who were generous with their help.

My greatest debt of gratitude is to my former supervisor, Giuseppe Da Prato, who has introduced me to the study of SPDE's with his unfailing advice and constant encouragement. It is mainly due to him that in the group of all his students (and former students) there has always been great collaboration and friendship, creating the best conditions for work.

Among the people who have been generous with their time and help, I would like to thank Ben Goldys and Bohdan Maslowski, who refereed my thesis and gave me many useful suggestions. I am also particularly indebted to Michael Röckner and Francesco Russo, who kindly invited me in Bielefeld and Paris, where I could take advantage of interesting discussions with many people and write part of these lecture notes. Finally, I would like to express my gratitude to Arnaud Debussche, Fausto Gozzi and Jerzy Zabczyk who always answered to all my questions with great willingness.

Sandra Cerrai

Contents

Introduction

In the first part of these notes we consider the following class of stochastic differential equations

$$d\xi(t) = b(\xi(t))\,dt + \sigma(\xi(t))\,dw(t), \quad \xi(0) = x \in \mathbb{R}^d, \tag{1}$$

where $w(t) = (w_1(t), \ldots, w_d(t))$ is a standard d-dimensional Brownian motion, the vector field $b : \mathbb{R}^d \to \mathbb{R}^d$ and the matrix valued function $\sigma : \mathbb{R}^d \to \mathcal{L}(\mathbb{R}^d)$ are smooth and have polynomial growth together with their derivatives and b enjoys some dissipativity conditions. As we shall see later on, we can consider vector fields b on $\mathbb{R}^d$ whose i-th component is given by

$$b_i(x) = p_i(x_i) + g_i(x), \quad x \in \mathbb{R}^d,$$

for a polynomial p_i of degree $2m+1$ having the leading coefficient strictly negative and for a smooth function g_i such that

$$\lim_{|x| \to +\infty} \frac{|D^\beta g_i(x)|}{1 + |x|^{2m-j}} = 0, \quad |\beta| = j.$$

As far as the diffusion term σ is concerned, we assume that its growth is controlled by the growth of b, that is

$$\sup_{x \in \mathbb{R}^d} \frac{\|D^\beta \sigma_i(x)\|}{1 + |x|^{k-j}} < \infty, \quad |\beta| = j,$$

for some $k \leq m$.

The existence and the uniqueness of solutions for this class of stochastic differential equations are known. Actually, in his book [83] Krylov considers a larger class of equations which includes the equation (1). By introducing the notion of *Euler solvability* for stochastic differential problems, Krylov proves existence and uniqueness results and gives some useful estimates for the p-th moments of the solutions. Here we give for completeness a self-contained proof of these results which uses different arguments and which adapts to the present situation of locally Lipschitz coefficients.

After we have proved the existence and the uniqueness of the solution $\xi(t; x)$ for the equation (1), we introduce the corresponding transition semigroup P_t and its generator, which is the second order elliptic operator

$$\mathcal{L}_0(x, D) = \frac{1}{2} \text{Tr} \left[\sigma(x)\sigma^*(x)D^2 \right] + \langle b(x), D \rangle, \quad x \in \mathbb{R}^d.$$

In the *first chapter* we study the elliptic and the parabolic problems associated with the operator $\mathcal{L}_0(x, D)$, namely

$$\lambda\varphi(x) - \mathcal{L}_0(x, D)\varphi(x) = f(x), \quad x \in \mathbb{R}^d, \tag{2}$$

with $\lambda > 0$, and

$$\begin{cases} \dfrac{\partial u}{\partial t}(t, x) = \mathcal{L}_0(x, D)u(t, x) + f(t, x), & t \geq 0 \quad x \in \mathbb{R}^d, \\[2mm] u(0, x) = \varphi(x), & x \in \mathbb{R}^d. \end{cases} \tag{3}$$

We show that both the problem (2) and the problem (3) admit a unique solution and prove global optimal regularity results for such solutions in the space of Hölder continuous functions (Schauder estimates). More precisely, we show that if in (2) the datum f is θ-Hölder continuous[1], for some $\theta \in (0, 1)$, then the unique solution φ is twice differentiable with its second derivative θ-Hölder continuous and it holds

$$\|\varphi\|_{C_b^{2+\theta}(\mathbb{R}^d)} \leq c \|f\|_{C_b^{\theta}(\mathbb{R}^d)},$$

for some constant c independent of f. In the same way, we show that if in (3) the function f is continuous in both variables and $f(t, \cdot)$ is θ-Hölder continuous for any $t \in [0, T]$, with $\sup_{t \in [0,T]} \|f(t, \cdot)\|_{C_b^{\theta}(\mathbb{R}^d)} < \infty$, and if the initial datum φ is twice differentiable with the second derivative θ-Hölder continuous, then the solution $u(t, \cdot)$ is twice differentiable and its second derivative is θ-Hölder continuous, for any $t \in [0, T]$. Besides it holds

$$\sup_{t \in [0,T]} \|u(t, \cdot)\|_{C_b^{2+\theta}(\mathbb{R}^d)} \leq c \left(\|\varphi\|_{C_b^{2+\theta}(\mathbb{R}^d)} + \sup_{t \in [0,T]} \|f(t, \cdot)\|_{C_b^{\theta}(\mathbb{R}^d)} \right),$$

with c independent of φ and f.

---

[1] If X is a separable Banach space, we denote by $B_b(X)$ the Banach space of bounded Borel functions from X into $\mathbb{R}$. If $k \geq 0$, we denote by $C_b^k(X)$ the Banach space of k-times differentiable functions from X into $\mathbb{R}$, which are uniformly continuous, together with their derivatives up to the k-th order. Finally, if $\theta \in (0, 1)$ we denote by $C_b^{k+\theta}(X)$ the Banach space of functions in $C_b^k(X)$ such that the k-th derivative is θ-Hölder continuous.

In the literature there exists a large number of works about boundary value problems for elliptic and parabolic equations in bounded domains of $\mathbb{R}^d$. They can be adapted to the case of unbounded domains, assuming the coefficients are bounded. In [77], Itô established the existence of a fundamental solution, provided $\det \sigma(x)\sigma^*(x) \equiv 1$ and the coefficients are smooth enough. In a series of papers by Aronson and Besala (see [3] and [6]) and Cannarsa and Vespri (see [15] and [16]) the case of unbounded coefficients in unbounded domains has been studied, but only in weighted spaces of functions and by adding a zero order term to the operator $\mathcal{L}_0$, in order to balance the unbounded part. In other more recent papers by Da Prato, Lunardi and Vespri (see [44], [92], [94] and [96]) properties of realization and optimal regularity results for the Ornstein-Uhlenbeck operator and some of its generalizations have been studied, both in the space of bounded continuous functions and in the space of square integrable functions with respect to the invariant measure. In [95], Lunardi could improve all these results, by considering the case of coefficients growing no more that exponentially and satisfying some dissipativity conditions. We remark that in all these papers only deterministic techniques are used.

Instead, here we use the probabilistic interpretation of $\mathcal{L}_0$ as the diffusion operator associated with the stochastic problem (1). The study of partial differential equations by probabilistic methods is classical by now. Starting with the books by Strook and Varadhan [114], Ethier and Kurtz [58] and others, many results have been proved about existence, uniqueness and regularity. These results have been extended in various aspects, including less growth restrictions and less regularity for the coefficients, as well as more degeneracy. To this purpose it is worthwhile to cite the interesting books by Friedlin [66] and Krylov [83].

The first preliminary step, in order to prove any existence and regularity result, is verifying that the semigroup P_t has a regularizing effect. Actually, we prove that if b and σ are of class C^k, then

$$\varphi \in B_b(\mathbb{R}^d) \Longrightarrow P_t\varphi \in C_b^k(\mathbb{R}^d).$$

Moreover, we prove that for any $0 \leq i \leq j \leq k$ it holds

$$\sup_{x \in \mathbb{R}^d} |D^j(P_t\varphi)(x)| \leq c\,(t \wedge 1)^{-(j-i)/2} \sup_{x \in \mathbb{R}^d} |D^i\varphi(x)|. \tag{4}$$

The main tool we use is the Bismut-Elworthy formula, which provides an expression for the derivatives of $P_t\varphi$ in terms of φ and not of its derivatives. We recall that in order to have such formulas we have preliminarily to prove that the solution $\xi(t;x)$ of (1) is k-times mean-square differentiable with respect to x. We would like to stress that this result is not trivial and requires some work, as we are dealing with coefficients which are not Lipschitz-continuous and we cannot apply the usual theorem of contractions depending on parameters. Actually, we have first to approximate the coefficients in order to get *a priori* estimates, then we pass to the limit. We also want to recall that there exist cases in which the mean-square differentiability with respect to the initial datum for the solutions of stochastic equations with regular coefficients is not true. In the book by Krylov [83, chapter 5] examples of such situations are provided. Finally, we would like to stress that, since the coefficients b and σ have polynomial growth, together with their derivatives, the estimates (4) are not trivial. In fact they follow from the crucial estimate

$$\sup_{x \in \mathbb{R}^d} \mathbb{E} |\xi(t; x)|^p \leq c \, (t \wedge 1)^{-\frac{p}{2m}}, \tag{5}$$

which is a consequence of the dissipativity conditions that we are assuming for b.

In the *chapter 2* we investigate the large time behaviour of the solution of the problem (1). More precisely we show that there exists a unique invariant measure μ for the system (1) that enjoys some further properties. We recall that the case of $\mathbb{R}^d$-valued processes with a drift term of gradient type having polynomial growth, perturbed by a Lipschitz term, has been studied by Kusuoka and Stroock in 1985 (see [86]).

The existence of an invariant measure easily follows from the boundedness of the second moment of the solution $\xi(t; x)$. The non-trivial part concerns its uniqueness which we prove by using the Khas'minskii and the Doob theorems. To this purpose we need to show that the semigroup P_t fulfills the *strong Feller* and the *irreducibility* properties. In the first chapter we show that if the coefficients are differentiable, then P_t is differentiable and in particular is strongly Feller. In this second chapter we show that P_t is strongly Feller, even if the coefficients are not so regular. As far as the irreducibility of P_t is concerned, we follow a general method introduced by Da Prato and Goldys in [42], which adapts to coefficients having any growth. As a by-product, from the Doob Theorem we have that the unique invariant measure μ is *strongly mixing* and a *strong law of large numbers* holds. Notice that due to a theorem proved by Stettner in [110], we have something more, namely $P_t\varphi$ converges to the equilibrium strongly in $L^2(\mathbb{R}^d, \mu)$. Moreover, proceeding again by local arguments, we prove that μ is absolute continuous with respect to the Lebesgue measure.

Finally, in the last section of the chapter we show that under stronger dissipativity conditions on the coefficients $P_t\varphi$ converges uniformly to the mean of φ with respect to the measure μ with the following rate of convergence

$$\|P_t\varphi - \langle \varphi, \mu \rangle\|_0 \leq c \, (t+1)^{-\frac{1}{m+1}}, \quad t \geq 0.$$

In some cases such a rate of convergence to equilibrium is of exponential type, that is there exists some $c > 0$ such that

$$\|P_t\varphi - \langle \varphi, \mu \rangle\|_0 \leq c \, e^{-ct} \|\varphi\|_0,$$

so that μ is exponentially mixing

In the *third chapter* we are showing that the diffusion operator $\mathcal{L}_0$ corresponding to the equation (1) generates an analytic semigroup in the space of continuous functions, even if the coefficients are unbounded and the diffusion term σ is possibly degenerate. This means that we are far from the classical framework in the study of the analyticity of semigroups generated by elliptic operators (see Lunardi [91] for a comprehensive overview on the more recent results in this field) under two respects: firstly, we are able to overcome the usual assumption of boundedness of coefficients; secondly our results adapt to a wide class of degenerate operators. Moreover, we do not use the classical deterministic techniques developed beginning from the works of Stewart [111] and [112]. Actually, since we regard our operators as the diffusions of suitable stochastic differential equations, we give an explicit probabilistic representation of the semigroup and thanks to that we obtain all estimates.

The problem of the analyticity of Markovian diffusion semigroups has been widely developed starting with the book by Stein [109] and later by the works of Devinatz [52], Liskevich and Perelmuter [90] and others, mainly in L^p spaces and in the symmetric case. As a matter of fact, our study of non-symmetric operators in the space of bounded uniformly continuous functions seems to be new in the probabilistic literature. As we recalled before, the case of unbounded coefficients in the whole space has been studied by Aronson and Besala in [3] and [6]. After them, Cannarsa and Vespri in [15] and [16] developed this argument and, by assuming the uniform ellipticity of the operators and giving suitable assumptions on the growth of the coefficients, they proved the generation of analytic semigroups in $L^p(\mathbb{R}^d)$ and in the space of continuous functions. Concerning the generation for degenerate operators, after the papers by Feller (see [60] and [61]) and the paper by Brezis, Rosenkrantz and Singer [8], several authors as Clément and Timmermans in [32], Campiti, Pallara and Metafune in [9] and [10] and Favini, J. Goldstein and Romanelli in [59] widely developed the case of ordinary differential operators with Ventcel's boundary conditions, both in L^p spaces and in spaces of continuous functions on a real interval I. In other papers by Baouendi and Gaulaouic and by Vespri (see [4] and [118]) the case of operators in bounded domains of $\mathbb{R}^d$ which are strongly elliptic everywhere but on the boundary is considered. The d-dimensional case is also considered in the recent work by Gozzi, Monte and Vespri [73].

In all these papers deterministic techniques are used and rather restrictive conditions are given on the way the coefficients vanish. Actually, they cannot vanish but in a negligible set and suitable assumptions are given on their behaviour near the zeros. Nevertheless, a differential operator having coefficients identically zero does generate an analytic semigroup in any functions space, so that the restrictions on the set where the coefficients vanish seem to be mainly related to the methods used in the proofs. Our aim here is to show how, by using completely different techniques, derived from the theory of stochastic differential equations, we can avoid such restrictions. The main step in the proof of these results is a generalization of the Bismut-Elworthy formula, which adapts to a larger number of situations than the classical one; namely, it works for a wide class of non-invertible diffusion terms and for many cases in which the mean-square differentiability with respect to the initial datum for the solution of the stochastic problem does not hold.

In the *second part* we approach the study of a Ginzburg-Landau class of reaction-diffusion systems perturbed by a random term, in a regular bounded domain $\mathcal{O}$ of $\mathbb{R}^d$, with $d \leq 3$,

$$
\begin{cases}
\dfrac{\partial u_k}{\partial t}(t,\xi) = \mathcal{A}_k(\xi, D)u_k(t,\xi) + f_k(\xi, u_1(t,\xi), \ldots, u_n(t,\xi)) + Q_k \dfrac{\partial^2 w_k}{\partial t\, \partial \xi}(t,\xi) \\[2mm]
u_k(0,\xi) = x_k(\xi), \quad t > 0, \ \ \xi \in \overline{\mathcal{O}} \\[2mm]
\mathcal{B}_k(\xi, D)u_k(t,\xi) = 0, \quad t > 0, \ \ \xi \in \partial\mathcal{O}, \qquad k = 1, \ldots, n.
\end{cases}
\tag{6}
$$

For each k, $\mathcal{A}_k$ is a uniformly elliptic second order operator with regular real coefficients and $\mathcal{B}_k$ is a linear operator acting on the boundary of $\mathcal{O}$. The reaction term $f = (f_1, \ldots, f_n) : \overline{\mathcal{O}} \times \mathbb{R}^n \to \mathbb{R}^n$ is continuous and in the first two chapters $f(\xi, \cdot)$ is assumed to be Lipschitz-continuous, uniformly with respect to $\xi \in \overline{\mathcal{O}}$, while in the other ones has polynomial growth. Finally, Q_k is a non-negative bounded linear operator from $L^2(\mathcal{O})$ into itself and $\partial^2 w_k/\partial t\, \partial \xi$ are independent space-time white noises.

The study of the existence and the uniqueness of solutions for reaction-diffusion equations with polynomial non-linearities, perturbed by a Gaussian random field, has been developed in several papers by Da Prato and Zabczyk (see [47] for a bibliography), Gyöngy and Pardoux [70], Manthey [98], Walsh [120] and others. In some of these works methods based on *comparison theorems* are used and in the others semigroups theory and dissipativity arguments. Here we are using semigroup techniques. To this purpose, it is important to notice that the maximum principle for systems is not true, in general, and then the dissipativity of the system in the space of continuous functions is not always true, but only under suitable assumptions. Moreover, since we are working with systems, the reaction term is not of gradient type, in general.

If F is the Nemytskii operator corresponding to the function f and if A is the realization in H of the differential operator $\mathcal{A} = (\mathcal{A}_1, \ldots, \mathcal{A}_n)$, with the boundary conditions $\mathcal{B} = (\mathcal{B}_1, \ldots, \mathcal{B}_n)$, the system (6) can be rewritten as the following abstract evolution problem

$$
du(t) = (Au(t) + F(u(t)))\, dt + Q\, dw(t), \quad u(0) = x.
\tag{7}
$$

In the *chapter 4* we consider the case where for any $\xi \in \overline{\mathcal{O}}$ the reaction term $f(\xi, \cdot)$ is Lipschitz-continuous. In this case F is Lipschitz-continuous in H and then for any initial datum $x \in H$ the system (6) admits a unique *mild* solution $u(\cdot; x)$ which belongs to $C([0,T] \times \overline{\mathcal{O}}; \mathbb{R}^n)$, $\mathbb{P}$-a.s. (for a proof see for example [47]).

In order to prove the regularizing effect of P_t, we need to study the differentiable dependence of the solution $u(\cdot; x)$ on initial data, which is strictly related to the regularity of coefficients. But the present situation of composition type non-linearities F is quite delicate, since F is never Fréchet differentiable in H, unless $f(\xi, \cdot)$ is linear for any $\xi \in \overline{\mathcal{O}}$. Actually, in order to prove the mean-square differentiability of $u(\cdot; x)$ with respect to x we cannot proceed as in the classical case of Fréchet differentiable coefficients. Nevertheless, we show that if for any $\xi \in \overline{\mathcal{O}}$ the function $f(\xi, \cdot)$ is of class C^j, for some $j \geq 4/d + 2$, then $u(t; x)$ is k-times differentiable, with $k < 4/d + 2$, and its derivatives satisfy suitable estimates which are uniform with respect to x.

Once proved the differentiable dependence of $u(t; x)$ on initial data, if the diffusion operator Q were invertible with a bounded inverse, we could prove that P_t has a smoothing effect by applying directly the infinite dimensional version of the Bismut-Elworthy formula due to Da Prato, Elworthy and Zabczyk in the case of additive noise (see [40]) and due to Peszat and Zabczyk in the case of multiplicative noise (see [104]). But here the diffusion term is not necessarily non-degenerate and then, in order to write a Bismut-Elworthy type formula for the derivatives of P_t, we have preliminarily to prove that the mean-square derivatives of $u(t; x)$ along any directions of H belong to the domain of Q^{-1}, for any $t > 0$, and the processes $Q^{-1} D^j u(t; x)(h_1, \ldots, h_j)$ are Itô integrable. Once proved this, we conclude that for any $t > 0$

$$\varphi \in B_b(H) \implies P_t \varphi \in C_b^k(H).$$

Furthermore, due to the crucial estimate

$$\sup_{x \in H} |u(t; x)|_H \leq \lambda(t) \, (t \wedge 1)^{-1/2m}, \qquad \mathbb{P} - \text{a.s.}$$

(for a suitable increasing process $\lambda(t)$ such that $\lambda(t) \in L^p(\Omega)$, for any $p \in [1, \infty)$ and $t \geq 0$), for any $0 \leq i \leq j \leq k$ we have

$$\sup_{x \in H} |D^j(P_t \varphi)(x)|_{\mathcal{L}^j(H)} \leq c(t) \, (t \wedge 1)^{-\frac{(j-i)(1+\epsilon)}{2}} \sup_{x \in H} |D^i \varphi(x)|, \qquad t > 0, \qquad (8)$$

where $c(t)$ is some continuous increasing function and ϵ is a constant depending on the degeneracy of Q, which can be taken strictly less than 1 when $d \leq 3$.

In the *chapter 5* we apply these results to the study of the existence and uniqueness of classical solutions for the elliptic and parabolic problems associated with the operator

$$\mathcal{L}(x, D) = \frac{1}{2} \text{Tr}\left[QQ^* D^2\right] + \langle Ax + F(x), D \rangle_H, \qquad x \in D(A).$$

Moreover, by using the estimates (8) and some generalizations to the infinite dimensional case of classical interpolation results, we prove Schauder estimates analogous to those proved in the first part for the finite dimensional case. This kind of problems has been studied by Cannarsa and Da Prato in [14] and other papers, in the linear case and in the case of bounded linear non-linearities satisfying some regularity properties. Such problems have also been studied by Zambotti in [121], under the assumption that the diffusion term is only Hölder continuous. Here we consider the case of unbounded non-linearities F, which are not Fréchet differentiable.

As $\mathcal{L}$ is the diffusion operator corresponding to the system (6), the solution of the problems

$$\lambda\psi(x) - \mathcal{L}(x, D)\psi(x) = \varphi(x), \quad x \in D(A), \tag{9}$$

with $\lambda > 0$, and

$$\begin{cases} \dfrac{\partial y}{\partial t}(t, x) = \mathcal{L}(x, D)y(t, x) + f(t, x), & t \geq 0 \quad x \in D(A), \\[3mm] y(0, x) = \varphi(x), \quad x \in H, \end{cases} \tag{10}$$

can be expressed in terms of P_t. Notice that here we are not assuming QQ^* to be of trace-class, so that in order to give a meaning to the equations above, we have to verify that the operator $QQ^*D^2(P_t\varphi)(x)$ is trace-class, for any $x \in H$ and $\varphi \in C_b(H)$. The proof of this fact is not trivial and requires some further assumptions on the operators A and Q.

In the *chapter 6* we assume that the reaction term $f(\xi, \cdot)$ has polynomial growth, for any $\xi \in \overline{\mathcal{O}}$ and we study the system (6) in the Banach space E of continuous functions instead of in the Hilbert space H of square integrable functions. The Nemytskii operator F corresponding to the function f is well defined and differentiable in E, but it is only locally Lipschitz-continuous. It is possible to show that for any initial datum $x \in E$, there exists a unique solution $u(t; x)$ which is an E-valued process. Then we can introduce the transition semigroup P_t^E corresponding to the system (6), regarded as a system in E.

We prove that if the function $f(\xi, \cdot)$ is of class C^k, for any fixed $\xi \in \overline{\mathcal{O}}$, then for any $t > 0$

$$\varphi \in B_b(E) \Longrightarrow P_t\varphi \in C_b^k(E).$$

Moreover we give a generalization of the Bismut-Elworthy formula to this non-Hilbertian situation. This allows us to prove estimates analogous to (8). We would like to stress that here we are dealing with non-Lipschitz coefficients and then, as in the finite dimensional case, we cannot apply the classical arguments based on the theorem of contractions depending on parameters, in order to get the mean-square differentiability of the solution with respect to initial data.

In the *chapter 7* we prove the regularizing properties of the semigroup P_t corresponding to the system (6), regarded as a system in H. The differentiability of transition semigroups in Hilbert spaces has been studied in several papers, in the linear and in the semilinear case, both with additive and with multiplicative noise (see [48] for a comprehensive bibliography). In the semilinear case the non-linear drift term F is always assumed to be Lipschitz-continuous and differentiable. Instead here we are dealing with drift terms F which are not even well defined in H. As far as we know, [40] is the only paper in which the case of functionals F which satisfy our assumptions is considered, but it only treats the case of dimension $d = 1$ and of covariance operator Q equal to identity (white noise in space and time). In such a paper it is verified that the semigroup P_t maps bounded Borel functions into continuous functions. Here we are able to improve this result in two respects: firstly we work with operators Q having an inverse not necessarily bounded, so that the case of dimension d greater than one can be covered, secondly we show that if $\varphi \in B_b(H)$ then $P_t \varphi$ is not only Lipschitz-continuous, but even belongs to $C_b^1(H)$.

Clearly, F is not well defined in H, nevertheless it can be proved that for any $x \in H$ the equation (6) has a unique *generalized* solution $u(\cdot; x)$. By using the results given in the previous chapter, we show that for any $t > 0$

$$\varphi \in B_b(H) \implies P_t \varphi \in C_b^1(H)$$

and the following estimate holds

$$\sup_{x \in H} |D(P_t \varphi)(x)|_H \le c \, (t \wedge 1)^{-\frac{1+\epsilon}{2}} \sup_{x \in H} |\varphi(x)|. \tag{11}$$

This result is proved by verifying that the first variation equation corresponding to (6) has a unique generalized solution $v(t; x, h)$ which belongs to the domain of Q^{-1} for any $t > 0$ and such that

$$\mathbb{E} \int_0^t |Q^{-1} v(s; x, h)|_H^2 \, ds < \infty.$$

This allows us to prove the following generalized Bismut-Elworthy formula

$$\langle D(P_t \varphi)(x), h \rangle_H = \frac{1}{t} \mathbb{E} \varphi(u(t; x)) \int_0^t \langle Q^{-1} v(s; x, h), dw(s) \rangle_H.$$

Notice that if the dimension d is less or equal to 3, we can assume that $\epsilon < 1$ and then the singularity arising in the right hand side of (11) is integrable near zero. Due to this fact, in the chapter 9 we will be able to prove the existence and the uniqueness of a mild solution for the Hamilton-Jacobi-Bellman equation corresponding to the problem (6)

In the *chapter 8* we approach the study of the asymptotic behaviour of the solution of the system (6), both in E and in H. As far as we know, the study of the ergodic properties of reaction-diffusion systems like (6) is not as large as the study of the existence and uniqueness of solutions. There are some works by Da Prato and Zabczyk (see [47]), dealing with reaction-diffusion equations and spin systems, where the authors mainly assume that $A + F$ is strictly dissipative and construct explicitly the invariant measure by a limiting procedure. We also recall the paper by [41] where compactness arguments are used. Here, with our methods we are able to enlarge considerably the class of reaction terms f considered and to apply our results also to systems of the following type

$$\begin{cases} \dfrac{\partial u}{\partial t} = c_1 \Delta u - u^3 + \lambda_1 u + \mu_1 v + \dfrac{\partial^2 w}{\partial \xi \partial t} \\[3mm] \dfrac{\partial v}{\partial t} = c_2 \Delta v - v^3 + \lambda_2 u + \mu_2 v + \dfrac{\partial^2 w}{\partial \xi \partial t}, \end{cases}$$

for $c_i > 0$ and λ_i, μ_i in $\mathbb{R}$ not necessarily positive.

We prove the existence of an invariant measure μ by using the Krylov-Bogoliubov theorem and the uniqueness by using the Khas'minskii and the Doob theorems. Notice that we first prove the existence of an invariant measure for P_t^E and then, as an easy consequence, we get the existence of an invariant measure for P_t. Hence we prove the uniqueness of the invariant measure for P_t, by showing that P_t is *strongly Feller* (even without assuming the differentiability of $f(\xi, \cdot)$) and *irreducible* in H. As a by-product, we get the uniqueness of the invariant measure for P_t^E. Moreover, since we can apply the Doob theorem, if μ is the unique invariant measure, then the time average of $\varphi(u(t; x))$ converges $\mathbb{P}$-a.s. to the space average of φ with respect to μ. Moreover, due to a theorem proved by Stettner in [110], we have that $P_t\varphi$ converges to equilibrium in $L^2(H, \mu)$, for any $\varphi \in L^2(H, \mu)$

In the last two chapters we apply the results proved in the chapter 7 to some stochastic optimal control problems both with finite and with infinite horizon. More precisely, in the *chapter 9* we study the associated Hamilton-Jacobi-Bellman equations and in the *chapter 10* we prove the verification theorems, that is we show that th solution of the Hamilton-Jacobi-Bellman problems can be identified with the value functions corresponding to some cost functionals.

The study of Hamilton-Jacobi-Bellman equations has been widely developed by many authors, mainly in finite dimension, but also in infinite dimension. Some of these authors have used the approach of viscosity solutions. For the finite dimensional case we refer to the paper by Crandall, Ishii and Lions [34] and to the book by Fleming and Soner [64] and for the infinite dimensional case we refer to the papers by Lions [89] and to the thesis of Swiech [115]. Other authors have studied regular solutions of second order Hamilton-Jacobi-Bellman equations and as far as the infinite dimension is concerned, we refer to the works by Barbu and Da Prato [5], Cannarsa and Da Prato [11] and [13], Gozzi [71] and [72], Haverneau [76], for the evolution case, and by Gozzi and Rouy [74] and Chow and Menaldi [9], for the stationary case. In these papers abstract semilinear stochastic problems are considered and the non-linear term f is assumed to be Lipschitz-continuous. Instead, here we are able to skip the condition of Lipschitz-continuity for f and we can consider the case of reaction terms which have polynomial growth (and hence are not well defined in H).

We consider the following infinite dimensional Hamilton-Jacobi-Bellman problems

$$\begin{cases} \dfrac{\partial y}{\partial t}(t,x) + \mathcal{L}(x,D)y(t,x) - K(Dy(t,x)) + g(x) = 0 \\[2mm] y(T,x) = \varphi(x), \end{cases} \tag{12}$$

and, for $\lambda > 0$,

$$\lambda\varphi(x) - \mathcal{L}(x,D)\varphi(x) + K(D\varphi(x)) = g(x), \tag{13}$$

where $\mathcal{L}$ is the differential operator defined by

$$\mathcal{L}(x,D)\psi(x) = \frac{1}{2}\mathrm{Tr}\left[QQ^*D^2\psi(x)\right] + \langle Ax + F(x), D\psi(x)\rangle_H.$$

Notice that here we are dealing with hamiltonians K which are not Lipschitz-continuous, so that the quadratic case can be covered. Moreover we find regular solutions of the Hamilton-Jacobi-Bellman equation even if the semigroup P_t does not map C^2 functions into C^2 functions. Due to these difficulties, our study is quite delicate and we have to proceed by several approximations. Actually, we have first to approximate the reaction term F by Lipschitz-continuous functionals F_α, in order to get C^2 regularity for the semigroup P_t^α associated with the approximating system

$$du(t) = (Au(t) + F_\alpha(u(t)))\,dt + Q\,dw(t), \qquad u(0) = x, \tag{14}$$

and then, in order to apply the usual Itô calculus, we have to approximate P_t^α by the semigroups $P_t^{\alpha,n}$ associated with the finite dimensional version of (14). Unfortunately, the direct approximation of the semigroup P_t by its Galerkin finite dimensional version P_t^n does not work and we have to proceed by these two double approximations. Moreover, we have to prove some a priori estimates in order to extend our results from the case of Lipschitz hamiltonians to the case of locally Lipschitz hamiltonians.

Concerning the parabolic case, thanks to the Itô formula and the variation of constants formula, we write the equation (12) in the mild form

$$y(t,x) = P_t\varphi(x) - \int_0^t P_{t-s}K(Dy(s,\cdot))(x)\,ds + \int_0^t P_{t-s}g(x)\,ds.$$

By using a fixed point argument we show that for any $\varphi, g \in C_b^1(H)$ there exists a unique differentiable solution $y(t,x)$ which is defined only in a small time interval $[0,T_0]$, as K is only locally Lipschitz-continuous. In order to get a global solution we prove an a priori estimate, whose proof requires some approximations.

As far as the elliptic case is concerned, we denote by L the weak generator of P_t and we show that $D(L) \subset C_b^1(H)$. Hence, by proceeding with suitable approximations, we show that when K is Lipschitz-continuous the operator

$$N(\varphi) = L\varphi - K(D\varphi)$$

is m-dissipativite and $D(\overline{N}) = D(L)$. This yields the existence and uniqueness of solutions of (13) for any $\lambda > 0$ and $g \in C_b(H)$. Then, we consider a locally Lipschitz hamiltonian K. We approximate it by a sequence of Lipschitz functions, we consider the problems associated with the approximating hamiltonians and, by a suitable *a-priori* estimate, we get that $\bar{N}$ is m-dissipative. In particular, for any $\lambda > 0$ and $g \in C_b(H)$ there exists a unique solution $\varphi \in D(\overline{N})$ to the problem

$$\lambda\varphi - \overline{N}(\varphi) = g.$$

Moreover, we show that there exists some $\lambda_0 > 0$ such that for any $g \in C_b^1(H)$ and $\lambda > \lambda_0$ there exists a unique solution to (13) in $D(L) \subset C_b^1(H)$. Unfortunately, here we are not able to prove any regularity of the elements of $D(\overline{N})$ and then we can not prove the existence of regular solutions for any $\lambda > 0$.

In the *last chapter* we are concerned with the study of the following *cost functionals*

$$J(t, x; z) = \mathbb{E}\varphi(u(T, t; x, z)) + \mathbb{E} \int_t^T [g(u(s, t; x, z)) + k(z(s))] \, ds, \qquad (15)$$

and

$$I(x; z) = \mathbb{E} \int_0^{+\infty} e^{-\lambda s} [g(u(s, 0; x, z)) + k(z(s))] \, ds, \qquad (16)$$

where $u(s, t; x, z)$ is the solution of the controlled system

$$du(s) = (Au(s) + F(u(s)) + z(s)) \, ds + Q \, dw(s), \qquad u(t) = x, \qquad (17)$$

φ and g are bounded Lipschitz-continuous functions from H into $\mathbb{R}$, $k : H \to (-\infty, +\infty]$ is a convex lower semi-continuous function and λ is a positive constant.

The *value functions* corresponding to the cost functionals (15) and (16) are defined respectively by

$$V(t, x) = \inf \left\{ J(t, x; z) \,;\, z \in L^2(\Omega; L^2(0, T; H)) \text{ adapted} \right\}$$

and

$$U(x) = \inf \left\{ I(x; z) \,;\, z \in L^2(\Omega; L^2(0, T; H)) \text{ adapted} \right\}.$$

Our aim is to prove that they are given respectively by the solution of the evolutionary Hamilton-Jacobi-Bellman equation (12) and of the stationary Hamilton-Jacobi-Bellman equation (13), where the hamiltonian K is the Legendre transform of k, that is

$$K(x) = \sup_{z \in H} \left\{ -\langle x, z \rangle_H - k(z) \right\}, \qquad x \in H.$$

We would like to remark that since the solutions of the problems (12) and (13) are only C^1, then the *closed loop* equations corresponding to such problems have coefficients which are only continuous. This means that they admit only martingale solutions which are not adapted to the original filtration in general and then do not provide admissible controls. Thus, we are not able to prove the existence of an optimal state and control, unless we consider the notion of *relaxed controls* as in the book of Fleming and Soner [64]. Nevertheless, we show that the functionals $J(t, x; z)$ and $I(x; z)$ can be approximated respectively by functionals $J_\alpha(t, x; z)$ and $I_\alpha(x; z)$ which have unique optimal states and controls.

Finally, in the last section we show that in the one dimensional case it is possible to solve the closed loop equation so that there exists a unique optimal state and a unique optimal control.

Notations

If X and Y are separable Banach spaces, we denote by $\mathcal{L}(X;Y)$ the Banach space of bounded linear operators A from X into Y, endowed with the norm

$$\|A\| = \sup_{|h|_X \leq 1} |Ah|_Y.$$

When $X = Y$, instead of $\mathcal{L}(X;Y)$ we write $\mathcal{L}(X)$. If $n \geq 1$ and $X_1,\ldots,X_n,Y$ are Banach spaces, we define $\mathcal{L}(X_1 \times \cdots \times X_n;Y)$ as the Banach space of bounded n-linear operators A from $X_1 \times \cdots \times X_n$ into Y, endowed with the norm

$$\|A\| = \sup_{|h_i|_{X_i} \leq 1} |A(h_1,\ldots,h_n)|_Y.$$

When $X_i = X$, for $i = 1,\ldots,n$ we write $\mathcal{L}^n(X;Y)$ instead of $\mathcal{L}(X \times \cdots \times X;Y)$.

If X is a Hilbert space, $\mathcal{L}^+(X)$ is the subset of $\mathcal{L}(X)$ consisting of symmetric non-negative operators. $\mathcal{L}_1(X)$ is the Banach space of all operators in $\mathcal{L}(X)$ which are of *trace class*, endowed with the norm

$$\|A\|_1 = \mathrm{Tr}\,[\sqrt{AA^\star}],$$

where $A^\star$ is the adjoint of A. The set of all symmetric non-negative operators of trace-class is denoted $\mathcal{L}_1^+(X)$. Finally, we denote by $\mathcal{L}_2(X)$ the Hilbert space of *Hilbert-Schmidt* operators, endowed with the scalar product

$$\langle A,B \rangle_2 = \mathrm{Tr}\,[AB^\star]$$

and the corresponding norm $\|A\|_2 = \sqrt{\langle A,A \rangle_2}$.

Functional spaces

If $\mathcal{O} \subseteq \mathbb{R}^d$ is an open set, we shall denote by $L^2(\mathcal{O};\mathbb{R}^n)$ the Hilbert space consisting of all functions $x = (x_1,\ldots,x_n) : \mathcal{O} \to \mathbb{R}^n$ such that $x_k \in L^2(\mathcal{O})$, for all $k = 1,\ldots,n$. The space $L^2(\mathcal{O};\mathbb{R}^n)$ is endowed with the scalar product

$$\langle x,y \rangle_{L^2(\mathcal{O};\mathbb{R}^n)} = \int_{\mathcal{O}} \langle x(\xi),y(\xi) \rangle \, d\xi = \sum_{k=1}^{n} \langle x_k,y_k \rangle_{L^2(\mathcal{O})}$$

and the associated norm $|\cdot|_{L^2(\mathcal{O};\mathbb{R}^n)}$. In a similar way, for any $p \geq 1$ we denote by $L^p(\mathcal{O};\mathbb{R}^n)$ the Banach space of all functions $x : \mathcal{O} \to \mathbb{R}^n$ such that $x_k \in L^p(\mathcal{O})$, for all $k = 1,\ldots,n$, endowed with the usual norm

$$|x|_p^p = \sum_{k=1}^{n} \int_{\mathcal{O}} |x_k(\xi)|^p \, d\xi = \sum_{k=1}^{n} |x_k|_{L^p(\mathcal{O})}^p.$$

For any integer s, the Sobolev space $W^{s,p}(\mathcal{O};\mathbb{R}^n)$ can be defined as the space of functions x such that the distributional derivatives $D^\alpha x_k$ (with $|\alpha| \le s$) are functions from $L^p(\mathcal{O})$, for each $k = 1,\ldots,n$. The norm in $W^{s,p}(\mathcal{O};\mathbb{R}^n)$ is defined by

$$|x|_{s,p} = \sum_{k=1}^n \sum_{|\alpha| \le s} |D^\alpha x|_p.$$

For any $s \in (0,1)$ and $p \ge 1$, $W^{s,p}(\mathcal{O};\mathbb{R}^n)$ is the set of all functions $x \in L^p(\mathcal{O};\mathbb{R}^n)$ such that

$$[x]_{s,p} = \sum_{k=1}^n [x_k]_{s,p} = \sum_{k=1}^n \int_{\mathcal{O}\times\mathcal{O}} \frac{|x_k(\xi) - x_k(\zeta)|^p}{|\xi - \zeta|^{sp+d}} \, d\xi \, d\zeta < \infty.$$

$W^{s,p}(\mathcal{O};\mathbb{R}^n)$, endowed with the norm $|x|_{s,p} = |x|_p + [x]_{s,p}$, is a Banach space.

We shall also consider the Banach space $C(\overline{\mathcal{O}};\mathbb{R}^n)$, with the *sup-norm*

$$|x|^2_{C(\overline{\mathcal{O}};\mathbb{R}^n)} = \sum_{k=1}^n \sup_{\xi \in \overline{\mathcal{O}}} |x_k(\xi)|^2$$

and the duality form on $C(\overline{\mathcal{O}};\mathbb{R}^n) \times (C(\overline{\mathcal{O}};\mathbb{R}^n))^\star$ denoted by $\langle \cdot, \cdot \rangle_{C(\overline{\mathcal{O}};\mathbb{R}^n)}$. If $\theta \in (0,1)$, we define $C^\theta(\overline{\mathcal{O}};\mathbb{R}^n)$ as the subspace of all functions $x \in C(\overline{\mathcal{O}};\mathbb{R}^n)$ such that

$$[x]_\theta = \sum_{k=1}^n [x_k]_\theta = \sum_{k=1}^n \sup_{\substack{\xi,\eta \in \overline{\mathcal{O}} \\ \xi \ne \eta}} \frac{|x_k(\xi) - x_k(\eta)|}{|\xi - \eta|^\theta} < \infty.$$

$C^\theta(\overline{\mathcal{O}};\mathbb{R}^n)$ is a Banach space, endowed with the norm $|x|_\theta = |x|_{C(\overline{\mathcal{O}};\mathbb{R}^n)} + [x]_\theta$.

Now we introduce some notations concerning functions defined on infinite dimensional spaces. If X and Y are any two separable Banach spaces, we denote by $B_b(X;Y)$ the Banach space of all bounded Borel functions $\varphi : X \to Y$, endowed with the supremum norm

$$\|\varphi\|_0^X = \sup_{x \in X} |\varphi(x)|_Y.$$

$C_b(X;Y)$ is the subspace of uniformly continuous functions. If $k \in \mathbb{N}$, we denote by $C_b^k(X;Y)$ the subspace of all k-times differentiable functions of $C_b(X;Y)$ which have bounded uniformly continuous derivatives up the k-th order. $C_b^k(X;Y)$ is a Banach space endowed with the norm

$$\|\varphi\|_k^X = \|\varphi\|_0^X + \sum_{i=1}^k \sup_{x \in X} |D^i\varphi(x)|_{\mathcal{L}^i(X;Y)},$$

where $\mathcal{L}^i(X;Y) = \mathcal{L}(X;\mathcal{L}^{i-1}(X;Y))$ and $\mathcal{L}^0(X;Y) = Y$.

If $\varphi \in C_b(X;Y)$ and $\theta \in (0,1)$ we set

$$[\varphi]_\theta^X = \sup_{\substack{x,y \in X \\ x \neq y}} \frac{|\varphi(x) - \varphi(y)|_Y}{|x - y|_X^\theta}.$$

Then the space $C_b^\theta(X;Y)$ of bounded θ-Hölder continuous functions with exponent θ is the set of functions $\varphi \in C_b(X;Y)$ such that $[\varphi]_\theta < \infty$. $C_b^\theta(X;Y)$, endowed with the norm

$$\|\varphi\|_{C_b^\theta(X;Y)} = \|\varphi\|_0^X + [\varphi]_\theta^X,$$

is a Banach space. Moreover, if $\theta > 1$ and $\theta \notin \mathbb{N}$, we define $C_b^\theta(X;Y)$ as the set of all functions $\varphi \in C_b^{[\theta]}(X;Y)$, such that

$$[D^{[\theta]}\varphi]_{\theta-[\theta]}^X = \sup_{\substack{x,y \in X \\ x \neq y}} \frac{|D^{[\theta]}\varphi(x) - D^{[\theta]}\varphi(y)|_{\mathcal{L}^{[\theta]}(X;Y)}}{|x - y|_X^{\theta-[\theta]}} < \infty$$

(here $[\theta]$ denotes the integer part of θ). $C_b^\theta(X;Y)$ is a Banach space endowed with the norm

$$\|\varphi\|_{C_b^\theta(X;Y)} = \|\varphi\|_{C_b^{[\theta]}(X;Y)} + [D^{[\theta]}\varphi]_{\theta-[\theta]}^X.$$

We denote by $C_b^{0,1}(X;Y)$ the Banach space of Lipschitz-continuous functions, endowed with the norm

$$\|\varphi\|_{0,1}^X = \|\varphi\|_0^X + [\varphi]_{0,1}^X,$$

where

$$[\varphi]_{0,1}^X = \sup_{\substack{x,y \in X \\ x \neq y}} \frac{|\varphi(x) - \varphi(y)|_Y}{|x - y|_X}.$$

If $k \in \mathbb{N}$, we denote by $C_b^{k,1}(X;Y)$ the set of all functions $\varphi \in C_b^k(X;Y)$, such that

$$[D^k\varphi]_{0,1}^X = \sup_{\substack{x,y \in X \\ x \neq y}} \frac{|D^k\varphi(x) - D^k\varphi(y)|_{\mathcal{L}^k(X;Y)}}{|x - y|_X} < \infty.$$

$C_b^{k,1}(X;Y)$ is a Banach space, endowed with the norm

$$\|\varphi\|_{C_b^{k,1}(X;Y)} = \|\varphi\|_{C_b^k(X;Y)} + [D^k\varphi]_{0,1}^X.$$

Finally, for any $\alpha \in [0,2)$ we say that a function $\varphi \in C_b(X;Y)$ belongs to the Zygmund space $C_b^\alpha(X;Y)$ if

$$[\varphi]_{C_b^\alpha(X;Y)} = \sup_{\substack{x,y \in E \\ x \neq y}} \frac{|\varphi(x) - 2\varphi\left(\frac{x+y}{2}\right) + \varphi(y)|_Y}{|x - y|_X^\alpha} < \infty.$$

In this case we set

$$\|\varphi\|_{C_b^\alpha(X;Y)} = \|\varphi\|_0^X + [\varphi]_{C_b^\alpha(X;Y)}.$$

It is possible to show that if $\alpha \neq 1$ then

$$C_b^\alpha(X;Y) = \mathcal{C}_b^\alpha(X;Y) \tag{18}$$

(for a proof see for example [117]).

In what follows, if $Y = \mathbb{R}$, we denote the spaces $B_b(X;Y)$, $C_b(X;Y)$, $C_b^\theta(X;Y)$, $C_b^{k,1}(X;Y)$, and $\mathcal{C}^\alpha(X;Y)$, respectively by $B_b(X)$, $C_b(X)$, $C_b^\theta(X)$, $C_b^{k,1}(X)$, and $\mathcal{C}^\alpha(X)$.

As well known, if $\dim H < \infty$ then the space $C_b^\infty(H)$ is dense in $C_b(H)$. In 1954 J. Kurtzweil in [85] proved that if $\dim H = \infty$, then for any $\varphi \in C_b(H)$ there exists a sequence $\{\varphi_n\} \subset C^\infty(H) \cap C_b(H)$ converging to φ in $C_b(H)$. However in 1973 A.S. Nemirovski and S.M. Semenov proved in [101] that $C_b^2(H)$ is not dense in $C_b(H)$, whereas $C_b^{1,1}(H)$ is.

Interpolation spaces

We recall here some elements of interpolation theory which will be used in what follows. For the proofs we refer to [117] and [91].

Let X, Y be two Banach spaces, with Y continuously embedded in X. We define

$$K(t, x, X, Y) = \inf \{\|a\|_X + t\|b\|_Y : a \in X \text{ and } b \in Y, \ x = a + b\},$$

and, for any $\theta \in (0, 1)$, we set

$$(X, Y)_{\theta,\infty} = \left\{ x \in X : \|x\|_{(X,Y)_{\theta,\infty}} = \sup_{t \in (0,1]} \frac{1}{t^\theta} K(t, x, X, Y) < \infty \right\}.$$

$(X, Y)_{\theta,\infty}$ is a Banach space with the norm $\|\cdot\|_{(X,Y)_{\theta,\infty}}$. Moreover, from the definition above it easily follows that if for any $t > 0$

$$x = a(t) + b(t),$$

with $a(t) \in X$ and $b(t) \in Y$ such that

$$\|a(t)\|_X \leq c\, t^\theta, \quad \|b(t)\|_Y \leq c\, t^{\theta-1},$$

then $x \in (X, Y)_{\theta,\infty}$. We conclude recalling the following useful result.

Theorem 0.0.1. *If X_i and Y_i, with $i = 1, 2$, are Banach spaces such that Y_i is continuously embedded into X_i and if T belongs to $L(X_1, X_2) \cap L(Y_1, Y_2)$, then T belongs to $L\left((X_1, Y_1)_{\theta,\infty}, (X_2, Y_2)_{\theta,\infty}\right)$, for every $\theta \in (0, 1)$. Moreover it holds*

$$\|T\|_{L\left((X_1,Y_1)_{\theta,\infty},(X_2,Y_2)_{\theta,\infty}\right)} \leq \left(\|T\|_{L(X_1,X_2)}\right)^{1-\theta} \left(\|T\|_{L(Y_1,Y_2)}\right)^\theta.$$

Remark In what follows we shall denote by c (without index) any positive constant appearing in inequalities, whose dependence on some parameters is not important. Such constant may change even in the same chain of inequalities. When we want to emphasize the dependence of the constant c on some parameters $p_1, \ldots, p_r$, we denote it by $c_{p_1, \ldots, p_r}$.

Similarly, we denote by $c(t)$ any continuous non-decreasing and non-negative function, whose explicit expression is not important. If we need to emphasize its dependence on some parameters $p_1, \ldots, p_r$, we denote it by $c_{p_1, \ldots, p_r}(t)$.

Part I

Finite dimension

Chapter 1

Kolmogorov equations in $\mathbb{R}^d$ with unbounded coefficients

In this chapter we are concerned with the following class of second order elliptic operators

$$\mathcal{L}_0(x, D) = \sum_{i,j=1}^{d} a_{ij}(x) D_{ij} + \sum_{i=1}^{d} b_i(x) D_i, \quad x \in \mathbb{R}^d.$$

The vector field $b = (b_1, \ldots, b_d) : \mathbb{R}^d \to \mathbb{R}^d$ is of class C^3 and the matrix $a(x) = [a_{ij}(x)]$ is symmetric, strictly positive and of class C^3, so that it can be written as

$$a(x) = \frac{1}{2}\sigma(x)\sigma^*(x), \quad x \in \mathbb{R}^d,$$

for some function $\sigma : \mathbb{R}^d \to \mathcal{L}(\mathbb{R}^d)$ of class C^3 (in fact, we can take $\sigma = \sqrt{a}$). Both b and σ are assumed to have polynomial growth and b enjoys some *dissipativity* conditions which will be described in more details later on. Our aim is to prove the existence and the uniqueness of solutions for the elliptic and the parabolic problems associated with the operator $\mathcal{L}_0$. Moreover, we want to give optimal regularity results in the space of bounded Hölder continuous functions (*Schauder type estimates*).

As known, $\mathcal{L}_0$ is the diffusion operator corresponding to the stochastic differential problem

$$d\xi(t) = b(\xi(t))\, dt + \sigma(\xi(t))\, dw(t), \quad \xi(0) = x, \tag{1.0.1}$$

where $w(t) = (w_1(t), \ldots, w_d(t))$ is a standard d-dimensional Brownian motion, defined on the stochastic basis $(\Omega, \mathcal{F}, \mathbb{P})$. This means that the semigroup P_t corresponding to the operator $\mathcal{L}_0$ is the transition Markov semigroup corresponding to the equation (1.0.1). More precisely, if $\xi(t; x)$ denotes the solution of problem (1.0.1) and if $\varphi : \mathbb{R}^d \to \mathbb{R}$ is a Borel bounded function, then for any $t \geq 0$

$$P_t\varphi(x) = \mathbb{E}\varphi(\xi(t; x)), \quad x \in \mathbb{R}^d.$$

In the section 1.2 we prove existence and uniqueness of solutions for the equation (1.0.1) and we show that for any $p \geq 1$ it holds

$$\sup_{t \geq 0} \mathbb{E}\,|\xi(t;x)|^p \leq c\,(1+|x|^p), \quad x \in \mathbb{R}^d.$$

Moreover, and this is the main result of the section which will be crucial in order to get all other results, we show that for any $p \geq 1$

$$\sup_{x \in \mathbb{R}^d} \mathbb{E}\,|\xi(t;x)|^p \leq c_p(t)(t \wedge 1)^{-\frac{p}{2m}}, \quad t > 0,$$

for a suitable constant $m > 0$, depending on the growth of b, and for some continuous increasing function $c(t)$.

In the section 1.3 we study the mean-square differentiability of $\xi(t;x)$ with respect to the initial datum $x \in \mathbb{R}^d$. The present situation is more complicated than the standard situation of Lipschitz-continuous coefficients. Actually, since the derivatives of b and σ are not bounded, we can not apply the classical theorem of contractions depending on parameters. This is the reason why we have to consider some approximating problems

$$d\xi(t) = b(\xi(t))\,dt + \sigma_n(\xi(t))\,dw(t), \quad \xi(0) = x.$$

We show that if b and σ are k-times differentiable, the solution $\xi_n(t;x)$ is k times mean-square differentiable with respect to $x \in \mathbb{R}^d$. Moreover we prove some estimates for the p-th moments of the derivative of $\xi_n(t;x)$, which are uniform with respect to $n \in \mathbb{N}$. Finally we show that for any fixed $t \geq 0$ the derivatives of $\xi_n(t;x)$ converge in $L^2(\Omega; \mathbb{R}^n)$ to the solutions of the variation equations corresponding to the problem (1.0.1). This implies that they are the mean-square derivatives of the process $\xi(t;x)$.

As shown in [17], in general the semigroup P_t is not strongly continuous in $C_b(\mathbb{R}^d)$. Nevertheless, in the section 1.4 we prove that P_t fulfills some regularity properties. Namely, we show that it is a weakly continuous semigroup (see the appendix B for the definition and main properties).

In the section 1.5 we show that P_t has a regularizing effect. Actually, we prove that if the coefficients are of class C^k then for any $\varphi \in B_b(\mathbb{R}^d)$ and $t > 0$ the function $P_t\varphi$ is k-times differentiable. Moreover we prove that for $j = 1, \ldots, k$ it holds

$$\sup_{x \in \mathbb{R}^d} \left| D^\beta (P_t\varphi)(x) \right| \leq c\,(t \wedge 1)^{-j/2} \sup_{x \in \mathbb{R}^d} |\varphi(x)|, \quad |\beta| = j. \tag{1.0.2}$$

In the section 1.6 we prove that there exists a unique solution both for the parabolic and for the elliptic problem associated with the operator $\mathcal{L}_0$. More precisely, we prove that for any $\lambda > 0$ and $f \in C_b(\mathbb{R}^d)$ the elliptic problem

$$\lambda\varphi(x) - \mathcal{L}_0(x,D)\varphi(x) = f(x), \quad x \in \mathbb{R}^d, \tag{1.0.3}$$

admits a unique solution $\varphi \in \cap_{p \geq 1} W^{2,p}_{\text{loc}}(\mathbb{R}^d) \cap C_b(\mathbb{R}^d)$. Similarly we prove that for any $\varphi \in C_b(\mathbb{R}^d)$ and $T > 0$ the parabolic problem

$$\begin{cases} \dfrac{\partial u}{\partial t}(t,x) & = \mathcal{L}_0(x,D)u(t,x), \quad t \in (0,T] \quad x \in \mathbb{R}^d, \\[2mm] u(0,x) & = \varphi(x), \quad x \in \mathbb{R}^d, \end{cases} \tag{1.0.4}$$

has a unique classical solution $u(t;x)$. To this purpose we give a *maximum principle* for $\mathcal{L}_0$ and we give a characterization of the domain of the weak generator $\mathcal{L}$ of the semigroup P_t.

In the last section, the section 1.7, we prove Schauder type estimates, both in the parabolic and in the elliptic case. By using the estimates (1.0.2) and a general method based on interpolation due to Lunardi (see [93]), we show that if in (1.0.3) f is θ-Hölder continuous, then the solution φ is twice differentiable and its second derivative is θ-Hölder continuous. Moreover

$$\|\varphi\|_{C^{2+\theta}_b(\mathbb{R}^d)} \leq c \|f\|_{C^\theta_b(\mathbb{R}^d)}.$$

Similarly, if in the non-homogeneous version of the problem (1.0.4)

$$\begin{cases} \dfrac{\partial u}{\partial t}(t,x) & = \mathcal{L}_0(x,D)u(t,x) + f(t,x), \quad t \in (0,T], \quad x \in \mathbb{R}^d, \\[2mm] u(0,x) & = \varphi(x), \quad x \in \mathbb{R}^d, \end{cases}$$

$f(t,\cdot)$ is θ-Hölder continuous, for any $t \in [0,T]$, and $\sup_{t \in [0,T]} \|f(t,\cdot)\|_{C^\theta_b(\mathbb{R}^d)} < \infty$, and if φ is twice differentiable with its second derivative θ-Hölder continuous, then the solution $u(t,\cdot)$ is twice differentiable and the second derivatives are θ-Hölder continuous, for any $t \in [0,T]$. Moreover, it holds

$$\sup_{t \in [0,T]} \|u(t,\cdot)\|_{C^{2+\theta}_b(\mathbb{R}^d)} \leq c \left(\|\varphi\|_{C^{2+\theta}_b(\mathbb{R}^d)} + \sup_{t \in [0,T]} \|f(t,\cdot)\|_{C^\theta_b(\mathbb{R}^d)} \right).$$

1.1 Assumptions

We recall that, as $a(x)$ is assumed to be strictly positive, then it can be written as

$$a(x) = \frac{1}{2}\sigma(x)\sigma^*(x), \quad x \in \mathbb{R}^d,$$

for some matrix valued function $\sigma : \mathbb{R}^d \to \mathcal{L}(\mathbb{R}^d)$ which has the same regularity as $a(x)$ (for a proof see for example [66] and [114]). Notice that since a is non-degenerate, all conditions on σ can be rephrased in terms of a, but for the sake of simplicity we prefer to proceed with σ.

Hypothesis 1.1.   *1. The vector field $b : \mathbb{R}^d \to \mathbb{R}^d$ is of class C^h, for some $h \geq 3$ and there exists $m \geq 0$ such that for any $j = 0, \ldots, h$*

$$\sup_{x \in \mathbb{R}^d} \frac{|D^\beta b(x)|}{1 + |x|^{2m+1-j}} < +\infty, \quad |\beta| = j.$$

2. The mapping $\sigma : \mathbb{R}^d \to \mathcal{L}(\mathbb{R}^d)$ is of class C^h and there exists $k \leq m$ such that for any $j = 0, \ldots, h$ it holds

$$\sup_{x \in \mathbb{R}^d} \frac{|D^\beta \sigma(x)|}{1 + |x|^{k-j}} < +\infty, \quad |\beta| = j.$$

3. For all $p > 0$ there exists $c_p \in \mathbb{R}$ such that

$$\langle Db(x)y, y \rangle + p \, \|D\sigma(x)y\|_2^2 \leq c_p |y|^2, \qquad x, y \in \mathbb{R}^d.$$

The condition 2 means that the function $a : \mathbb{R}^d \to \mathcal{L}(\mathbb{R}^d)$ is of class C^h, has polynomial growth together with all its derivatives and is strictly positive, that is there exists $\nu > 0$ such that

$$\inf_{x \in \mathbb{R}^d} \sum_{i,j=1}^{d} a_{ij}(x) h_i h_j \geq \nu |h|^2, \quad h \in \mathbb{R}^d.$$

Moreover, the condition 3. implies that for any $x, y \in \mathbb{R}^d$

$$\langle b(x) - b(y), x - y \rangle + p \, \|\sigma(x) - \sigma(y)\|_2^2 \leq c_p |x - y|^2. \tag{1.1.1}$$

In what follow, if $m > 0$ we shall also assume the following dissipativity condition for the drift term b.

Hypothesis 1.2. *There exist $a > 0$ and $\gamma, c \geq 0$ such that for any $x, h \in \mathbb{R}^d$ it holds*

$$\langle b(x + h) - b(x), h \rangle \leq -a|h|^{2m+2} + c \left(|x|^\gamma + 1 \right).$$

Remark 1.1.1. Let p_i be any polynomial of degree $2m + 1$, having the leading coefficient strictly negative, that is

$$p_i(t) = -c_i t^{2m+1} + \sum_{j=0}^{2m} c_{ij} t^j, \quad t \in \mathbb{R},$$

for some $c_i > 0$ and $c_{ij} \in \mathbb{R}$. Moreover, let $g_i : \mathbb{R}^d \to \mathbb{R}$ be any function of class C^h such that

$$\lim_{|x| \to +\infty} \frac{|D^\beta g_i(x)|}{1 + |x|^{2m}} = 0, \quad |\beta| = 1.$$

and for any $j \geq 2$

$$\sup_{x \in \mathbb{R}^d} \frac{|D^\beta g_i(x)|}{1 + |x|^{2m+1-j}} < \infty, \qquad |\beta| = j.$$

If we define for any $i = 1, \ldots, d$

$$b_i(x) = p_i(x_i) + g_i(x), \quad x \in \mathbb{R}^d,$$

it is immediate to check that the vector field $b = (b_1, \ldots, b_d) : \mathbb{R}^d \to \mathbb{R}^d$ fulfills the conditions of the Hypothesis 1.1 and 1.2.

In order to have the smoothing effect of the transition semigroup, we will need the following non-degeneracy condition for the diffusion term σ.

Hypothesis 1.3. *For any $x \in \mathbb{R}^d$ there exists $\sigma^{-1}(x) \in \mathcal{L}(\mathbb{R}^d)$ and*

$$\sup_{x \in \mathbb{R}^d} \|\sigma^{-1}(x)\| < \infty.$$

1.2 The solution of the associated SDE

With the notations introduced in the previous section, the operator $\mathcal{L}_0$ can be written as

$$\mathcal{L}_0(x, D) = \frac{1}{2} \mathrm{Tr} \left[\sigma(x)\sigma^\star(x)D^2 \right] + \langle b(x), D \rangle, \quad x \in \mathbb{R}^d.$$

$\mathcal{L}_0$ is the diffusion operator associated with the stochastic differential equation

$$d\xi(t) = b(\xi(t))\, dt + \sigma(\xi(t))\, dw(t), \qquad \xi(0) = x, \tag{1.2.1}$$

where $w(t) = (w_1(t), \ldots, w_d(t))$ is a standard d-dimensional Brownian motion defined on the stochastic basis $(\Omega, \mathcal{F}, \mathbb{P})$. Our aim is to show that the problem (1.2.1) admits a unique solution.

For any $n \in \mathbb{N}$, let $\sigma_n : \mathbb{R}^d \to \mathcal{L}(\mathbb{R}^d)$ be a matrix valued function of class C^h such that

$$\sigma_n(x) = \begin{cases} \sigma(x) & \text{if } |x| \leq n \\[2mm] \sigma\left((n+1)x/|x|\right) & \text{if } |x| \geq n+1. \end{cases}$$

It is easy to check that σ_n has bounded derivatives, for any $n \in \mathbb{N}$, and in particular is Lipschitz-continuous. Moreover, for any $j = 0, \ldots, h$ it holds

$$\sup_{n \in \mathbb{N}} \sup_{x \in \mathbb{R}^d} \frac{|D^\beta \sigma_n(x)|}{1 + |x|^{k-j}} < \infty, \qquad |\beta| = j. \tag{1.2.2}$$

We remark that due to the Hypothesis 1.1-3. it is possible to choose σ_n in such a way that for any $p \geq 0$

$$\langle Db(x)y, y \rangle + p \|D\sigma_n(x)y\|_2^2 \leq c_p |y|^2, \quad x, y \in \mathbb{R}^d, \tag{1.2.3}$$

for some constant c_p independent of $n \in \mathbb{N}$.

If we show that for any $n \in \mathbb{N}$ the approximating problem

$$d\xi(t) = b(\xi(t))dt + \sigma_n(\xi(t))dw(t), \qquad \xi(0) = x \qquad (1.2.4)$$

admits a unique solution $\xi_n(\cdot; x) \in L^2(\Omega; C([0, +\infty); \mathbb{R}^d))$, then we can fix $\Gamma_x \in \mathcal{F}$ such that $\mathbb{P}(\Gamma_x) = 0$ and such that for any $n \in \mathbb{N}$

$$\xi_n(\cdot; x)(\omega) \in C([0, +\infty); \mathbb{R}^d), \qquad \omega \in \Gamma_x^C.$$

For any $x \in \mathbb{R}^d$ and $n \in \mathbb{N}$ we can define the *stopping time* $\tau_n^x : \Omega \to \mathbb{R} \cup \{+\infty\}$ in the following way

$$\tau_n^x(\omega) = \begin{cases} \inf\{t \geq 0 \; : \; |\xi_n(t; x)(\omega)| \geq n\} & \text{if } \omega \in \Gamma_x^C \\ 0 & \text{if } \omega \in \Gamma_x \end{cases}$$

with the usual convention that if $\omega \in \Gamma_x^C$ and $|\xi_n(t; x)(\omega)| < n$ for any $t \geq 0$ then $\tau_n^x(\omega) = +\infty$. As $\{\tau_n^x\}_{n \in \mathbb{N}}$ is a non-decreasing sequence, we can define for any $\omega \in \Omega$

$$\tau^x(\omega) = \lim_{n \to +\infty} \tau_n^x(\omega), \qquad \omega \in \Omega.$$

If we show that for any $T > 0$ and $x \in \mathbb{R}^d$ there exists a constant $c_T(x) > 0$ independent of n such that

$$\mathbb{E} \sup_{t \in [0,T]} |\xi_n(t; x)|^2 \leq c_T(x), \qquad (1.2.5)$$

then it follows

$$\mathbb{P}(\tau^x = +\infty) = 1. \qquad (1.2.6)$$

Actually, by using the Chebischev inequality

$$\mathbb{P}\left(\sup_{t \in [0,T]} |\xi_n(t; x)|^2 \leq n \right) = 1 - \mathbb{P}\left(\sup_{t \in [0,T]} |\xi_n(t; x)|^2 > n \right)$$

$$\geq 1 - \frac{1}{n^2} \mathbb{E}\left(\sup_{t \in [0,T]} |\xi_n(t; x)|^2 \geq 1 - \frac{c_T(x)}{n^2}, \right.$$

and then

$$\mathbb{P}\left(\tau_n^x \geq T\right) \geq 1 - \frac{c_T(x)}{n^2}.$$

Therefore

$$\lim_{n \to +\infty} \mathbb{P}\left(\tau_n^x \geq T\right) = 1$$

and (1.2.6) follows from the arbitrariness of T.

Finally, since for any $t \leq \tau_n^x$

$$\xi_n(t;x) = \xi_{n+1}(t;x), \quad \mathbb{P}-\text{a.e.,}$$

by setting

$$\xi(t;x) = \xi_n(t \wedge \tau_n^x;x)$$

for all (t,ω) such that $t \leq \tau^x(\omega)$, the existence and the uniqueness of solutions for (1.2.1) follow.

We denote by $\mathcal{P}_\infty$ the σ-algebra of subsets of $[0,\infty) \times \Omega$ generated by sets of the form $\{0\} \times F$, with $F \in \mathcal{F}_0$, and of the form $(s,t] \times F$, with $0 \leq s < t < \infty$ and $F \in \mathcal{F}_s$. This σ-algebra is called the *predictable σ-algebra*. For any $T > 0$ the restriction of $\mathcal{P}_\infty$ to $[0,T] \times \Omega$ is denoted by $\mathcal{P}_T$. Any measurable mapping from $([0,\infty) \times \Omega, \mathcal{P}_\infty)$ (or $([0,T) \times \Omega; \mathcal{P}_T)$) into $(E; \mathcal{E})$ is said a *predictable process*. Next, we denote by $\mathcal{H}_2(T,\mathbb{R}^d)$ the set of all predictable processes belonging to $C([0,T]; L^2(\Omega; \mathbb{R}^d))$. $\mathcal{H}_2(T,\mathbb{R}^d)$ is a closed subspace of $C([0,T]; L^2(\Omega; \mathbb{R}^d))$, endowed with the norm

$$\|\xi\|_{\mathcal{H}_2(T,\mathbb{R}^d)}^2 = \sup_{t \in [0,T]} \mathbb{E}\,|\xi(t)|^2.$$

We recall that the following result holds

Lemma 1.2.1. *For any $m \geq 1$ and for any process $\xi \in \mathcal{H}_2(T,\mathbb{R}^d)$ we have*

$$\mathbb{E} \sup_{s \in [0,t]} \left| \int_0^s \xi(r)dw(r) \right|^{2m} \leq c_m\, \mathbb{E} \left(\int_0^s |\xi(r)|^2\, dr \right)^m, \quad t \in [0,T],$$

for some constant c_m.

When $m = 1$ the proof easily follows from the maximal martingale inequality. When $m > 1$ it is based on the maximal martingale inequality and Itô's formula applied to the process

$$\eta(t) = \int_0^t \xi(s)\, dw(s)$$

and to the function $f(x) = |x|^{2m}$. Notice that this result holds in infinite dimension, as well (for a proof in this more general case, see [47, Lemma 7.21]).

Proposition 1.2.2. *Assume that the Hypothesis 1.1 holds. Then for any $n > 0$ and $T > 0$ the approximating problem (1.2.4) admits a unique solution $\xi_n(\cdot;x)$ in $\mathcal{H}_2(T,\mathbb{R}^d)$.*

Proof. By proceeding pathwise it is not difficult to show that for any fixed $\eta \in \mathcal{H}_2(T,\mathbb{R}^d)$ the problem

$$d\zeta(t) = b(\zeta(t))\, dt + \sigma_n(\eta(t))\, dw(t), \quad \zeta(0) = x$$

admits an unique solution $\Gamma_n(\eta) \in \mathcal{H}_2(T, \mathbb{R}^d)$. Actually, since σ_n is bounded, the stochastic integral

$$w_n^\eta(t) = \int_0^t \sigma_n(\eta(s)) \, dw(s)$$

is well defined and, due to the Lemma 1.2.1, for any $p \geq 1$ it holds

$$\mathbb{E}|w_n^\eta(t)|^p \leq c_{n,p} \, t^{p/2}, \quad t \geq 0. \tag{1.2.7}$$

By setting $\theta_n^\eta(t) = \zeta(t) - w_n^\eta(t)$, we have

$$\frac{d}{dt}\theta_n^\eta(t) = b(\theta_n^\eta(t) + w_n^\eta(t)), \quad \theta_n^\eta(0) = x. \tag{1.2.8}$$

Since b is continuous, the problem (1.2.8) admits a local solution and by standard arguments such a solution can be extended to the whole interval $[0, T]$, due to the dissipativity of b (see the Hypothesis 1.3) and to (1.2.7). The uniqueness follows from the Hypothesis 1.3.

Whence, if we show that

$$\Gamma_n : \mathcal{H}_2(T_0, \mathbb{R}^d) \to \mathcal{H}_2(T_0, \mathbb{R}^d)$$

is a contraction for T_0 sufficiently small, then its fixed point is the unique solution of (1.2.4).

For any $\eta_1, \eta_2 \in \mathcal{H}_2(T, \mathbb{R}^d)$, we have

$$\Gamma_n(\eta_1)(t) - \Gamma_n(\eta_2)(t) =$$

$$\int_0^t \left(b\left(\Gamma_n(\eta_1)(s)\right) - b\left(\Gamma_n(\eta_2)(s)\right)\right) ds + \int_0^t \left(\sigma_n(\eta_1(s)) - \sigma_n(\eta_2(s))\right) dw(s).$$

Due to Itô's formula, by taking the expectation we have

$$\mathbb{E}|\Gamma_n(\eta_1)(t) - \Gamma_n(\eta_2)(t)|^2$$

$$= 2\,\mathbb{E} \int_0^t \left\langle b\left(\Gamma_n(\eta_1)(s)\right) - b\left(\Gamma_n(\eta_2)(s)\right), \Gamma_n(\eta_1)(s) - \Gamma_n(\eta_2)(s)\right\rangle ds$$

$$+ \mathbb{E} \int_0^t \|\sigma_n(\eta_1(s)) - \sigma_n(\eta_2(s))\|_2^2 \, ds.$$

By using the Hypothesis 1.1-3. (in the version (1.1.1)) and the Lipschitz property of σ_n, this implies that

$$\mathbb{E}|\Gamma_n(\eta_1)(t) - \Gamma_n(\eta_2)(t)|^2$$

$$\leq 2c \int_0^t \mathbb{E}|\Gamma_n(\eta_1)(s) - \Gamma_n(\eta_2)(s)|^2 \, ds + M_n^2 \int_0^t \mathbb{E}|\eta_1(s) - \eta_2(s)|^2 \, ds,$$

so that, by the Gronwall lemma

$$\mathbb{E}|\Gamma_n(\eta_1)(t) - \Gamma_n(\eta_2)(t)|^2 \leq M_n^2 \int_0^t e^{2c(t-s)} \mathbb{E}|\eta_1(s) - \eta_2(s)|^2 \, ds.$$

In particular we have

$$\sup_{t \in [0,T]} \mathbb{E}|\Gamma_n(\eta_1)(t) - \Gamma_n(\eta_2)(t)|^2 \leq M_n^2 e^{2cT} T \sup_{t \in [0,T]} \mathbb{E}|\eta_1(t) - \eta_2(t)|^2.$$

Hence, by choosing T_0 such that

$$M_n \sqrt{T_0} e^{cT_0} < 1,$$

Γ_n is a contraction in $\mathcal{H}_2(T_0, \mathbb{R}^d)$. Moreover, as T_0 depends only on c and M_n, we can proceed in the same way in $[T_0, 2T_0]$ and so on, and we get the existence of a unique solution for (1.2.4) which is defined in the whole interval $[0, T]$. $\qquad\square$

In order to conclude the proof of the existence and uniqueness of the solution $\xi(t; x)$, we have to verify (1.2.5). To this purpose we prove a preliminary result.

Lemma 1.2.3. *Under the Hypothesis 1.1, for any $p \geq 1$ we have*

$$\mathbb{E}|\xi_n(t; x)|^p \leq c_p(t) (|x|^p + 1), \quad x \in \mathbb{R}^d, \tag{1.2.9}$$

where $c_p(t)$ is an increasing continuous function independent of n.

Proof. If we prove (1.2.9) for $p \geq 4$, then the case $p \in [1, 4)$ follows from the Hölder inequality. We set $p = 2q$, with $q \geq 2$.

For $r > 0$ and $x \in \mathbb{R}^d$ we define

$$\sigma_{n,r}^x = \inf \{ t \geq 0 : |\xi_n(t; x)| \geq r \},$$

with $\sigma_{n,r}^x = +\infty$ if $|\xi_n(t; x)| < r$ for any $t \geq 0$. Since $\xi_n(\cdot; x) \in \mathcal{H}_2(T, \mathbb{R}^d)$, for all $n \in \mathbb{N}$ and $t \geq 0$ we have

$$\mathbb{E}|\xi_n(t; x)|^2 < \infty.$$

Then, if we define $\sigma_n^x = \lim_{r \to +\infty} \sigma_{n,r}^x$, we have

$$\mathbb{P}(\sigma_n^x = +\infty) = 1. \tag{1.2.10}$$

We apply Itô's formula to the process $t \mapsto \xi_n(t; x)$ and to the function $f(x) = |x|^{2q}$ and we get

$$|\xi_n(t \wedge \sigma_{n,r}^x)|^{2q} = |x|^{2q} + 2q \int_0^{t \wedge \sigma_{n,r}^x} |\xi_n(s)|^{2(q-1)} \left(\langle \xi_n(s), b(\xi_n(s)) \rangle + \frac{1}{2} \mathrm{Tr}\, [a_n(s)] \right) ds$$

$$+ 2q(q-1) \int_0^{t \wedge \sigma_{n,r}^x} |\xi_n(s)|^{2(q-2)} \mathrm{Tr}\, [a_n(s) (\xi_n(s) \otimes \xi_n(s))] \, ds$$

$$+ 2q \int_0^{t \wedge \sigma_{n,r}^x} |\xi_n(s)|^{2(q-1)} \langle \xi_n(s), \sigma_n(\xi_n(s)) \, dw(s) \rangle,$$

where for any $n \in \mathbb{N}$ and $t \geq 0$ we have defined

$$a_n(t; x) = \sigma_n(\xi_n(t; x)) \sigma_n^*(\xi_n(t; x)), \qquad x \in \mathbb{R}^d.$$

Now, by taking the expectation we get

$$\mathbb{E}|\xi_n(t \wedge \sigma_{n,r}^x)|^{2q} = |x|^{2q}$$

$$+ 2q \, \mathbb{E} \int_0^t \chi_{\{\sigma_{n,r}^x \geq s\}} |\xi_n(s)|^{2(q-1)} \left(\langle \xi_n(s), b(\xi_n(s)) \rangle + \frac{1}{2} \mathrm{Tr}\,[a_n(s)] \right) ds \qquad (1.2.11)$$

$$+ 2q(q-1) \, \mathbb{E} \int_0^t \chi_{\{\sigma_{n,r}^x \geq s\}} |\xi_n(s)|^{2(q-2)} \mathrm{Tr}\,[a_n(s) \xi_n(s) \otimes \xi_n(s)] \, ds.$$

Due to the Hypothesis 1.1-3. this easily yields

$$\mathbb{E}|\xi_n(t \wedge \sigma_{n,r}^x)|^{2q} \leq |x|^{2q} + c_q \, \mathbb{E} \int_0^t \chi_{\{\sigma_{n,r}^x \geq s\}} \left(|\xi_n(s \wedge \sigma_{n,r}^x)|^{2q} + 1 \right) ds$$

$$\leq |x|^{2q} + c_q \int_0^t \left(\mathbb{E}|\xi_n(s \wedge \sigma_{n,r}^x)|^{2q} + 1 \right) ds$$

and from the Gronwall lemma it follows

$$\mathbb{E}|\xi_n(t \wedge \sigma_{n,r}^x; x)|^{2q} \leq c_q \, e^{c_q t} \left(|x|^{2q} + 1 \right).$$

Now, due to (1.2.10), we have that for any fixed $t \geq 0$ and $x \in \mathbb{R}^d$

$$\lim_{r \to +\infty} |\xi_n(t \wedge \sigma_{n,r}^x; x)|^{2q} = |\xi_n(t; x)|^{2q}, \qquad \mathbb{P} - \text{a.s.}$$

and then, from Fatou's lemma we get (1.2.9). $\qquad\qquad\qquad\qquad\qquad\qquad \square$

Proposition 1.2.4. *Assume that the Hypothesis 1.1 holds. Then for any $T > 0$ we have that $\xi_n(\cdot; x) \in L^2(\Omega; C([0, T]; \mathbb{R}^d))$ and*

$$\mathbb{E} \sup_{s \in [0,t]} |\xi_n(s; x)|^2 \leq c(t) \left(|x|^{4m+2} + 1 \right), \qquad t \in [0, T], \qquad (1.2.12)$$

for a continuous increasing function $c(t)$ not depending on $n \in \mathbb{N}$.

Proof. We have

$$\xi_n(s; x) = x + \int_0^s b(\xi_n(r; x)) \, dr + \int_0^s \sigma_n(\xi_n(r; x)) \, dw(r).$$

Then

$$|\xi_n(s;x)|^2 \le 3|x|^2 + 3 \left| \int_0^s b(\xi_n(r;x)) \, dr \right|^2 + 3 \left| \int_0^s \sigma_n(\xi_n(r;x)) \, dw(r) \right|^2$$

$$\le 3|x|^2 + cs \int_0^s \left(|\xi_n(r;x)|^{2(2m+1)} + 1 \right) dr + 3 \left| \int_0^s \sigma_n(\xi_n(r;x)) \, dw(r) \right|^2,$$

so that

$$\sup_{s \in [0,t]} |\xi_n(s;x)|^2 \le c|x|^2 + ct \int_0^t \left(|\xi_n(r;x)|^{2(2m+1)} + 1 \right) dr$$

$$+ c \sup_{s \in [0,t]} \left| \int_0^s \sigma_n(\xi_n(r;x)) \, dw(r) \right|^2, \qquad \mathbb{P} - \text{a.s.}$$

Now, by taking the expectation, due to the maximal martingale inequality and to the growth conditions on σ_n, according to the Lemma 1.2.1 we have

$$\mathbb{E} \sup_{s \in [0,t]} |\xi_n(s;x)|^2 \le c|x|^2 + ct \int_0^t \left(\mathbb{E}|\xi_n(r;x)|^{2(2m+1)} + 1 \right) dr$$

$$+ c \mathbb{E} \left| \int_0^t \sigma_n(\xi_n(r;x)) \, dw(r) \right|^2$$

$$\le c|x|^2 + ct \int_0^t \left(\mathbb{E}|\xi_n(r;x)|^{2(2m+1)} + 1 \right) dr + c \int_0^t \left(\mathbb{E}|\xi_n(r;x)|^{2k} + 1 \right) dr.$$

Thus, thanks to the previous lemma, we can conclude that

$$\mathbb{E} \sup_{s \in [0,t]} |\xi_n(t;x)|^2 \le c \left(|x|^2 + t^2 c_{4m+2}(t)(|x|^{4m+2} + 1) + t c_{2k}(t)(|x|^{2k} + 1) + t^2 + t \right)$$

and (1.2.12) follows. The continuity of the trajectories is obtained in a similar way. $\qquad \square$

Now, we can conclude with the existence and uniqueness of solutions for the equation (1.2.1).

Theorem 1.2.5. *Under the Hypothesis 1.1 there exists a unique d-dimensional predictable process $\xi(t;x)$, having continuous trajectories, which satisfies the equation (1.2.1). Moreover, for any $p \ge 1$ we have*

$$\mathbb{E} \sup_{s \in [0,t]} |\xi(s;x)|^p \le c_p(t)(|x|^p + 1), \tag{1.2.13}$$

where $c_p(t)$ is a suitable increasing function.

The estimate that we are going to prove will play an essential rôle in order to prove any further estimate. In the proof we will need the following lemma.

Lemma 1.2.6. *Let $u(t; u_0)$ be the solution of the Cauchy problem*

$$u'(t) = a\left(c - u^{1+\delta}(t)\right), \qquad u(0) = u_0, \tag{1.2.14}$$

for some $a, c, \delta > 0$. Then for any $t > 0$ it holds

$$\sup_{u_0 \geq 0} |u(t; u_0)| \leq c(t)\, t^{-\frac{1}{\delta}}, \tag{1.2.15}$$

where $c(t)$ is a continuous function, increasing with respect to t and depending only on δ, a and c.

Proof. We first recall a classical comparison argument in the theory of ordinary differential equations. Let $I \subseteq \mathbb{R}$ be an interval and $f, g : I \times \mathbb{R} \to \mathbb{R}$ continuous and locally Lipschitz-continuous with respect to the second variable. Let $u, v : I \to \mathbb{R}$ be two differentiable functions such that

$$u'(t) \leq f(t, u(t)), \qquad v'(t) \leq g(t, v(t)), \quad t \in I.$$

Then, if $u(t_0) \geq v(t_0)$, for some $t_0 \in I$, we have $u(t) \geq v(t)$, for any $t \leq t_0$.

Now, we fix $t_0 > 0$ and $u_{t_0} > (2c)^{\frac{1}{1+\delta}}$. The Cauchy problem

$$u'(t) = c\left(1 - u^{1+\delta}(t)\right), \qquad u(t_0) = u_{t_0} \tag{1.2.16}$$

admits a solution $u(t; u_{t_0})$ defined in some maximal interval $I = (\alpha, \beta)$ and it is clear that $u(t; u_{t_0}) > (2c)^{\frac{1}{1+\delta}}$, for every $t \in (\alpha, t_0)$. Then $u^{1+\delta}(t; u_{t_0}) > 2c$ and

$$u'(t) \leq -\frac{a}{2} u^{1+\delta}(t), \quad t \in (a, t_0].$$

By using the comparison argument quoted before, with

$$f(t, y) = c\left(1 - y^{1+\delta}\right), \qquad g(t, y) = -\frac{a}{2} y^{1+\delta},$$

we get

$$u(t; u_{t_0}) \geq u_{t_0}\left(1 - \frac{\delta a}{2} u_{t_0}^{\delta}(t_0 - t)\right)^{-\frac{1}{\delta}}, \qquad t \in (a, t_0].$$

By choosing

$$u_{t_0} > \max\left\{(2c)^{\frac{1}{1+\delta}}, \left(\frac{2}{\delta a}\right)^{\frac{1}{\delta}} t_0^{-\frac{1}{\delta}}\right\} = c(t_0) t_0^{-\frac{1}{\delta}},$$

it follows that $u(t; u_{t_0})$ is not defined for $t \leq 0$, so that $\alpha > 0$. Therefore, if we come back to the equation (1.2.16), by uniqueness we get that the solution $u(t; u_0)$ of the problem (1.2.14) satisfies the estimate (1.2.15). $\qquad\square$

Proposition 1.2.7. *Assume that the Hypotheses 1.1 and 1.2 hold. Then for any $p \geq 1$ and $n \in \mathbb{N}$*

$$\sup_{x \in \mathbb{R}^d} \mathbb{E}|\xi_n(t;x)|^p \leq c_p(t)(t \wedge 1)^{-\frac{p}{2m}}, \tag{1.2.17}$$

where $c_p(t)$ is a continuous function, which is increasing with respect to t and independent of $n \in \mathbb{N}$. In particular, we have

$$\sup_{x \in \mathbb{R}^d} \mathbb{E}|\xi(t;x)|^p \leq c_p(t)(t \wedge 1)^{-\frac{p}{2m}}. \tag{1.2.18}$$

Proof. Due to (1.2.11) and to the Hypothesis 1.2, by using the Young inequality it easily follows that for any $r > 0$ and $p \geq 4$

$$\frac{d}{dt}\mathbb{E}|\xi_n(t \wedge \sigma^x_{n,r};x)|^p \leq -a\,\mathbb{E}|\xi_n(t \wedge \sigma^x_{n,r};x)|^{p+2m} + c\,\mathbb{E}|\xi_n(t \wedge \sigma^x_{n,r};x)|^{p+2(k-1)} + c.$$

By using again the Young inequality, since $2(k-1)$ is assumed to be less or equal to $2m$, this implies that

$$\frac{d}{dt}\mathbb{E}|\xi_n(t \wedge \sigma^x_{n,r};x)|^p \leq -\frac{a}{2}\mathbb{E}|\xi_n(t \wedge \sigma^x_{n,r};x)|^{p+2m} + c.$$

Then, from the Hölder inequality we get

$$\frac{d}{dt}\mathbb{E}|\xi_n(t \wedge \sigma^x_{n,r};x)|^p \leq -\frac{a}{2}\left(\mathbb{E}|\xi_n(t \wedge \sigma^x_{n,r};x)|^p\right)^{\frac{p+2m}{p}} + c$$

and by comparison it follows

$$\mathbb{E}|\xi_n(t \wedge \sigma^x_{n,r};x)|^p \leq u(t;|x|^p), \quad t \geq 0,$$

where $u(t;|x|^p)$ is the solution of the problem

$$u'(t) = c - \frac{a}{2}u^{1+\frac{2m}{p}}(t), \quad u(0) = |x|^p.$$

Therefore, from the estimate (1.2.15) we get

$$\sup_{x \in \mathbb{R}^d} \mathbb{E}|\xi_n(t \wedge \sigma^x_{n,r};x)|^p \leq c_p(t)(t \wedge 1)^{-\frac{p}{2m}}.$$

Now, by taking the limit as r goes to infinity, from Fatou's lemma (1.2.17) follows. $\square$

Remark 1.2.8. If the matrix σ does not depend on $x \in \mathbb{R}^d$, the stochastic equation corresponding to the operator $\mathcal{L}_0$ has an additive noise and then we can proceed pathwise in our estimates. Then for any $p \geq 1$ and $t > 0$ we get the stronger estimate

$$\sup_{x \in \mathbb{R}^d} |\xi(t;x)|^p \leq c_p(t)t^{-\frac{p}{2m}}, \quad \mathbb{P} - a.s.$$

where $c_p(t)$ is an increasing process having finite moments of any order.

1.3   Estimates for the derivatives of the solution

If the coefficients σ and b in the equation (1.2.1) are Lipschitz-continuous and of class C^h, with $h \geq 1$, by using the *contraction theorem of contractions depending on parameters* (see Appendix C for a generalization), it is possible to prove that the solution $\xi(t;x)$ of the problem (1.2.1) is mean-square differentiable with respect to the initial datum x, up to the h-th order. In particular, for any fixed $t \geq 0$ the mapping

$$\mathbb{R}^d \to L^2(\Omega; \mathbb{R}^d), \qquad x \mapsto \xi(t;x)$$

is h-times differentiable. Moreover the derivatives are themselves solutions of suitable stochastic differential equations which one obtains by differentiating formally the coefficients of the equation (1.2.1). Such equations are known as the *variation equations* corresponding to the problem (1.2.1).

In particular, since $\xi(t;x)$ is h-times mean-square differentiable with respect to x, for any $\varphi \in C_b^h(\mathbb{R}^d)$ it is possible to differentiate h times with respect to $x \in \mathbb{R}^d$ under the integral sign in

$$P_t\varphi(x) = \mathbb{E}\varphi(\xi(t;x)),$$

so that $P_t\varphi \in C_b^h(\mathbb{R}^d)$. For instance, for the first derivative we have

$$\langle D(P_t\varphi)(x), h \rangle = \mathbb{E}\langle D\varphi(\xi(t;x)), D\xi(t;x)h \rangle.$$

Similar formulas can be obtained for higher order derivatives.

Instead, the present situation is more delicate as the coefficients b and σ have unbounded first derivative and then we can not apply directly the classical theory. This is the reason why we have to proceed approximating the coefficients by Lipschitz-continuous functions.

According to the Hypothesis 1.3 there exists $c \in \mathbb{R}$ such that the mapping $g : \mathbb{R}^d \to \mathbb{R}^d$, $x \mapsto g(x) = b(x) - cx$, is dissipative. Then, thanks to the Proposition A.2.1 we can introduce the Yosida approximations of g, by setting for any $\alpha > 0$

$$g_\alpha(x) = g(J_\alpha(x)), \qquad x \in \mathbb{R}^d,$$

where

$$J_\alpha(x) = (I - \alpha g)^{-1}(x), \qquad x \in \mathbb{R}^d.$$

As proved in the Proposition A.2.2, for any $\alpha > 0$ the function g_α is dissipative and Lipschitz-continuous. As shown at the end of the Appendix A, since we are assuming that g is of class C^h, we have that J_α is of class C^h, and then g_α is of class C^h. Moreover, since

$$g_\alpha(x) = (J_\alpha(x) - x)/\alpha, \qquad x \in \mathbb{R}^d,$$

we easily get an explicit expression for the derivatives of g_α in terms of the derivatives of J_α. Whence, due to (A.2.2) and to (A.2.3), (A.2.4) and (A.2.5), for any $j =$

$0, \ldots, h$ and $R > 0$ we have

$$\lim_{\alpha \to 0} \sup_{|x| \leq R} |D^j g_\alpha(x) - D^j g(x)| = 0. \qquad (1.3.1)$$

Finally, due to (A.2.1) we have

$$\sup_{x \in \mathbb{R}^d} |D J_\alpha(x)| \leq c,$$

for some constant c independent of α. By using (A.2.3), (A.2.4) and (A.2.5) this implies that for any $j = 0, \ldots, h$ and $x \in \mathbb{R}^d$

$$|D^j g_\alpha(x)| \leq c |D^j g(x)|, \qquad x \in \mathbb{R}^d. \qquad (1.3.2)$$

Now, for any $\alpha > 0$ we define

$$b_\alpha(x) = g_\alpha(x) + c x.$$

It is easy to check that b_α is of class C^h, is Lipschitz-continuous and there exists $c \in \mathbb{R}$ independent of α such that for any $x, y \in \mathbb{R}^d$

$$\langle D b_\alpha(x) y, y \rangle \leq c |y|^2, \qquad \alpha > 0. \qquad (1.3.3)$$

Moreover, from (1.3.1), for any $R > 0$ we have

$$\lim_{\alpha \to 0} \sup_{|x| \leq R} |D^j b_\alpha(x) - D^j b(x)| = 0 \qquad (1.3.4)$$

and, from (1.3.2) for $j = 0, \ldots, h$ we have

$$\sup_{\alpha > 0} \sup_{x \in \mathbb{R}^d} \frac{|D^\beta b_\alpha(x)|}{1 + |x|^{2m+1-j}} < \infty, \qquad |\beta| = j. \qquad (1.3.5)$$

In what follows, for any $\alpha > 0$ and $n \in \mathbb{N}$ we shall consider the approximating problem

$$d\xi(t) = b_\alpha(\xi(t)) \, dt + \sigma_n(\xi(t)) \, dw(t), \qquad \xi(0) = x. \qquad (1.3.6)$$

As the coefficients b_α and σ_n are Lipschitz-continuous, the equation (1.3.6) admits a unique solution $\xi_{n,\alpha}(t; x)$ and for any $p \geq 1$

$$\mathbb{E}|\xi_{n,\alpha}(t; x)|^p \leq c_{n,p}(t) (1 + |x|^p), \qquad t \geq 0, \qquad (1.3.7)$$

for an increasing function $c_{n,p}(t)$, not depending on α. The proof of (1.3.7) is completely analogous to the proof of the Lemma 1.2.3, so we do not repeat it. We only remark that it is based on the fact that the estimate (1.3.3) is independent of α and σ_n is bounded.

Due to (1.3.4), with $j = 0$, we have the following approximation result.

Lemma 1.3.1. *For any $n \in \mathbb{N}$ and $p \geq 1$ it holds*

$$\lim_{\alpha \to 0} \mathbb{E} |\xi_{n,\alpha}(t; x) - \xi_n(t; x)|^p = 0. \tag{1.3.8}$$

Proof. For the sake of brevity, we prove it only for $p = 2$. The case $p > 2$ is analogous with more complicate notations; the case $p \in [1, 2)$ follows as usual from the Hölder inequality. We set

$$v_{n,\alpha}(t; x) = \xi_{n,\alpha}(t; x) - \xi_n(t; x).$$

We have

$$dv_{n,\alpha}(t) = [b_\alpha(\xi_{n,\alpha}(t)) - b(\xi_n(t))] \, dt + [\sigma_n(\xi_{n,\alpha}(t)) - \sigma_n(\xi_n(t))] \, dw(t), \quad v_{n,\alpha}(0) = 0.$$

Then, if we apply Itô's formula and take first the expectation and then the derivative with respect to t, recalling that σ_n is Lipschitz-continuous we get

$$\frac{d}{dt} \mathbb{E} |v_{n,\alpha}(t)|^2 = 2 \mathbb{E} \langle b_\alpha(\xi_{n,\alpha}(t; x)) - b(\xi_n(t; x)), v_{n,\alpha}(t) \rangle$$

$$+ \mathbb{E} \|\sigma_n(\xi_{n,\alpha}(t; x)) - \sigma_n(\xi_n(t; x))\|_2^2 \leq 2 \mathbb{E} \langle b_\alpha(\xi_{n,\alpha}(t; x)) - b_\alpha(\xi_n(t; x)), v_{n,\alpha}(t) \rangle$$

$$+ 2 \mathbb{E} \langle b_\alpha(\xi_n(t; x)) - b(\xi_n(t; x)), v_{n,\alpha}(t) \rangle + c_n \mathbb{E} |v_{n,\alpha}(t)|^2.$$

Then, due to (1.3.3), by using the Young inequality we have

$$\frac{d}{dt} \mathbb{E} |v_{n,\alpha}(t)|^2 \leq c_n \mathbb{E} |v_{n,\alpha}(t)|^2 + c_n \mathbb{E} |b_\alpha(\xi_n(t; x)) - b(\xi_n(t; x))|^2,$$

and from the Gronwall lemma we can conclude that

$$\mathbb{E} |v_{n,\alpha}(t)|^2 \leq c_n \int_0^t e^{c_n(t-s)} \mathbb{E} |b_\alpha(\xi_n(s; x)) - b(\xi_n(s; x))|^2 \, ds.$$

Now, thanks to (1.3.5) we have

$$\mathbb{E} |b_\alpha(\xi_n(t; x)) - b(\xi_n(t; x))|^2 \leq c \left(1 + \mathbb{E} |\xi_n(t; x)|^{4m+2}\right)$$

and then, as (1.3.4) holds, by using the Lemma 1.2.3 and the dominated convergence theorem it follows that

$$\lim_{\alpha \to 0} \mathbb{E} |v_{n,\alpha}(t)|^2 = 0.$$

$\square$

Since b_α and σ_n are of class C^h and Lipschitz-continuous, then $\xi_{n,\alpha}(t; x)$ is h-times mean-square differentiable with respect to $x \in \mathbb{R}^d$. In the next subsections we are showing that the solution $\xi(t; x)$ of the equation (1.2.1) is h-times mean-square differentiable, as well.

1.3.1 First derivative

For any $n \in \mathbb{N}$ and $x, h \in \mathbb{R}^d$, we consider the problem

$$d\eta(t) = Db(\xi_n(t;x))\eta(t)dt + D\sigma_n(\xi_n(t;x))\eta(t)dw(t), \qquad \eta(0) = h. \qquad (1.3.9)$$

The equation (1.3.9) is known as the *first variation* equation corresponding to the problem (1.2.4). The following result holds true.

Proposition 1.3.2. *Under the Hypotheses 1.1 and 1.2 for any $n \in \mathbb{N}$ the problem (1.3.9) admits a unique solution $\eta_n^h(\cdot;x) \in C([0,+\infty); L^2(\Omega; \mathbb{R}^d))$ such that for any $p \geq 1$*

$$\sup_{x \in \mathbb{R}^d} \mathbb{E}|\eta_n^h(t;x)|^p \leq f_{1,p}(t)|h|^p, \quad t \geq 0, \qquad (1.3.10)$$

for a suitable increasing continuous function $f_{1,p}(t)$, not depending on n. Moreover, for any $p \geq 1$ it holds

$$\lim_{\substack{\alpha \to 0 \\ h \in \mathbb{R}^d \\ h \neq 0}} \sup \frac{\mathbb{E}|D\xi_{n,\alpha}(t;x)h - \eta_n^h(t;x)|^p}{|h|^p} = 0. \qquad (1.3.11)$$

Proof. We fix $n \in \mathbb{N}$, $T > 0$ and $x \in \mathbb{R}^d$. For any $\eta \in C([0,T]; L^2(\Omega; \mathbb{R}^d))$, we define

$$\Gamma_n(\eta)(t) = h + \int_0^t Db(\xi_n(s;x))\eta(s)\,ds + \int_0^t D\sigma_n(\xi_n(s;x))\eta(s)\,dw(s), \quad t \in [0,T].$$

By using the growth conditions on $D\sigma_n$ and the estimate (1.2.9), it is not difficult to show that Γ_n is a contraction in $C([0,T_0]; L^2(\Omega; \mathbb{R}^d))$, for T_0 sufficiently small. Then there exists a unique fixed point for Γ_n, which is the unique solution $\eta_n^h(t;x)$ of the problem (1.3.9) in $C([0,T_0]; L^2(\Omega; \mathbb{R}^d))$. As we can repeat the same arguments in the intervals $[T_0, 2T_0]$, $[2T_0, 3T_0]$ and so on, we have a unique global solution $\eta_n^h(t;x)$ in $C([0,T]; L^2(\Omega; \mathbb{R}^d))$, for any $T > 0$.

Now, let us prove (1.3.10). By using Itô's formula and by proceeding as in the proof of the Proposition 1.2.7, for any $p \geq 4$ we have

$$\frac{d}{dt}\mathbb{E}|\eta_n^h(t)|^p = p\,\mathbb{E}|\eta_n^h(t)|^{p-2} \left\langle \eta_n^h(t), Db(\xi_n(t;x))\eta_n^h(t) \right\rangle$$

$$+\frac{p(p-2)}{2}\mathbb{E}|\eta_n^h(t)|^{p-4}\mathrm{Tr}\left[a_n^h(t;x)\left(\eta_n^h(t) \otimes \eta_n^h(t)\right)\right] + \frac{p}{2}\mathbb{E}|\eta_n^h(t)|^{p-2}\mathrm{Tr}\left[a_n^h(t;x)\right]$$

where

$$a_n^h(t;x) = \left(D\sigma_n(\xi_n(t;x))\eta_n^h(t)\right)\left(D\sigma_n(\xi_n(t;x))\eta_n^h(t)\right)^\star.$$

Then from (1.2.3) we get

$$\frac{d}{dt}\mathbb{E}|\eta_n^h(t)|^p \leq c_p\,\mathbb{E}|\eta_n^h(t)|^p,$$

for some constant c_p not depending on n. By using the Gronwall lemma it follows

$$\mathbb{E}|\eta_n^h(t))|^p \le |h|^p \exp(c_p t) = f_{1,p}(t)|h|^p, \quad t \ge 0,$$

which implies (1.3.10).

Next, we prove (1.3.11) only for $p = 2$, as the other cases are analogous. For any $\alpha > 0$ and $n \in \mathbb{N}$ we denote $D\xi_{n,\alpha}(t;x)h = \eta_{n,\alpha}^h(t;x)$ and we set

$$v_{n,\alpha}(t;x) = \eta_{n,\alpha}^h(t;x) - \eta_n^h(t;x).$$

By applying Itô's formula and taking first the expectation and then the derivative with respect to t, we get

$$\frac{d}{dt}\mathbb{E}|v_{n,\alpha}(t)|^2 = 2\left\langle Db_\alpha(\xi_{n,\alpha}(t;x))\eta_{n,\alpha}^h(t) - Db(\xi_n(t;x))\eta_n^h(t), v_{n,\alpha}(t)\right\rangle$$

$$+\mathbb{E}\|D\sigma_n(\xi_{n,\alpha}(t;x))\eta_{n,\alpha}^h(t) - D\sigma_n(\xi_n(t;x))\eta_n^h(t)\|_2^2$$

and then

$$\frac{d}{dt}\mathbb{E}|v_{n,\alpha}(t)|^2 \le c\left(\langle Db_\alpha(\xi_{n,\alpha}(t;x))v_{n,\alpha}(t), v_{n,\alpha}(t)\rangle\right.$$

$$+\left\langle (Db_\alpha(\xi_{n,\alpha}(t;x)) - Db_\alpha(\xi_n(t;x)))\eta_n^h(t), v_{n,\alpha}(t)\right\rangle$$

$$+\left\langle (Db_\alpha(\xi_n(t;x)) - Db(\xi_n(t;x)))\eta_n^h(t), v_{n,\alpha}(t)\right\rangle$$

$$\left. + \mathbb{E}\|D\sigma_n(\xi_{n,\alpha}(t;x))v_{n,\alpha}(t)\|_2^2 + \mathbb{E}\|(D\sigma_n(\xi_{n,\alpha}(t)) - D\sigma_n(\xi_n(t;x)))\eta_n^h(t)\|_2^2\right).$$

Therefore, recalling that σ_n has bounded derivatives up to the h-th order and that (1.3.3) holds, due to the Young inequality we obtain

$$\frac{d}{dt}\mathbb{E}|v_{n,\alpha}(t)|^2 \le c_n\left(\mathbb{E}|v_{n,\alpha}(t)|^2 + \mathbb{E}\left|(Db_\alpha(\xi_{n,\alpha}(t;x)) - Db_\alpha(\xi_n(t;x)))\eta_n^h(t)\right|^2\right.$$

$$\left. + \mathbb{E}\left|(Db_\alpha(\xi_n(t;x)) - Db(\xi_n(t;x)))\eta_n^h(t)\right|^2 + \mathbb{E}\left|\xi_{n,\alpha}(t;x) - \xi_n(t;x)\right|^2|\eta_n^h(t)|^2\right)$$

$$= c_n\mathbb{E}|v_{n,\alpha}(t)|^2 + \sum_{i=1}^{3} I_i^h(t;n,\alpha).$$

Recalling that $v_{n,\alpha}(t) = D\xi_{n,\alpha}(t;x)h - \eta_n^h(t;x)$, by the Gronwall lemma this implies that

$$\mathbb{E}|D\xi_{n,\alpha}(t;x)h - \eta_n^h(t;x)|^2 \le \int_0^t e^{c_n(t-s)}\sum_{i=1}^{3} I_i^h(s;n,\alpha)\,ds.$$

We have

$$I_1^h(t; n, \alpha)$$

$$= c_n \, \mathbb{E} \left| \int_0^1 D^2 b_\alpha(\theta \xi_{n,\alpha}(t; x) + (1 - \theta) \xi_n(t; x))(\xi_{n,\alpha}(t; x) - \xi_n(t; x)) \, d\theta \, \eta_n^h(t) \right|^2$$

and due to (1.3.5) we have

$$I_1^h(t; n, \alpha) \leq c_n \left(\mathbb{E} |\eta_n^h(t)|^4 \right)^{1/2} \left(\mathbb{E} \, |\xi_{n,\alpha}(t; x) - \xi_n(t; x)|^8 \right)^{1/4}$$

$$\left(\mathbb{E} \left(1 + |\xi_{n,\alpha}(t; x)|^{4m-2} + |\xi_n(t; x)|^{4m-2} \right)^4 \right)^{1/4}.$$

Therefore, from (1.2.9) and (1.3.7), and from the Lemma 1.3.1 and the estimate (1.3.10), thanks to the dominated convergence theorem we get

$$\lim_{\substack{\alpha \to 0 \\ \substack{h \in \mathbb{R}^d \\ h \neq 0}}} \sup |h|^{-2} \int_0^t e^{c_n(t-s)} I_1^h(s; n, \alpha) \, ds = 0.$$

Analogous arguments can be used for the other terms $I_i^h(t; n, \alpha)$, $i = 1, 2$, and by taking into accounts of (1.2.9), (1.3.4) and (1.3.7) and the estimate (1.3.10) the same conclusion holds. $\qquad\square$

1.3.2 Higher order derivatives

For any $n \in \mathbb{N}$ and $x, h_1, h_2 \in \mathbb{R}^d$, we consider the problem

$$d\zeta(t) = Db(\xi_n(t; x))\zeta(t)dt + D\sigma_n(\xi_n(t; x))\zeta(t) \, dw(t) + d\rho_n^{h_1 h_2}(t; x), \quad \zeta(0) = 0, \tag{1.3.12}$$

where $\rho_n^{h_1 h_2}(t; x)$ is the process defined by

$$\rho_n^{h_1 h_2}(t) = \int_0^t D^2 b(\xi_n(s; x))(\eta_n^{h_1}(s; x), \eta_n^{h_2}(s; x)) \, ds$$

$$+ \int_0^t D^2 \sigma_n(\xi_n(s; x))(\eta_n^{h_1}(s; x), \eta_n^{h_2}(s; x)) \, dw(s).$$

The equation (1.3.12) is the *second variation equation* corresponding to the equation (1.2.4).

Proposition 1.3.3. *Assume that the Hypotheses 1.1, 1.2 and 1.3 hold. Then the problem (1.3.12) has a unique solution $\zeta_n^{h_1 h_2}(\cdot; x) \in C([0, +\infty); L^2(\Omega; \mathbb{R}^d))$ such that for any $p \geq 1$*

$$\sup_{x \in \mathbb{R}^d} \mathbb{E} |\zeta_n^{h_1 h_2}(t; x)|^p \leq f_{2,p}(t) |h_1|^p |h_2|^p, \quad t \geq 0, \tag{1.3.13}$$

for an increasing continuous function $f_{2,p}(t)$ not depending on n. Moreover, for any $p \geq 1$ we have

$$\lim_{a \to 0} \sup_{\substack{h_1, h_2 \in \mathbb{R}^d \\ h_1, h_2 \neq 0}} |h_1|^{-p} |h_2|^{-p} \mathbb{E} |D^2 \xi_{n,a}(t; x)(h_1, h_2) - \zeta_n^{h_1 h_2}(t; x)|^p = 0. \qquad (1.3.14)$$

Proof. Fix $n \in \mathbb{N}$ and $T > 0$ and define for any $\zeta \in C([0, T]; L^2(\Omega; \mathbb{R}^d))$

$$\Gamma_n(\zeta)(t) = \int_0^t Db(\xi_n(s; x))\zeta(s)\, ds + \int_0^t D\sigma_n(\xi_n(s; x))\zeta(s)\, dw(s) + \int_0^t d\rho_n^{h_1 h_2}(s).$$

By using the boundedness of the derivatives of σ_n and (1.2.3), together with the estimate (1.3.10), we have that Γ_n is a contraction on $C([0, T_0]; L^2(\Omega; \mathbb{R}^d))$, for T_0 sufficiently small. Then, by proceeding as in the proof of the Proposition 1.3.2, we get the existence and the uniqueness of the solution $\zeta_n^{h_1 h_2}(\cdot; x) \in C([0, +\infty); L^2(\Omega; \mathbb{R}^d))$ for the second variation equation (1.3.12).

As in the proof of the Proposition 1.3.2, in order to obtain the estimate (1.3.13) we apply Itô's formula to the process $|\zeta_n^{h_1 h_2}(t; x)|^{2q}$. Thus, by taking the expectation and differentiating with respect to t we get

$$\frac{d}{dt} \mathbb{E} |\zeta_n^{h_1 h_2}(t)|^{2q} = 2q \, \mathbb{E} |\zeta_n^{h_1 h_2}(t)|^{2(q-1)} \left\langle \zeta_n^{h_1 h_2}(t), Db(\xi_n(t; x))\zeta_n^{h_1 h_2}(t) \right\rangle$$

$$+ 2q \, \mathbb{E} |\zeta_n^{h_1 h_2}(t)|^{2(q-1)} \left\langle \zeta_n^{h_1 h_2}(t), D^2 b(\xi_n(t; x))(\eta_n^{h_1}(t; x), \eta_n^{h_2}(t; x)) \right\rangle$$

$$+ 2q(q-1) \, \mathbb{E} |\zeta_n^{h_1 h_2}(t)|^{2(q-2)} \mathrm{Tr} \left[a_n^{h_1 h_2}(t; x) \left(\zeta_n^{h_1 h_2}(t) \otimes \zeta_n^{h_1 h_2}(t) \right) \right]$$

$$+ q(q-1) \, \mathbb{E} |\zeta_n^{h_1 h_2}(t)|^{2(q-1)} \mathrm{Tr} \left[a_n^{h_1 h_2}(t; x) \right],$$

with $a_n^{h_1 h_2}(t; x)$ defined by

$$a_n^{h_1 h_2}(t; x) = \left(D\sigma_n(\xi_n(t; x))\zeta_n^{h_1 h_2}(t; x) + D^2 \sigma_n(\xi_n(t; x))(\eta_n^{h_1}(t; x), \eta_n^{h_2}(t; x)) \right)$$

$$\left(D\sigma_n(\xi_n(t; x))\zeta_n^{h_1 h_2}(t; x) + D^2 \sigma_n(\xi_n(t; x))(\eta_n^{h_1}(t; x), \eta_n^{h_2}(t; x)) \right)^*.$$

From (1.2.2) we have

$$\mathrm{Tr} \left[a_n^{h_1 h_2}(t; x) \right]$$

$$\leq c \left\| D\sigma_n(\xi_n(t; x))\zeta_n^{h_1 h_2}(t; x) \right\|_2^2 + c \left(1 + |\xi_n(t; x)|^{2(k-2)} \right) \left(|\eta_n^{h_1}(t; x)| \, |\eta_n^{h_2}(t; x)| \right)^2.$$

Hence, by using (1.2.3) and the growth condition for $D^2 b$, it follows

$$\frac{d}{dt}\mathbb{E}|\zeta_n^{h_1 h_2}(t)|^{2q} \leq c\,\mathbb{E}|\zeta_n^{h_1 h_2}(t)|^{2q}$$

$$+c\,\mathbb{E}|\zeta_n^{h_1 h_2}(t)|^{2q-1}\left(1+|\xi_n(t;x)|^{2m-1}\right)|\eta_n^{h_1}(t;x)|\,|\eta_n^{h_2}(t;x)|$$

$$+c\,\mathbb{E}|\zeta_n^{h_1 h_2}(t)|^{2(q-1)}\left(1+|\xi_n(t;x)|^{2(k-2)}\right)|\eta_n^{h_1}(t;x)|^2\,|\eta_n^{h_2}(t;x)|^2,$$

so that we have

$$\frac{d}{dt}\mathbb{E}|\zeta_n^{h_1 h_2}(t)|^{2q} \leq c\,\mathbb{E}|\zeta_n^{h_1 h_2}(t)|^{2q}$$

$$+c\left(\mathbb{E}|\zeta_n^{h_1 h_2}(t)|^{2q}\right)^{1-\frac{1}{2q}}\left(\mathbb{E}\left(|\eta_n^{h_1}(t;x)|\,|\eta_n^{h_2}(t;x)|\right)^{2q}\left(1+|\xi_n(t;x)|^{2m-1}\right)^{2q}\right)^{\frac{1}{2q}}$$

$$+c\left(\mathbb{E}|\zeta_n^{h_1 h_2}(t;x)|^{2q}\right)^{1-\frac{1}{q}}\left(\mathbb{E}\left(|\eta_n^{h_1}(t;x)|\,|\eta_n^{h_2}(t;x)|\right)^{2q}\left(1+|\xi_n(t;x)|^{2(k-2)}\right)^{q}\right)^{\frac{1}{q}}.$$

Thanks to the Young inequality, this implies

$$\frac{d}{dt}\mathbb{E}|\zeta_n^{h_1 h_2}(t)|^{2q} \leq c\,\mathbb{E}|\zeta_n^{h_1 h_2}(t)|^{2q}+\mathbb{E}\left(|\eta_n^{h_1}(t;x)|\,|\eta_n^{h_2}(t;x)|\right)^{2q}\left(1+|\xi_n(t;x)|^{2m-1}\right)^{2q}$$

$$+\mathbb{E}\left(|\eta_n^{h_1}(t;x)|\,|\eta_n^{h_2}(t;x)|\right)^{2q}\left(1+|\xi_n(t;x)|^{2(k-2)}\right)^{q}.$$

Now, by using the estimates (1.2.17) and (1.3.10) and the Hölder inequality it is easy to check that for any $\alpha, \beta, \gamma \geq 1$ and $h_1, \ldots, h_k \in \mathbb{R}^d$

$$\sup_{x\in\mathbb{R}^d}\mathbb{E}\,(1+|\xi_n(t;x)|^{\alpha})^{\beta}\left(|\eta_n^{h_1}(t;x)|\cdots|\eta_n^{h_k}(t;x)|\right)^{\gamma} \leq c\,(t\wedge 1)^{-\frac{\alpha\beta}{2m}}\bar{f}_{\gamma}(t)\prod_{i=1}^{k}|h_i|^{\gamma},$$

where $\bar{f}_{\gamma} : \mathbb{R}^+ \to \mathbb{R}^+$ is an increasing continuous function depending only on α, β and γ. Therefore, since $k \leq m$, we get

$$\frac{d}{dt}\mathbb{E}|\zeta_n^{h_1 h_2}(t;x)|^{p} \leq c\,\mathbb{E}|\zeta_n^{h_1 h_2}(t;x)|^{p} + c\left(1+(t\wedge 1)^{\frac{1}{2m}-1}\bar{f}_p(t)\right)|h_1|^p|h_2|^p,$$

and then, by setting

$$f_{2,p}(t) = \int_0^t e^{c(t-s)}\left(1+(s\wedge 1)^{\frac{1}{2m}-1}\bar{f}_p(s)\right)ds, \quad t > 0,$$

we have (1.3.13).

Finally, we prove (1.3.14). As in the case of the first derivative equation, we prove it for $p = 2$. For any $\alpha > 0$ and $n \in \mathbb{N}$ we set $D^2\xi_{n,\alpha}(t;x)(h_1,h_2) = \zeta_{n,\alpha}^{h_1 h_2}(t;x)$ and we define

$$v_{n,\alpha}(t;x) = \zeta_{n,\alpha}^{h_1 h_2}(t;x) - \zeta_n^{h_1 h_2}(t;x).$$

By applying Itô's formula to $|v_{n,\alpha}(t)|^2$ and by taking the expectation and the derivative with respect to t we have

$$\frac{d}{dt}\mathbb{E}|v_{n,\alpha}(t)|^2 = 2\,\mathbb{E}\Big\langle Db_\alpha(\xi_{n,\alpha}(t))\zeta_{n,\alpha}^{h_1 h_2}(t) - Db(\xi_n(t))\zeta_n^{h_1 h_2}(t), v_{n,\alpha}(t)\Big\rangle$$

$$+2\,\mathbb{E}\Big\langle D^2 b_\alpha(\xi_{n,\alpha}(t))(\eta_{n,\alpha}^{h_1}(t), \eta_{n,\alpha}^{h_2}(t)) - D^2 b(\xi_n(t))(\eta_n^{h_1}(t), \eta_n^{h_2}(t)), v_{n,\alpha}(t)\Big\rangle$$

$$+\mathbb{E}\,\Big\|\Big(D\sigma_n(\xi_{n,\alpha}(t))\zeta_{n,\alpha}^{h_1 h_2}(t) - D\sigma_n(\xi_n(t))\zeta_n^{h_1 h_2}(t)\Big)$$

$$+ \Big(D^2 \sigma_n(\xi_{n,\alpha}(t))(\eta_{n,\alpha}^{h_1}(t), \eta_{n,\alpha}^{h_2}(t)) - D^2 \sigma_n(\xi_n(t))(\eta_n^{h_1}(t), \eta_n^{h_2}(t))\Big)\Big\|_2^2$$

$$= I_1(t;n,\alpha) + I_2(t;n,\alpha) + I_3(t;n,\alpha).$$

Now, we estimate the three terms in the right hand side. Concerning $I_1(t;n,\alpha)$ we have

$$I_1(t;n,\alpha) = 2\,\mathbb{E}\langle Db_\alpha(\xi_{n,\alpha}(t))v_{n,\alpha}(t), v_{n,\alpha}(t)\rangle$$

$$+2\,\mathbb{E}\Big\langle (Db_\alpha(\xi_{n,\alpha}(t)) - Db_\alpha(\xi_n(t)))\,\zeta_n^{h_1 h_2}(t), v_{n,\alpha}(t)\Big\rangle$$

$$+2\,\mathbb{E}\Big\langle (Db_\alpha(\xi_n(t)) - Db(\xi_n(t)))\zeta_n^{h_1 h_2}(t), v_{n,\alpha}(t)\Big\rangle.$$

By using (1.3.3) and the Young inequality, this yields

$$|I_1(t;n,\alpha)| \le c\,\mathbb{E}|v_{n,\alpha}(t)|^2 + c\,\mathbb{E}\,\Big|(Db_\alpha(\xi_{n,\alpha}(t)) - Db_\alpha(\xi_n(t)))\,\zeta_n^{h_1 h_2}(t)\Big|^2$$

$$+c\,\mathbb{E}\,\Big|(Db_\alpha(\xi_n(t)) - Db(\xi_n(t)))\,\zeta_n^{h_1 h_2}(t)\Big|^2$$

$$= c\,\mathbb{E}|v_{n,\alpha}(t)|^2 + J_1(t;n,\alpha) + J_2(t;n,\alpha).$$

For the third term we easily have

$$|I_3(t;n,\alpha)|$$

$$\le c_n\,\mathbb{E}|v_{n,\alpha}(t)|^2 + c\,\mathbb{E}\|(D\sigma_n(\xi_{n,\alpha}(t)) - D\sigma_n(\xi_n(t)))\,\zeta_n^{h_1 h_2}(t)\|_2^2$$

$$+c\,\mathbb{E}\|D^2\sigma_n(\xi_{n,\alpha}(t))(\eta_{n,\alpha}^{h_1}(t), \eta_{n,\alpha}^{h_2}(t)) - D^2\sigma_n(\xi_n(t))(\eta_n^{h_1}(t), \eta_n^{h_2}(t))\|_2^2$$

$$= c_n\,\mathbb{E}|v_{n,\alpha}(t)|^2 + J_3(t;n,a).$$

Therefore, collecting all terms, from the Gronwall lemma we obtain

$$\mathbb{E}|v_{n,\alpha}(t)|^2 \le \int_0^t e^{c_n(t-s)} \left(I_2(s;n,\alpha) + \sum_{i=1}^3 J_i(s;n,\alpha) \right) ds,$$

and we can conclude, as in the proof of the previous proposition, by showing that

$$\lim_{\alpha \to 0} \sup_{\substack{h_1,h_2 \in \mathbb{R}^d \\ h_1,h_2 \ne 0}} |h_1|^{-2}|h_2|^{-2} \int_0^t e^{c_n(t-s)} I_2(s;n,\alpha)\, ds = 0$$

and for any $i = 1,2,3$

$$\lim_{\alpha \to 0} \sup_{\substack{h_1,h_2 \in \mathbb{R}^d \\ h_1,h_2 \ne 0}} |h_1|^{-2}|h_2|^{-2} \int_0^t e^{c_n(t-s)} J_i(s;n,\alpha)\, ds = 0.$$

In fact, the limits above follow from the dominated convergence theorem, by using the estimates (1.2.2), (1.2.9), (1.3.3), (1.3.4), (1.3.7), (1.3.10), (1.3.13) and the limits (1.3.8) and (1.3.11). □

Finally, for any $n \in \mathbb{N}$ and $x, h_i \in \mathbb{R}^d$, $i = 1,2,3$, we consider the problem

$$\begin{cases} d\theta(t) = Db(\xi_n(t;x))\theta(t)\, dt + D\sigma_n(\xi_n(t;x))\theta(t)\, dw(t) \\[2mm] \qquad\quad + d\rho_{n,1}^{h_1h_2h_3}(t;x) + d\rho_{n,2}^{h_1h_2h_3}(t;x), \\[2mm] \theta(0) = 0, \end{cases} \qquad (1.3.15)$$

where the processes $\rho_{n,1}^{h_1h_2h_3}(t;x)$ and $\rho_{n,2}^{h_1h_2h_3}(t;x)$ are null at zero and are respectively defined by

$$d\rho_{n,1}^{h_1h_2h_3}(t;x) = D^3b(\xi_n(t;x))(\eta_n^{h_1}(t;x), \eta_n^{h_2}(t;x), \eta_n^{h_3}(t;x))\, dt$$

$$+ D^3\sigma_n(\xi_n(t;x))(\eta_n^{h_1}(t;x), \eta_n^{h_2}(t;x), \eta_n^{h_3}(t;x))\, dw(t)$$

and

$$d\rho_{n,2}^{h_1h_2h_3}(t;x) = \frac{1}{4} \sum_{\pi \in S_3} D^2b(\xi_n(t;x)) \left(\eta_n^{h_{\pi(1)}}(t;x), \zeta_n^{h_{\pi(2)}h_{\pi(3)}}(t;x) \right) dt$$

$$+ \frac{1}{4} \sum_{\pi \in S_3} D^2\sigma_n(\xi(t;x)) \left(\eta_n^{h_{\pi(1)}}(t,x), \zeta_n^{h_{\pi(2)}h_{\pi(3)}}(t;x) \right) dw(t)$$

(here S_3 denotes the set of all permutations of a set of three elements). As for the first two variation equations, we have the following result for the third variation equation.

Proposition 1.3.4. *Assume that the Hypotheses 1.1, 1.2 and 1.3 hold. Then, for all* $x, h_i \in \mathbb{R}^d$, $i = 1, 2, 3$, *the problem (1.3.15) admits a unique solution* $\theta_n^{h_1 h_2 h_3}(\cdot; x) \in C([0, +\infty); L^2(\Omega; \mathbb{R}^d))$ *such that for any* $p \geq 1$

$$\sup_{x \in \mathbb{R}^d} \mathbb{E} \left| \theta_n^{h_1 h_2 h_3}(t; x) \right|^p \leq f_{3,p}(t) \prod_{i=1}^{3} |h_i|^p, \quad t \geq 0,$$

where $f_{3,p}(t)$ *is an increasing continuous function, not depending on* n. *Moreover, for any* $p \geq 1$ *it holds*

$$\lim_{\alpha \to 0} \sup_{\substack{h_i \in \mathbb{R}^d \\ h_i \neq 0}} \mathbb{E} |D^3 \xi_{n,\alpha}(t; x)(h_1, h_2, h_3) - \theta_n^{h_1 h_2 h_3}(t; x)|^p \prod_{i=1}^{3} |h_i|^{-p} = 0.$$

Analogous results hold for higher order derivatives, up to the order h.

1.3.3   Conclusion

Thanks to the results proved in the previous subsections, we are now able to show the differentiability of the process $\xi(t; x)$ with respect to $x \in \mathbb{R}^d$. As a preliminary step we prove the mean-square differentiability of $\xi_n(t; x)$.

Proposition 1.3.5. *If the Hypotheses 1.1 and 1.2 are satisfied, then the solution* $\xi_n(t; x)$ *of the problem (1.2.4) is* h-*times mean-square differentiable with respect to* $x \in \mathbb{R}^d$. *Moreover, the derivatives of* $\xi_n(t; x)$ *coincide with the solutions of the variation equations corresponding to the equation (1.2.4). In particular, for any* $p \geq 1$ *and* $j = 1, \ldots, h$ *it holds*

$$\sup_{x \in \mathbb{R}^d} \mathbb{E} |D^j \xi_n(t; x)|^p \leq f_{j,p}(t), \tag{1.3.16}$$

for suitable increasing functions $f_{j,p}(t)$, *not depending on* n.

Proof. We prove the existence of the first mean-square derivative. The existence of higher order mean-square derivatives can be proved in an identical way.

Let us fix $\alpha > 0$ and $n \in \mathbb{N}$. Since $\xi_{n,\alpha}(t; x)$ is mean-square differentiable, for any $x, h \in \mathbb{R}^d$ and $t \geq 0$ we have

$$\xi_{n,\alpha}(t; x + h) - \xi_{n,\alpha}(t; x)$$

$$= D\xi_{n,\alpha}(t; x)h + \int_0^1 \left(D\xi_{n,\alpha}(t; x + \theta h) - D\xi_{n,\alpha}(t; x) \right) h \, d\theta.$$

Notice that the equality above has to be intended in $L^2(\Omega; \mathbb{R}^d)$. Therefore, if we take the limits as α goes to zero in $L^2(\Omega; \mathbb{R}^d)$, due to the Lemma 1.3.1 and the Proposition 1.3.2 we have

$$\xi_n(t; x + h) - \xi_n(t; x) = \eta_n^h(t; x) + \int_0^1 \left(\eta_n^h(t; x + \theta h) - \eta_n^h(t; x) \right) d\theta.$$

Now, since b and σ_n are continuous, together with their derivatives up to the third order, by using the same arguments used in the previous subsections it is possible to show that the solution $\xi_n(t;x)$ of the equation (1.2.4) and the solutions of the corresponding variation equations are mean-square continuous with respect to $x \in \mathbb{R}^d$. In particular, the process $\eta_n^h(t;x)$ is mean-square continuous with respect to x and from (1.3.10) and (1.3.11) it follows that

$$\lim_{|h| \to 0} \frac{1}{|h|^2} \mathbb{E} \left| \int_0^1 \left(\eta_n^h(t; x + \theta h) - \eta_n^h(t; x) \right) d\theta \right|^2 = 0.$$

Then $\xi_n(t;x)$ is mean-square differentiable and $D\xi_n(t;x)h = \eta_n^h(t;x)$. $\square$

Theorem 1.3.6. *Assume that the Hypotheses 1.1 and 1.2 hold. Then the solution $\xi(t;x)$ of the problem (1.2.1) is h times mean-square differentiable and for any $j = 1, \ldots, h$ it holds*

$$\sup_{x \in \mathbb{R}^d} \mathbb{E} |D^j \xi(t;x)|^p \leq f_{j,p}(t), \tag{1.3.17}$$

for suitable continuous increasing functions $f_{j,p}(t)$.

Proof. We recall that for any $n \in \mathbb{N}$ we have

$$\xi(t;x) = \xi_n(t \wedge \tau_n^x; x), \quad \mathbb{P} - \text{a.s.}$$

In particular, due to (1.2.6) for any $t \geq 0$ and $x \in \mathbb{R}^d$ we have

$$\lim_{n \to +\infty} \xi_n(t;x) = \xi(t;x), \quad \mathbb{P} - \text{a.s.}$$

The process $D\xi_n(t;x)h$ is the solution of the equation

$$d\eta(t) = Db(\xi_n(t;x))\eta(t)\,dt + D\sigma_n(\xi_n(t;x))\eta(t)\,dw(t), \quad \eta(0) = h$$

and then it is immediate to check that for any $t \leq \tau_n^x$

$$D\xi_n(t;x)h = D\xi_{n+1}(t;x)h, \quad \mathbb{P} - \text{a.s.}$$

This means that the process

$$\eta^h(t;x) = D\xi_n(t \wedge \tau_n^x; x)h$$

is well defined, does not depend on h and is the solution of the first variation equation corresponding to the problem (1.2.1). Moreover, for any $t \geq 0$ and $x \in \mathbb{R}^d$ it holds

$$\lim_{n \to +\infty} D\xi_n(t;x)h = \eta^h(t;x), \quad \mathbb{P} - \text{a.s.}$$

and then, since for any $p \geq 2$

$$\mathbb{E}|D\xi_n(t;x)h|^p \leq f_{1,p}(t)|h|^p,$$

we get

$$\lim_{n \to +\infty} \mathbb{E}|D\xi_n(t;x)h - \eta^h(t;x)|^2 = 0.$$

Exactly in the same way we can check that for any $k \leq h$ the k-th variation equation corresponding to (1.2.1) admits a unique solution, which we denote by $\eta^{h_1,\dots,h_k}(t;x)$. Furthermore it is possible to show that

$$\lim_{n \to +\infty} \mathbb{E}|D^k\xi_n(t;x)(h_1,\dots,h_k) - \eta^{h_1,\dots,h_k}(t;x)|^2 = 0.$$

Therefore, by proceeding as in the proof of the previous propositions, we can conclude that $\xi(t;x)$ is h-times differentiable and its derivatives coincide with the processes $\eta^{h_1,\dots,h_k}(t;x)$. Now, the estimates (1.3.17) follow from (1.3.16), by using Fatou's lemma. □

1.4   The transition semigroup

For any $\varphi \in B_b(\mathbb{R}^d)$ and $t \geq 0$ we set

$$P_t\varphi(x) = \mathbb{E}\varphi(\xi(t;x)) = \int_{\mathbb{R}^d} \varphi(y)\, P_t(x,dy), \quad x \in \mathbb{R}^d,$$

where $P_t(x,dy)$, $t \geq 0$ and $x \in \mathbb{R}^d$, is the transition probabilities family corresponding to the equation (1.2.1). Since the Chapman-Kolmogorov equation holds true for $P_t(x,dy)$, P_t defines a semigroup of contractions on $B_b(\mathbb{R}^d)$.

We remark that due to the Theorem 1.3.6 the semigroup P_t maps $C_b(\mathbb{R}^d)$ into itself, but in general it is not strongly continuous in $C_b(\mathbb{R}^d)$ (for some counterexamples see [17] and [44]). Nevertheless, P_t enjoys some regularity properties. Namely, as we are going to prove in the next proposition, P_t is a weakly continuous semigroup (for the definition and main properties see the appendix B).

Proposition 1.4.1. *Under the Hypothesis 1.1, the semigroup P_t is weakly continuous on $C_b(\mathbb{R}^d)$.*

Proof. We prove the property 1 in the Definition B.1.2, that is we show that for any $\varphi \in C_b(\mathbb{R}^d)$ and $T > 0$ the family of functions $\{\, P_t\varphi;\ t \in [0,T]\,\}$ is equi-uniformly continuous. Due to the estimate (1.3.17), with $j = 1$, for any $x,y \in \mathbb{R}^d$ we have

$$\mathbb{E}|\xi(t;x) - \xi(t;y)| = \mathbb{E}\left|\int_0^1 D\xi(t;\theta x + (1-\theta)y)\, d\theta\right| |x-y| \leq f_{1,1}(t)|x-y|,$$

so that, if $\varphi \in C_b^{0,1}(\mathbb{R}^d)$

$$|P_t\varphi(x) - P_t\varphi(y)| \leq \|\varphi\|_{0,1}\, \mathbb{E}|\xi(t;x) - \xi(t;y)| \leq \|\varphi\|_{0,1} f_{1,1}(t)|x-y|.$$

Now, if $\varphi \in C_b(\mathbb{R}^d)$, for any $\epsilon > 0$ there exists $\varphi_\epsilon \in C_b^{0,1}(\mathbb{R}^d)$ such that $\|\varphi - \varphi_\epsilon\|_0 < \epsilon$ and then, since P_t is a contraction, we easily get

$$|P_t\varphi(x) - P_t\varphi(y)| \le 2\epsilon + \|\varphi_\epsilon\|_{0,1} f_{1,1}(t)|x - y|.$$

This means that $P_t\varphi \in C_b(\mathbb{R}^d)$ and the family of functions $\{P_t\varphi : t \in [0,T]\}$ is equi-uniformly continuous, for any fixed $T > 0$.

Next, let us prove the property 2 For any $\varphi \in C_b(\mathbb{R}^d)$ and $x \in \mathbb{R}^d$ we have

$$P_t\varphi(x) - \varphi(x) = \mathbb{E}\left(\varphi\left(x + \int_0^t b(\xi(s;x))\, ds + \int_0^t \sigma(\xi(s;x))\, dw(s)\right) - \varphi(x)\right).$$

Then, if $\varphi \in C_b^{0,1}(\mathbb{R}^d)$, due the growth conditions on b and σ it follows

$$|P_t\varphi(x) - \varphi(x)| \le \|\varphi\|_{0,1} \left(\mathbb{E}\int_0^t |b(\xi(s;x))|\, ds + \left(\mathbb{E}\int_0^t \|\sigma(\xi(s;x))\|_2^2\, ds\right)^{1/2}\right)$$

$$\le c\|\varphi\|_{0,1} \left(\int_0^t \left(1 + \mathbb{E}|\xi(s;x)|^{2m+1}\right) ds + \left(\int_0^t \left(1 + \mathbb{E}|\xi(s;x)|^{2k}\right) ds\right)^{1/2}\right).$$

By using (1.2.13), this implies that for any $R > 0$

$$\sup_{|x| \le R} |P_t\varphi(x) - \varphi(x)| \le c_R \|\varphi\|_{0,1}\, t \left(1 + c_{2m+1}(t) + \sqrt{c_{2k}(t)}\right),$$

so that, as P_t is a contraction, for any $\varphi \in C_b^{0,1}(\mathbb{R}^d)$ we get

$$\lim_{t \to 0} \sup_{|x| \le R} |P_t\varphi(x) - \varphi(x)| = 0, \tag{1.4.1}$$

that is $P_t\varphi$ is $\mathcal{K}$-convergent to φ, as t goes to zero (see the Definition B.1.1 for the notion of $\mathcal{K}$-convergence). Finally, since $C_b^{0,1}(\mathbb{R}^d)$ is dense in $C_b(\mathbb{R}^d)$ and uniform convergence implies $\mathcal{K}$-convergence, by proceeding as above we conclude that (1.4.1) holds for any $\varphi \in C_b(\mathbb{R}^d)$.

Concerning the property 3. it is sufficient to show that for any $R, T > 0$ the family of probability measures

$$\{P_t(x, dy) ; \ t \in [0,T], \ |x| \le R\}$$

is *tight*. Actually, assume that for any $\epsilon > 0$ there exists $c_\epsilon > 0$ such that for any $t \in [0,T]$ and $|x| \le R$

$$P_t(x, B_{c_\epsilon}(0)) \ge 1 - \epsilon, \tag{1.4.2}$$

where $B_{c_\epsilon}(0)$ is the closed ball of centre 0 and radius c_ϵ. If we fix any sequence $\{\varphi_n\} \subset C_b(\mathbb{R}^d)$ which is $\mathcal{K}$-convergent to some $\varphi \in C_b(\mathbb{R}^d)$, there exists $n_\epsilon \in \mathbb{N}$ such that

$$\sup_{|y| \le c_\epsilon} |\varphi_n(y) - \varphi(y)| \le \epsilon, \qquad n \ge n_\epsilon$$

and $L = \sup_{n \in \mathbb{N}} \|\varphi_n\|_0 < \infty$. Thus, for any $n \geq n_\epsilon$ we get

$$\sup_{|x| \leq R} |P_t \varphi_n(x) - P_t \varphi(x)| \leq \sup_{|x| \leq R} \int_{|y| \leq c_\epsilon} |\varphi_n(y) - \varphi(y)| P_t(x, dy)$$

$$+ \sup_{|x| \leq R} \int_{|y| > c_\epsilon} |\varphi_n(y) - \varphi(y)| P_t(x, dy) \leq \epsilon + (L + \|\varphi\|_0) \epsilon$$

and this implies the property 3. Thereby it remains to prove (1.4.2). But this immediately follows from the estimate (1.2.13), as for any $c_\epsilon > 0$

$$P_t(x, B_{c_\epsilon}(0)) = \mathbb{P}(|\xi(t; x)| \leq c_\epsilon) \leq \frac{1}{c_\epsilon} \mathbb{E}|\xi(t; x)| \leq \frac{1}{c_\epsilon} c(T)(R + 1).$$

Actually, it is sufficient to take $c_\epsilon > c(T)(R + 1)$.

Finally, P_t is a contraction semigroup on $C_b(\mathbb{R}^d)$ and then the property 4. in the Definition B.1.2 is satisfied with $M = 1$ and $\omega = 0$.

$\square$

An important consequence of the previous proposition is

Corollary 1.4.2. *For any $\varphi \in C_b(\mathbb{R}^d)$ the function $u : [0, +\infty) \times \mathbb{R}^d \to \mathbb{R}$ defined by $u(t, x) = P_t \varphi(x)$ is continuous.*

1.5   The derivatives of the semigroup

In this section we want to show that under the hypothesis of non-degeneracy for σ the semigroup P_t has a regularizing effect. More precisely, we want to show that if the coefficients b and σ are of class C^h, $P_t \varphi$ is h-times differentiable with bounded derivatives, for any $\varphi \in B_b(\mathbb{R}^d)$ and $t > 0$.

Notice that in general if $\varphi \in C_b^k(\mathbb{R}^d)$ and the solution $\xi(t; x)$ is k-times mean-square differentiable, then $P_t \varphi$ is k-times differentiable. Moreover an explicit formula for the derivatives of $P_t \varphi$ can be given, in terms of the derivatives of φ and of $\xi(t; x)$.

Here we need something more. Actually, we want to give an expression for the derivative of $P_t \varphi$ in terms of φ only, and not of its derivatives. To this purpose, we recall the *Bismut-Elworthy formula* (see [7] and [57] for a proof). Such a formula, which applies when $\xi(t; x)$ is mean-square differentiable and the diffusion term is non-degenerate, for any $\varphi \in C_b(\mathbb{R}^d)$, $t > 0$ and $x, h \in \mathbb{R}^d$ reads

$$\langle D(P_t \varphi)(x), h \rangle = \frac{1}{t} \mathbb{E} \varphi(\xi(t; x)) \int_0^t \langle \sigma^{-1}(\xi(s; x)) D\xi(s; x)h, dw(s) \rangle.$$

Proposition 1.5.1. *Assume the Hypotheses 1.1, 1.2 and 1.3. Then, for any $\varphi \in B_b(\mathbb{R}^d)$ and $t > 0$ the function $P_t \varphi$ is differentiable and it holds*

$$\|D(P_t \varphi)\|_0 \leq c (t \wedge 1)^{-1/2} \|\varphi\|_0. \tag{1.5.1}$$

Proof. By using the Bismut-Elworthy formula, for any $\varphi \in C_b(\mathbb{R}^d)$ and $h \in \mathbb{R}^d$ we have

$$\langle D(P_t\varphi)(x), h \rangle = \frac{1}{t}\mathbb{E}\,\varphi(\xi(t;x)) \int_0^t \langle \sigma^{-1}(\xi(s;x))D\xi(s;x)h, dw(s) \rangle .$$

Therefore, from the Hypothesis 1.3 and from (1.3.17), with $j = 1$ and $p = 2$, it follows

$$|\langle D(P_t\varphi)(x), h \rangle| \leq \frac{\|\varphi\|_0}{t}\mathbb{E}\left| \int_0^t \langle \sigma^{-1}(\xi(s;x))D\xi(s;x)h, dw(s) \rangle \right|$$

$$\leq \frac{\|\varphi\|_0}{t}\left(\mathbb{E}\int_0^t |\sigma^{-1}(\xi(s;x))D\xi(s;x)h|^2\,ds \right)^{1/2} \leq c\,\|\varphi\|_0 t^{-1/2}\sqrt{f_{1,2}(t)}|h|,$$

so that for $t \leq 1$ we get

$$\|D(P_t\varphi)\|_0 \leq c\,\|\varphi\|_0\, t^{-1/2}.$$

Now, if $t > 1$ we have

$$\|D(P_t\varphi)\|_0 = \|D(P_1(P_{t-1}\varphi))\|_0 \leq c\,\|P_{t-1}\varphi\|_0 \leq c\,\|\varphi\|_0$$

and then (1.5.1) follows for $\varphi \in C_b(\mathbb{R}^d)$ and $t > 0$.

In particular, for arbitrary $\varphi \in C_b(\mathbb{R}^d)$ and $t > 0$ we have

$$|P_t\varphi(x) - P_t\varphi(y)| \leq c\,(t \wedge 1)^{-1/2}\|\varphi\|_0|x - y|, \quad x, y \in \mathbb{R}^d.$$

This means that

$$\mathrm{Var}\,(P_t(x, \cdot) - P_t(y, \cdot)) \leq c\,(t \wedge 1)^{-1/2}|x - y|, \quad x, y \in \mathbb{R}^d$$

and $P_t\varphi$ is Lipschitz-continuous, for any $\varphi \in B_b(\mathbb{R}^d)$. Due to the semigroup law we can conclude that $P_t\varphi \in C_b(\mathbb{R}^d)$ for any $\varphi \in B_b(\mathbb{R}^d)$ and (1.5.1) holds. $\square$

The Bismut-Elworthy formula for the second derivative of $P_t\varphi$, along the directions $h_1, h_2 \in \mathbb{R}^d$ at any point $x \in \mathbb{R}^d$, for every $\varphi \in C_b(\mathbb{R}^d)$ gives

$$D^2(P_t\varphi)(x)(h_1, h_2)$$

$$= \frac{2}{t}\mathbb{E}\,\langle D(P_{t/2}\varphi)(\xi(t/2;x)), D\xi(t/2;x)h_1 \rangle \int_0^{t/2} \langle \sigma^{-1}(\xi(s;x))D\xi(s;x)h_2, dw(s) \rangle$$

$$- \frac{2}{t}\mathbb{E}\int_0^{t/2} \langle D(P_{t-s}\varphi)(\xi(s;x)), [D\sigma(\xi(s;x))D\xi(s;x)h_1]\,\sigma^{-1}(\xi(s;x))D\xi(s;x)h_2 \rangle\,ds$$

$$+ \frac{2}{t}\mathbb{E}\int_0^{t/2} \langle D(P_{t-s}\varphi)(\xi(s;x)), D^2\xi(s;x)(h_1, h_2) \rangle\,ds.$$

$$(1.5.2)$$

By proceeding in the same way, it is possible to prove that the following formula holds for the third derivative (here, for the sake of brevity, we omit to write the dependence of the variable $x \in \mathbb{R}^d$ for $\xi(t;x)$ and its derivatives).

$$D^3(P_t\varphi)(x)(h_1, h_2, h_3) = \frac{2}{t}\,\mathbb{E}\left(D^2(P_{t/2}\varphi)(\xi(t/2))\,(D\xi(t/2)h_1, D\xi(t/2)h_2)\right.$$

$$+ \left\langle D(P_{t/2}\varphi)(\xi(t/2)), D^2\xi(t/2)(h_1, h_2)\right\rangle\right) \int_0^{t/2} \left\langle \sigma^{-1}(\xi(s))D\xi(s)h_3, dw(s)\right\rangle$$

$$+ I_1(t; h_1, h_2) + I_1(t; h_2, h_1) + I_2(t; h_1, h_2) + I_2(t; h_2, h_1)$$

$$+ \frac{2}{t}\,\mathbb{E}\int_0^{t/2} \left\langle D(P_{t-s}\varphi)(\xi(s)), D^3\xi(s)(h_1, h_2, h_3)\right\rangle\,ds$$

$$- \frac{2}{t}\,\mathbb{E}\int_0^{t/2} \left\langle D(P_{t-s}\varphi)(\xi(s)), I_3(s; h_1, h_2)\sigma^{-1}(\xi(s))D\xi(s)h_3\right\rangle\,ds,$$

$$\tag{1.5.3}$$

where for any $t > 0$ and $h, k \in \mathbb{R}^d$ we have

$$I_1(t; h, k) = \frac{2}{t}\,\mathbb{E}\int_0^{t/2} D^2(P_{t-s}\varphi)(\xi(s; x))(D^2\xi(s)(h, k), D\xi(s)h_3)\,ds$$

and

$$I_2(t; h, k)$$

$$= -\frac{2}{t}\,\mathbb{E}\int_0^{t/2} D^2(P_{t-s}\varphi)(\xi(s))\left(D\xi(s)h, [D\sigma(\xi(s))D\xi(s)k]\,\sigma^{-1}(\xi(s))D\xi(s)h_3\right)\,ds.$$

and, finally,

$$I_3(t; h, k) = D^2\sigma(\xi(s))\,(D\xi(s)h, D\xi(s)k) + D\sigma(\xi(s))D^2\xi(s)(h, k).$$

Theorem 1.5.2. *Assume the Hypotheses 1.1, 1.2 and 1.3 and let $\varphi \in B_b(\mathbb{R}^d)$. Then $P_t\varphi$ is h-times differentiable and for any $t > 0$ it holds*

$$\|D^j(P_t\varphi)\|_0 \le c\,(t \wedge 1)^{-j/2}\|\varphi\|_0, \qquad j = 1, \ldots, h. \tag{1.5.4}$$

Proof. We only prove twice differentiability of $P_t\varphi$, as the higher order differentiability follows from analogous arguments.

Due to the Bismut-Elworthy formula (1.5.2) for the second derivative, it is easy to check that $P_t\varphi$ is twice differentiable, for any $\varphi \in C_b(\mathbb{R}^d)$. Moreover, due to the previous proposition, if $\varphi \in B_b(\mathbb{R}^d)$ then $P_t\varphi$ is Lipschitz-continuous, for every

$t > 0$, so that $P_t\varphi = P_{t/2}(P_{t/2}\varphi)$ is twice differentiable. As far as the estimate (1.5.4) is concerned, we rewrite the formula (1.5.2) as

$$D^2(P_t\varphi)(x)(h_1, h_2) = \frac{2}{t}(J_1(t; x) + J_2(t; x) + J_3(t; x)).$$

For the fist term we have

$$|J_1(t; x)|^2 \leq \|D(P_t\varphi)\|_0^2 \, \mathbb{E}\,|D\xi(t/2; x)h_1|^2 \int_0^{t/2} \mathbb{E}\,|\sigma^{-1}(\xi(s; x))D\xi(s; x)h_2|^2 \, ds.$$

Then, due to (1.3.17) and (1.5.1) we have

$$|J_1(t; x)| \leq c(t \wedge 1)^{-1/2}\sqrt{f_{1,2}(t)} \, t^{1/2}\|\varphi\|_0|h_1||h_2| \leq c(t) \, \|\varphi\|_0|h_1||h_2|. \qquad (1.5.5)$$

For the second term, due to the Hypothesis 1.1 we have

$$|J_2(t; x)| \leq c \int_0^{t/2} \|D(P_{t-s}\varphi)\|_0 \, \mathbb{E}\left(|D\xi(s; x)h_1||D\xi(s; x)h_2|\left(1 + |\xi(s; x)|^{k-1}\right)\right) ds$$

and then, thanks to (1.3.17) and (1.5.1), by using the Hölder inequality we have

$$|J_2(t; x)| \leq c \int_0^{t/2} ((t - s) \wedge 1)^{-1/2}\left(1 + \mathbb{E}|\xi(s; x)|^{2(k-1)}\right)^{1/2} \sqrt{f_{1,4}(s)} \, ds\|\varphi\|_0|h_1||h_2|.$$

Thanks to the estimate (1.2.18) this yields

$$|J_2(t; x)| \leq c \int_0^{t/2} ((t - s) \wedge 1)^{-1/2}\sqrt{f_{1,4}(s)}\left(1 + s^{-\frac{k-1}{2m}}\right) ds\|\varphi\|_0|h_1||h_2|$$

$$= c(t)\|\varphi\|_0|h_1||h_2|.$$

$$(1.5.6)$$

Finally, for the third term we have

$$|J_3(t; x)| \leq \int_0^{t/2} \|D(P_{t-s}\varphi)\|_0 \, \mathbb{E}\,|D^2\xi(s; x)(h_1, h_2)| \, ds$$

and from (1.3.17) and (1.5.1) it follows

$$|J_3(t; x)| \leq \int_0^{t/2} ((t - s) \wedge 1)^{-1/2}f_{2,1}(s) \, ds\|\varphi\|_0|h_1||h_2| \leq c(t)\|\varphi\|_0|h_1||h_2|. \quad (1.5.7)$$

From (1.5.5), (1.5.6) and (1.5.7) the estimate (1.5.4) follows, by using as in the proof of the Proposition 1.5.1 the semigroup law. $\qquad\qquad\qquad\square$

1.6   Existence and uniqueness of solutions

1.6.1   The parabolic case

Let us consider the parabolic problem associated with the operator $\mathcal{L}_0$

$$\begin{cases} \dfrac{\partial u}{\partial t}(t,x) = \mathcal{L}_0(x,D)u(t,x), & t > 0, \ x \in \mathbb{R}^d, \\[2mm] u(0,x) = \varphi(x) & x \in \mathbb{R}^d. \end{cases} \qquad (1.6.1)$$

Our aim now is proving the existence and uniqueness of solutions.

Definition 1.6.1. *A function* $u : [0,+\infty) \times \mathbb{R}^d \to \mathbb{R}$ *is a classical solution of (1.6.1) if*

1. u is continuous on $[0,+\infty) \times \mathbb{R}^d$,

2. for any $t > 0$, $u(t,\cdot) \in C_b^2(\mathbb{R}^d)$,

3. for any $x \in \mathbb{R}^d$, $u(\cdot,x) \in C^1((0,+\infty))$,

4. u_t, $D_x u$ and $D_x^2 u$ are continuous from $(0,+\infty) \times \mathbb{R}^d$ into $\mathbb{R}$,

5. u verifies the problem (1.6.1).

Theorem 1.6.2. *Under the Hypotheses 1.1, 1.2 and 1.3 for any $\varphi \in C_b(\mathbb{R}^d)$ the function $u : [0,+\infty) \times \mathbb{R}^d \to \mathbb{R}$ defined by $u(t,x) = P_t\varphi(x)$ is the unique classical solution of the problem (1.6.1).*

Proof. Existence: In the section 1.5 we have seen that for any $\varphi \in C_b(\mathbb{R}^d)$ and $t > 0$ the function $P_t\varphi \in C_b^2(\mathbb{R}^d)$. Then we can apply Itô's formula to the regular function $P_t\varphi$, with $t > 0$, and for any $h > 0$ we get

$$P_{t+h}\varphi(x) = P_h(P_t\varphi)(x) = P_t\varphi(x)$$

$$+ \mathbb{E} \int_0^h \left(\frac{1}{2}\mathrm{Tr}\left[a(\xi(s;x))D^2(P_t\varphi)(\xi(s;x))\right] + \langle b(\xi(s;x)), D(P_t\varphi)(\xi(s;x)) \rangle \right) ds.$$

As the estimate (1.2.13) for the momenta of $\xi(t;x)$ holds and $\xi(t;x)$ has continuous trajectories, from the dominated convergence theorem we have

$$\frac{\partial^+ u}{\partial t}(t,x) = \lim_{h \to 0^+} \frac{u(t+h,x) - u(t,x)}{h} = \lim_{h \to 0^+} \frac{P_{t+h}\varphi(x) - P_t\varphi(x)}{h}$$

$$= \frac{1}{2}\mathrm{Tr}\left[a(x)D^2(P_t\varphi)(x)\right] + \langle b(x), D(P_t\varphi)(x) \rangle$$

$$= \frac{1}{2}\mathrm{Tr}\left[a(x)D^2u(t,x)\right] + \langle b(x), Du(t,x) \rangle .$$

It is not difficult to check that the right hand side in the above equality is continuous and then for every $x \in \mathbb{R}^n$ the function $t \mapsto \frac{\partial^+ u}{\partial t}(t, x)$ is continuous. Moreover, since u is continuous, we conclude that for any $x \in \mathbb{R}^d$ the function $u(\cdot, x)$ is continuously differentiable in $(0, +\infty)$ and u is a classical solution of the problem (1.6.1).

Uniqueness: Let v be a classical solution of the problem (1.6.1) and let $t > 0$ be fixed. By applying Itô's formula to the process $v(t - s, \xi(s; , x))$, $s \in [0, t - \epsilon]$, with $\epsilon > 0$, we get

$$v(\epsilon, \xi(t - \epsilon; x)) = v(t, x) - \int_0^{t-\epsilon} \frac{\partial v}{\partial t}(t - s, \xi(s; x)) \, ds$$

$$+ \int_0^{t-\epsilon} \left(\frac{1}{2} \mathrm{Tr} \left[a(\xi(s; x)) D^2 v(t - s, \xi(s; x)) \right] + \langle Dv(t - s, \xi(s; x)), b(\xi(s; x)) \rangle \right) \, ds$$

$$+ \int_0^{t-\epsilon} \langle Dv(t - s, \xi(s; x)), \sigma(\xi(s; x)) \, dw(s) \rangle .$$

By taking the expectation and recalling that v is a solution of (1.6.1), it follows that

$$\mathbb{E} v(\epsilon, \xi(t - \epsilon; x)) = v(t, x). \tag{1.6.2}$$

Therefore, as v is continuous and bounded on $[0, +\infty) \times \mathbb{R}^d$ and $\xi(t; x)$ has continuous trajectories, we can take the limit in the left hand side of (1.6.2) for ϵ going to zero and we have

$$\mathbb{E} \varphi(\xi(t; x)) = v(t, x).$$

$\square$

Remark 1.6.3. If the initial function φ belongs to $C_b^2(\mathbb{R}^d)$, by proceeding in the same way as before, we can show that $u(t, x) = P_t \varphi(x)$ is the unique strict solution of the problem (1.6.1), that is u is smooth up to time $t = 0$.

1.6.2 The elliptic case

By using techniques similar to those used in [44] and [96], we want to describe the domain of the infinitesimal generator of P_t in $C_b(\mathbb{R}^d)$. First of all, we prove a *maximum principle* for the operator $\mathcal{L}_0$.

Lemma 1.6.4. *Let us fix $\varphi \in \cap_{p \geq 1} W_{loc}^{2,p}(\mathbb{R}^d) \cap C_b(\mathbb{R}^d)$ and $\lambda > 0$. If we assume that $\lambda \varphi - \mathcal{L}_0 \varphi \in C_b(\mathbb{R}^d)$, then we have*

$$\|\varphi\|_0 \leq \frac{1}{\lambda} \|\lambda \varphi - \mathcal{L}_0 \varphi\|_0. \tag{1.6.3}$$

Proof. Let $\mathcal{A}$ be any second order differential operator

$$\mathcal{A}(x, D) = \sum_{i,j=1}^d \alpha_{ij}(x) D_{ij} + \sum_{i=1}^d \beta_i(x) D_i, \quad x \in \mathbb{R}^d,$$

having bounded uniformly continuous coefficients α_{ij} and β_i. If the function φ belongs to $\cap_{p \geq 1} W_{loc}^{2,p}(\mathbb{R}^d) \cap C_b(\mathbb{R}^d)$ and $\mathcal{A}\varphi$ is in $C_b(\mathbb{R}^d)$, then at any relative maximum (resp. minimum) point x_0 it holds $\mathcal{A}\varphi(x_0) \leq 0$ (resp. $\mathcal{A}\varphi(x_0) \geq 0$), for a proof see [91, Proposition 3.1.10]. As the proof depends on local properties of coefficients only, then it works also in the present case, where unbounded coefficients are considered.

Therefore, if we assume that there exists $x_0 \in \mathbb{R}^d$ such that $|\varphi(x_0)| = \|\varphi\|_0$, if x_0 is a maximum point, we have $\mathcal{L}_0\varphi(x_0) \leq 0$ and if x_0 is a minimum point, we have $\mathcal{L}_0\varphi(x) \leq 0$, so that the statement follows.

Thus, assume that $|\varphi(x)| \neq \|\varphi\|_0$, for any $x \in \mathbb{R}^d$ and $\|\varphi\|_0 = \sup_{x \in \mathbb{R}^d} \varphi(x)$ (the case $\|\varphi\|_0 = \inf_{x \in \mathbb{R}^d} \varphi(x)$ is analogous). For any $n \in \mathbb{N}$ we define

$$\varphi_n(x) = \varphi(x) - \frac{1}{n}\left(|x|^2 + k\right),$$

where k is a non-negative constant, independent of n, to be determined later on. We have $\varphi_n(x) \leq \varphi(x) \leq \|\varphi\|_0$, for any $x \in \mathbb{R}^d$. Moreover, φ_n attains its maximum at a certain point $y_n \in \mathbb{R}^d$. For each $\epsilon > 0$ there exists $R_\epsilon > 0$ such that

$$\sup_{|x| \leq R_\epsilon} \varphi(x) \geq \|\varphi\|_0 - \epsilon,$$

and in correspondence to such R_ϵ we can choose n_0 such that

$$\sup_{|x| \leq R_\epsilon} \frac{1}{n}\left(|x|^2 + k\right) \leq \epsilon, \qquad n \geq n_0.$$

This implies that for any $n \geq n_0$

$$\sup_{x \in \mathbb{R}^d} \varphi_n(x) \geq \sup_{|x| \leq R_\epsilon} \varphi_n(x) \geq \|\varphi\|_0 - 2\epsilon,$$

so that we have

$$\lim_{n \to +\infty} \varphi_n(y_n) = \lim_{n \to +\infty} \max_{x \in \mathbb{R}^d} \varphi_n(x) = \|\varphi\|_0. \tag{1.6.4}$$

Now, setting $f = \lambda\varphi - \mathcal{L}_0\varphi$, we have

$$\lambda\varphi_n(x) - \mathcal{L}_0\varphi_n(x) = f(x) - \frac{1}{n}\left[\lambda\left(|x|^2 + k\right) - \mathcal{L}_0\left(|x|^2 + k\right)\right]$$

$$= f(x) - \frac{1}{n}\left[\lambda\left(|x|^2 + k\right) - \|\sigma(x)\|^2 - 2\langle b(x), x\rangle\right].$$

Due to the Hypotheses 1.1 and 1.2 it is easy to verify that for a suitable non-negative constant c it holds $\|\sigma(x)\|^2 + 2\langle b(x), x\rangle \le c$ and then

$$\lambda\varphi_n - \mathcal{L}_0\varphi_n \le f - \frac{1}{n}(\lambda k - c).$$

Therefore if we choose $k = c/\lambda$ this implies

$$\varphi_n(y_n) \le \frac{1}{\lambda}\left(\mathcal{L}_0\varphi_n(y_n) + f(y_n)\right) \le \frac{1}{\lambda}\|f\|_0$$

and by taking the limit as $n \to +\infty$ (1.6.3) follows. $\qquad\square$

The previous lemma makes possible to give a characterization of the domain of the weak generator $\mathcal{L}$ of the semigroup P_t (for the definition and main properties see the section B.2). We recall that for any $\lambda > 0$

$$D(\mathcal{L}) = R(\lambda, \mathcal{L})(C_b(\mathbb{R}^d)),$$

where

$$R(\lambda, \mathcal{L})\varphi(x) = \int_0^{+\infty} e^{-\lambda t} P_t\varphi(x)\, dt.$$

Moreover, as we have already seen, $P_t\varphi \in C_b^2(\mathbb{R}^d)$, for any $\varphi \in C_b^2(\mathbb{R}^d)$ and $\|P_t\varphi\|_2 \le c\|\varphi\|_2$ for any $t \ge 0$, so that it is possible to show that

$$\varphi \in C_b^2(\mathbb{R}^d) \Rightarrow R(\lambda, \mathcal{L})\varphi \in C_b^2(\mathbb{R}^d), \quad \lambda > 0. \tag{1.6.5}$$

We will prove this fact in the step 1 of the proof of the Theorem 5.4.3 in a more difficult framework. Actually, there we will consider the infinite dimensional case and we will prove that if φ is only once differentiable, then $R(\lambda, \mathcal{L})\varphi$ is twice differentiable.

Proposition 1.6.5. *It holds*

$$\begin{cases} D(\mathcal{L}) = \left\{\varphi \in \cap_{p\ge 1} W_{loc}^{2,p}(\mathbb{R}^d) \cap C_b(\mathbb{R}^d) : \mathcal{L}_0\varphi \in C_b(\mathbb{R}^d)\right\} \\[2mm] \mathcal{L}\varphi = \mathcal{L}_0\varphi, \quad \varphi \in D(\mathcal{L}). \end{cases}$$

Proof. Let $\varphi = R(\lambda, \mathcal{L})f$, for some $\lambda > 0$ and $f \in C_b(\mathbb{R}^d)$, and let $\{f_k\}$ be a sequence in $C_b^2(\mathbb{R}^d)$ converging uniformly to f. If we set $\varphi_k = R(\lambda, \mathcal{L})f_k$, due to (1.6.5) $\varphi_k \in C_b^2(\mathbb{R}^d)$. Besides, as the operator $R(\lambda, \mathcal{L})$ is in $\mathcal{L}(C_b(\mathbb{R}^d))$, the sequence $\{\varphi_k\}$ converges to φ in $C_b(\mathbb{R}^d)$ and according to the Remark 1.6.3 for any $k \in \mathbb{N}$ we have

$$\lambda\varphi_k - \mathcal{L}_0\varphi_k = \lambda\varphi_k - \mathcal{L}\varphi_k = f_k.$$

Now, due to general local regularity results for elliptic equations, the sequence $\{\varphi_k\}$ converges to φ in $W^{2,p}(K)$, for any compact set $K \subset \mathbb{R}^d$ and for any $p \ge 1$, so that $\varphi \in W^{2,p}_{loc}(\mathbb{R}^d)$, for any $p \geq 1$. Finally, as $\lambda\varphi - \mathcal{L}\varphi = f$ and $\mathcal{L}_0\varphi_k$ converges uniformly to $\mathcal{L}\varphi$, it follows that $\mathcal{L}_0\varphi \in C_b(\mathbb{R}^d)$ and $\mathcal{L}_0\varphi = \mathcal{L}\varphi$.

Conversely, let $\varphi \in \cap_{p\geq 1}W^{2,p}_{loc}(\mathbb{R}^d) \cap C_b(\mathbb{R}^d)$ be such that $\mathcal{L}_0\varphi \in C_b(\mathbb{R}^d)$. By setting $f = \lambda\varphi - \mathcal{L}_0\varphi$ and $g = R(\lambda, \mathcal{L})f$, if we show that $g = \varphi$ then our statement is completely proved. Thanks to the first part of the proof it holds

$$\mathcal{L}_0 g = \mathcal{L}g = \mathcal{L}R(\lambda, \mathcal{L})f = \lambda g - f,$$

which implies that

$$\lambda(g - f) - \mathcal{L}_0(g - f) = 0.$$

Therefore, from the maximum principle proved in the Lemma 1.6.4 we get that $g = f$. □

According to the previous proposition, we can easily prove the following result of existence and uniqueness of solutions for the elliptic equation associated with the operator $\mathcal{L}_0$.

Theorem 1.6.6. *For any $f \in C_b(\mathbb{R}^d)$ and $\lambda > 0$ there exists a unique solution $\varphi \in \cap_{p\geq 1}W^{2,p}_{loc}(\mathbb{R}^d) \cap C_b(\mathbb{R}^d)$ to the problem $\lambda\varphi - \mathcal{L}_0(x, D)\varphi = f$ which is given by*

$$\varphi(x) = R(\lambda, \mathcal{L})f(x) = \int_0^{+\infty} e^{-\lambda t} P_t f(x)\, dt, \qquad x \in \mathbb{R}^d.$$

Proof. We set $\varphi = R(\lambda, \mathcal{L})f \in D(\mathcal{L})$. By the previous proposition we have that $\varphi \in \cap_{p\geq 1}W^{2,p}_{loc}(\mathbb{R}^d) \cap C_b(\mathbb{R}^d)$ and $\mathcal{L}_0\varphi = \mathcal{L}\varphi$, so that

$$\lambda\varphi - \mathcal{L}_0\varphi = \lambda\varphi - \mathcal{L}\varphi = \lambda R(\lambda, \mathcal{L})f - \mathcal{L}R(\lambda, \mathcal{L})f = f.$$

The uniqueness is a consequence of the Lemma 1.6.4. □

1.7 Schauder estimates

By following a procedure similar to the one introduced by Lunardi in [94], we can prove Hölder optimal regularity properties of the solutions of the parabolic and the elliptic problems associated with $\mathcal{L}_0$. To this purpose we quote a classical interpolation result, whose proof can be found for example in [117].

Proposition 1.7.1. *For $0 \leq \theta_1 < \theta_2$ and $0 < \sigma < 1$ it holds*

$$\left(C_b^{\theta_1}(\mathbb{R}^d), C_b^{\theta_2}(\mathbb{R}^d)\right)_{\sigma,\infty} = C_b^{\theta_1+\sigma(\theta_2-\theta_1)}(\mathbb{R}^d).$$

By using this proposition we can give an estimate of the norm of P_t as an operator acting on spaces of Hölder continuous functions.

Proposition 1.7.2. *Let $\alpha \in (0,3]$ and $\theta \in (0,1)$, with $\theta < \alpha$. Then for any $t > 0$ it holds*

$$\|P_t\|_{\mathcal{L}(C_b^\theta(\mathbb{R}^d), C_b^\alpha(\mathbb{R}^d))} \le c\,(t \wedge 1)^{-\frac{\alpha-\theta}{2}}.$$

Proof. Step 1: For any $0 < \sigma < 1$ and $t > 0$ we have

$$\|P_t\|_{\mathcal{L}(C_b^\sigma(\mathbb{R}^d), C_b^1(\mathbb{R}^d))} \le c\,(t \wedge 1)^{-\frac{1-\sigma}{2}}. \tag{1.7.1}$$

Indeed, if $\varphi \in C_b^1(\mathbb{R}^d)$ and $x, h \in \mathbb{R}^d$, we have

$$\langle D(P_t\varphi)(x), h \rangle = \mathbb{E}\langle D\varphi(\xi(t;x)), D\xi(t;x)h \rangle.$$

Then, due to (1.3.17) with $p = j = 1$, if $t \le 1$

$$\|D(P_t\varphi)\|_0 \le f_{1,1}(t)\,\|D\varphi\|_0 \le f_{1,1}(1)\,\|\varphi\|_1.$$

On the other hand, as shown in the Proposition 1.5.1 if $t > 1$

$$\|D(P_t\varphi)\|_0 \le c\,\|\varphi\|_0 \le c\,\|\varphi\|_1,$$

so that for any $t \ge 0$

$$\|P_t\|_{\mathcal{L}(C_b^1(\mathbb{R}^d), C_b^1(\mathbb{R}^d))} \le c. \tag{1.7.2}$$

Moreover, by using once more the Proposition 1.5.1, we get

$$\|P_t\|_{\mathcal{L}(C_b(\mathbb{R}^d), C_b^1(\mathbb{R}^d))} \le c\,(t \wedge 1)^{-1/2}$$

and by interpolation (1.7.1) follows for any $\sigma \in (0,1)$.

Step 2: Let $\theta \in (0,1)$ and $\alpha \in (\theta, 1]$. Then, for any $t > 0$ it holds

$$\|P_t\|_{\mathcal{L}(C_b^\theta(\mathbb{R}^d), C_b^\alpha(\mathbb{R}^d))} \le c\,(t \wedge 1)^{-\frac{\alpha-\theta}{2}}.$$

From the first step, with $\sigma = \frac{\theta}{\alpha}$, we have that P_t is bounded from $C_b^{\frac{\theta}{\alpha}}(\mathbb{R}^d)$ into $C_b^1(\mathbb{R}^d)$, for $t > 0$. Then by interpolation we have that

$$P_t : (C_b(\mathbb{R}^d), C_b^{\theta/\alpha}(\mathbb{R}^d))_{\alpha,\infty} = C_b^\theta(\mathbb{R}^d) \to (C_b(\mathbb{R}^d), C_b^1(\mathbb{R}^d))_{\alpha,\infty} = C_b^\alpha(\mathbb{R}^d)$$

is bounded, and due to the previous step

$$\|P_t\|_{\mathcal{L}(C_b^\theta(\mathbb{R}^d), C_b^\alpha(\mathbb{R}^d))}$$

$$\le \left(\|P_t\|_{\mathcal{L}(C_b(\mathbb{R}^d), C_b(\mathbb{R}^d))}\right)^{1-\alpha} \left(\|P_t\|_{\mathcal{L}(C_b^{\theta/\alpha}(\mathbb{R}^d), C_b^1(\mathbb{R}^d))}\right)^{\alpha} \le c\,(t \wedge 1)^{-\frac{\alpha-\theta}{2}}.$$

Step 3: Let $\sigma \in (0,1)$. Then for any $t > 0$ we have

$$\|P_t\|_{\mathcal{L}(C_b^\sigma(\mathbb{R}^d), C_b^3(\mathbb{R}^d))} \le c\,(t \wedge 1)^{-\frac{3-\sigma}{2}}.$$

Thanks to the formula (1.5.3) we have that

$$D^3(P_t\varphi)(x)(h_1, h_2, h_3) = \frac{2}{t} \sum_{k=1}^{6} J_k(t; x, h_1, h_2, h_3)$$

$$+\frac{2}{t} \mathbb{E} \left[D^2(P_{t/2}\varphi)(\xi(t/2)) \left(D\xi(t/2)h_1, D\xi(t/2)h_2 \right) \right.$$

$$+ \left. \langle D(P_{t/2}\varphi)(\xi(t/2)), D^2\xi(t/2)(h_1, h_2) \rangle \right] \int_0^{t/2} \langle \sigma^{-1}(\xi(s))D\xi(s)h_3, dw(s) \rangle$$

By using (1.2.18), (1.3.17) and (1.5.4), it is immediate to check that for any $k = 1, \ldots, 6$ it holds

$$\sup_{x \in \mathbb{R}^d} |J_k(t; x, h_1, h_2, h_3)| \leq c(t)|h_1|_H|h_2|_H|h_3|_H.$$

Then, if we fix $\varphi \in C_b^1(\mathbb{R}^d)$, by using again (1.2.18), (1.3.17) and (1.5.4), after some some computations we get for any $t > 0$

$$\|D^3(P_t\varphi)\|_0 \leq \frac{1}{t} c(t)\|\varphi\|_1.$$

In particular, if $t > 1$ we have

$$\|D^3(P_t\varphi)\|_0 = \|D^3(P_1(P_{t-1}\varphi))\|_0 \leq c(1) \|P_{t-1}\varphi\|_1 \leq c \|\varphi\|_1,$$

last inequality following from (1.7.2). This implies that for any $t > 0$

$$\|P_t\|_{\mathcal{L}(C_b^1(\mathbb{R}^d), C_b^3(\mathbb{R}^d))} \leq c (t \wedge 1)^{-1}.$$

Moreover, thanks to (1.5.4) it holds

$$\|P_t\|_{\mathcal{L}(C_b(\mathbb{R}^d), C_b^3(\mathbb{R}^d))} \leq c (t \wedge 1)^{-3/2},$$

so that by interpolation we conclude that for any $t > 0$

$$\|P_t\|_{\mathcal{L}(C_b^\sigma(\mathbb{R}^d), C_b^3(\mathbb{R}^d))} \leq c (t \wedge 1)^{-\frac{3-\sigma}{2}}.$$

Step 4: Let $\theta \in (0, 1)$ and $\alpha \in (1, 3]$. Then for any $t > 0$ we have

$$\|P_t\|_{\mathcal{L}(C_b^\theta(\mathbb{R}^d), C_b^\alpha(\mathbb{R}^d))} \leq c (t \wedge 1)^{-\frac{\alpha-\theta}{2}}.$$

Indeed, by the first step P_t is bounded from $C_b^\theta(\mathbb{R}^d)$ into $C_b^1(\mathbb{R}^d)$ and by the third step P_t is bounded from $C_b^\theta(\mathbb{R}^d)$ into $C_b^3(\mathbb{R}^d)$. Then the semigroup P_t is bounded from $C_b^\theta(\mathbb{R}^d)$ into $(C_b^1(\mathbb{R}^d), C_b^3(\mathbb{R}^d))_{\frac{\alpha-1}{2},\infty} = C_b^\alpha(\mathbb{R}^d)$ and it holds

$$\|P_t\|_{\mathcal{L}(C_b^\theta(\mathbb{R}^d), C_b^\alpha(\mathbb{R}^d))}$$

$$\leq \left(\|P_t\|_{\mathcal{L}(C_b^\theta(\mathbb{R}^d), C_b^1(\mathbb{R}^d))} \right)^{\frac{3-\alpha}{2}} \left(\|P_t\|_{\mathcal{L}(C_b^\theta(\mathbb{R}^d), C_b^3(\mathbb{R}^d))} \right)^{\frac{\alpha-1}{2}} \leq c (t \wedge 1)^{-\frac{\alpha-\theta}{2}}.$$

$\square$

Before proving the Schauder estimates for the elliptic and the parabolic problem, we recall a lemma, whose proof can be found in [95, Lemma 3.5]

Lemma 1.7.3. *Let $\theta \in (0,3)$ not integer and let $I \subset \mathbb{R}$ be an interval. Assume that $h : I \to C_b^\theta(\mathbb{R}^d)$ is such that for any $x \in \mathbb{R}^d$ the mapping $t \mapsto h(t)(x)$ is continuous and*

$$\|h(t)\|_{C_b^\theta(\mathbb{R}^d)} \le \lambda(t), \qquad t \in I,$$

for some function $\lambda \in L^1(I)$. Then the function

$$g(x) = \int_I h(t)(x)\, dt, \qquad x \in \mathbb{R}^d,$$

belongs to $C_b^\theta(\mathbb{R}^d)$ and $\|g\|_{C_b^\theta(\mathbb{R}^d)} \le \|\lambda\|_{L^1(I)}$.

Theorem 1.7.4. *Let $f \in C_b^\theta(\mathbb{R}^d)$, with $\theta \in (0,1)$ and let $\lambda > 0$. If φ is the solution of the elliptic equation*

$$\lambda\varphi(x) - \mathcal{L}_0(x,D)\varphi(x) = f(x), \qquad x \in \mathbb{R}^d,$$

then $\varphi \in C_b^{2+\theta}(\mathbb{R}^d)$ and

$$\|\varphi\|_{C_b^{2+\theta}(\mathbb{R}^d)} \le c\|f\|_{C_b^\theta(\mathbb{R}^d)}. \tag{1.7.3}$$

Proof. As proved in the Theorem 1.6.6, the unique solution of (1.7.3) is given by

$$\varphi(x) = R(\lambda, \mathcal{L})f(x) = \int_0^{+\infty} e^{-\lambda t} P_t f(x)\, dt, \quad x \in \mathbb{R}^d.$$

For any $t \in (0,1]$ we set

$$\varphi(x) = a_t(x) + b_t(x),$$

where for any $x \in \mathbb{R}^d$

$$\begin{cases} a_t(x) = \displaystyle\int_t^{+\infty} e^{-\lambda s} P_s f(x)\, ds \\[2mm] b_t(x) = \displaystyle\int_0^t e^{-\lambda s} P_s f(x)\, ds. \end{cases}$$

Then, by using the estimates proved in the Proposition 1.7.2 and the Lemma 1.7.3, it is immediate to check that $a_t \in C_b^{2+\alpha}(\mathbb{R}^d)$ and $b_t \in C_b^\alpha(\mathbb{R}^d)$, for any $\alpha \in (\theta,1)$ and $\lambda > 0$, and

$$\|a_t)\|_{C_b^{2+\alpha}(\mathbb{R}^d)} \le c\, t^{-(\alpha-\theta)/2}\|f\|_{C_b^\theta(\mathbb{R}^d)},$$

$$\|b_t\|_{C_b^\alpha(\mathbb{R}^d)} \le c\, t^{1-(\alpha-\theta)/2}\|f\|_{C_b^\theta(\mathbb{R}^d)}.$$

Due to the general interpolation results quoted in the introduction, this yields

$$\varphi \in \left(C_b^\alpha(\mathbb{R}^d), C_b^{2+\alpha}(\mathbb{R}^d)\right)_{1-\frac{\alpha-\theta}{2},\infty} = C_b^{2+\theta}(\mathbb{R}^d)$$

and the estimate (1.7.3) follows. $\qquad\square$

We conclude with the Schauder estimates in the parabolic case.

Theorem 1.7.5. *Let us fix $\alpha, \theta \in (0,1)$, with $\alpha < \theta/2$, and let $f : [0, +\infty) \times \mathbb{R}^d \to \mathbb{R}$ be a continuous function such that $f(t, \cdot) \in C_b^\theta(\mathbb{R}^d)$, for any $t \in (0, +\infty)$, and*

$$\sup_{t \in (0,T]} t^\alpha \|f(t, \cdot)\|_{C_b^\theta(\mathbb{R}^d)} < \infty,$$

for $T > 0$. Then, for any $\varphi \in C_b(\mathbb{R}^d)$ there exists a unique bounded classical solution $u : [0, +\infty) \times \mathbb{R}^d \to \mathbb{R}$ of the non-homogeneous problem

$$\begin{cases} \dfrac{\partial u}{\partial t}(t, x) = \mathcal{L}_0(x, D)u(t, x) + f(t, x), & t > 0 \quad x \in \mathbb{R}^d, \\[2mm] u(0, x) = \varphi(x), & x \in \mathbb{R}^d, \end{cases} \qquad (1.7.4)$$

which is given by the variation of constant formula

$$u(t, x) = P_t\varphi(x) + \int_0^t P_{t-s}f(s, \cdot)(x)\,ds.$$

The function u belongs to $C([0,T] \times \mathbb{R}^d) \cap C^{1,2}((0,T] \times \mathbb{R}^d)$, $u(t, \cdot) \in C_b^{2+\theta-2\alpha}(\mathbb{R}^d)$, for any $t > 0$, and

$$\|u(t, \cdot)\|_0 + \sup_{t \in [0,T]} t^{1+\theta/2}\|u(t, \cdot)\|_{C_b^{2+\theta-\alpha}(\mathbb{R}^d)} \leq c_T \left(\|\varphi\|_0 + \sup_{t \in (0,T]} t^\alpha \|f(t, \cdot)\|_{C_b^\theta(\mathbb{R}^d)} \right),$$

$$(1.7.5)$$

for a constant c_T independent both of f and φ.

Moreover, if $\varphi \in C_b^{2+\theta}(\mathbb{R}^d)$ and $\alpha = 0$, we have that $u(t, \cdot) \in C_b^{2+\theta}(\mathbb{R}^d)$, for any $t \geq 0$, and

$$\sup_{t \in [0,T]} \|u(t, \cdot)\|_{C_b^{2+\theta}(\mathbb{R}^d)} \leq c_T \left(\|\varphi\|_{C_b^{2+\theta}(\mathbb{R}^d)} + \sup_{t \in (0,T]} \|f(t, \cdot)\|_{C_b^\theta(\mathbb{R}^d)} \right).$$

Proof. The proof of this theorem is based on the Theorem 1.6.2 and on the Proposition 1.7.2 and can be found in [95]; we repeat it for the readers convenience.

In the Theorem 1.6.2 we have already seen that the function $v(t, x) = P_t\varphi(x)$ belongs to $C([0, T] \times \mathbb{R}^d) \cap C^{1,2}([0, T] \times \mathbb{R}^d)$ and

$$\frac{\partial v}{\partial t}(t, x) = \mathcal{L}_0 v(t, x), \qquad t > 0, \ x \in \mathbb{R}^d. \qquad (1.7.6)$$

Moreover, due to the Proposition 1.7.2

$$\|v(t, \cdot)\|_{C_b^{2+\theta}(\mathbb{R}^d)} \leq c\, t^{-(1+\theta/2)} \|\varphi\|_0$$

and if $\varphi \in C_b^\theta(\mathbb{R}^d)$ for any $R > 0$

$$\sup_{|x| \leq R} \left| \frac{\partial v}{\partial t}(t, x) \right| = \sup_{|x| \leq R} |\mathcal{L}_0 v(t, x)| \leq c_R t^{-1+\theta/2} \|\varphi\|_{C_b^\theta(\mathbb{R}^d)}. \qquad (1.7.7)$$

Concerning the other term

$$g(t, x) = \int_0^t P_{t-s} f(s, \cdot)(x) \, ds,$$

as the mapping

$$\{(t, s) \, ; \, 0 \leq s \leq t \leq T\} \times \mathbb{R}^d \to \mathbb{R}, \qquad (t, s, x) \mapsto P_{t-s} f(s, \cdot)(x),$$

is continuous and bounded, the function g is well defined and continuous in $[0, T] \times \mathbb{R}^d$. Now, for any $0 < r < 1$ and for any $t \in [0, T]$, we have that $g(t, \cdot) = a_t(r) + b_t(r)$, where

$$a_t(r)(x) = \begin{cases} \displaystyle\int_0^{t-r} P_{t-s} f(s, \cdot)(x) \, ds, & \text{if } r < t \\[4mm] 0 & \text{if } r \geq t, \end{cases} \qquad x \in \mathbb{R}^d,$$

$$b_t(r)(x) = \begin{cases} \displaystyle\int_{t-r}^t P_{t-s} f(s, \cdot)(x) \, ds, & \text{if } r < t \\[4mm] \displaystyle\int_0^t P_{t-s} f(s, \cdot)(x) \, ds & \text{if } r \geq t, \end{cases} \qquad x \in \mathbb{R}^d.$$

Once fixed $\beta \in (\theta, 1)$, due to the Proposition 1.7.2 we have

$$\|P_{t-s} f(s, \cdot)\|_{C_b^{2+\beta}(\mathbb{R}^d)} \leq c \, (t - s)^{-1-(\beta-\theta)/2} s^{-\alpha} \sup_{s \in [0,t]} s^\alpha \|f(s, \cdot)\|_{C_b^\theta(\mathbb{R}^d)}.$$

Then, by using the Lemma 1.7.3 we have that $a_t(r) \in C_b^{2+\beta}(\mathbb{R}^d)$ and

$$\|a_t(r)\|_{C_b^{2+\beta}(\mathbb{R}^d)} \leq c r^{-\alpha-(\beta-\theta)/2} \sup_{t \in [0,T]} t^\alpha \|f(t, \cdot)\|_{C_b^\theta(\mathbb{R}^d)}.$$

In a completely analogous way it is possible to show that $b_t(r) \in C_b^\beta(\mathbb{R}^d)$ and

$$\|b_t(r)\|_{C_b^\beta(\mathbb{R}^d)} \leq c r^{1-\alpha-(\beta-\theta)/2} \sup_{t \in [0,T]} t^\alpha \|f(t, \cdot)\|_{C_b^\theta(\mathbb{R}^d)}.$$

Therefore, by using the characterization of interpolation spaces given in the introduction and the Proposition 1.7.1, we have

$$g(t, \cdot) \in \left(C_b^\beta(\mathbb{R}^d), C_b^{2+\beta}(\mathbb{R}^d) \right)_{1-\alpha-\frac{\beta-\theta}{2}, \infty} = C_b^{2+\theta-2\alpha}(\mathbb{R}^d)$$

and

$$\|g(t, \cdot)\|_{C_b^{2+\theta-2\alpha}(\mathbb{R}^d)} \le c \sup_{t \in [0,T]} t^\alpha \|f(t, \cdot)\|_{C_b^\theta(\mathbb{R}^d)},$$

where c is independent of f. This implies that the mapping $t \mapsto g(t, \cdot)$ is bounded in $[\epsilon, T]$ with values in $C_b^{2+\theta-2\alpha}(\mathbb{R}^d)$, for any $\epsilon > 0$, and continuous from $[0, T]$ with values in $C(K)$, for any $K \subset \mathbb{R}^d$ compact. Hence, if $\alpha < \theta/2$, by interpolation we get that $g \in C([\epsilon, T]; C^2(K))$, for any compact set K, and then v, Dv and D^2v are continuous and bounded on $[\epsilon, T] \times \mathbb{R}^d$.

As seen in the Theorem 1.6.2, the mapping $t \mapsto P_{t-s}f(s, \cdot)(x)$ is continuously differentiable in $(s, T]$, for any $s \in [0, T]$ and $x \in \mathbb{R}^d$, and due to (1.7.6) and (1.7.7) its derivative belongs to $L^1(s, T]$. This implies that

$$\frac{\partial g}{\partial t}(t, x) = \int_0^t \mathcal{L}_0 P_{t-s}f(s, \cdot)(x)\, ds + f(t, x).$$

Thus as the derivative D^β is a closed operator in $C(K)$, for any compact $K \subset \mathbb{R}^d$, we conclude that

$$\frac{\partial g}{\partial t}(t, x) = \mathcal{L}_0 \int_0^t P_{t-s}f(s, \cdot)(x)\, ds + f(t, x),$$

and then the function $u(t, x) = P_t\varphi(x) + g(t, x)$ is a solution of the problem (1.7.4), has the required regularity and fulfills the estimate (1.7.5).

In order to prove the second part of the theorem we have only to check that P_t maps $C_b^{2+\theta}(\mathbb{R}^d)$ into itself and for any $\varphi \in C_b^{2+\theta}(\mathbb{R}^d)$

$$\|P_t\varphi\|_{C_b^{2+\theta}(\mathbb{R}^d)} \le c \|\varphi\|_{C_b^{2+\theta}(\mathbb{R}^d)}. \tag{1.7.8}$$

Indeed, by differentiating twice with respect to $x \in \mathbb{R}^d$ in $P_t\varphi(x) = \mathbb{E}\varphi(\xi(t; x))$, for any $x, h_1, h_2 \in \mathbb{R}^d$ we have

$$D^2(P_t\varphi)(x)(h_1, h_2) = \mathbb{E}D^2\varphi(\xi(t; x))(D\xi(t; x)h_1, D\xi(t; x)h_2)$$

$$+\mathbb{E}\left\langle D\varphi(\xi(t; x)), D^2\xi(t; x)(h_1, h_2)\right\rangle.$$

Hence, if we fix $x, y \in \mathbb{R}^d$, we get

$$D^2(P_t\varphi)(x)(h_1, h_2) - D^2(P_t\varphi)(y)(h_1, h_2) = I_1(t; x, y) + I_2(t; x, y),$$

where

$$I_1(t; x, y) = \mathbb{E}\left[D^2\varphi(\xi(t; x)) - D^2\varphi(\xi(t; y))\right](D\xi(t; x)h_1, D\xi(t; x)h_2)$$

$$+\mathbb{E}D^2\varphi(\xi(t; y))(D\xi(t; x)h_1, D\xi(t; x)h_2 - D\xi(t; y)h_2)$$

$$+\mathbb{E}D^2\varphi(\xi(t; y))(D\xi(t; x)h_1 - D\xi(t; y)h_1, D\xi(t; y)h_2)$$

and

$$I_2(t; x, y) = \mathbb{E} \left\langle D\varphi(\xi(t; x)) - D\varphi(\xi(t; y)), D^2\xi(t; x)(h_1, h_2) \right\rangle$$

$$+\mathbb{E} \left\langle D\varphi(\xi(t; y)), D^2\xi(t; x)(h_1, h_2) - D^2\xi(t; y)(h_1, h_2) \right\rangle.$$

We have

$$|I_1| \leq [D^2\varphi]_\theta \left(\mathbb{E}|\xi(t; x) - \xi(t; y)|^{2\theta} \right)^{1/2} \left(\mathbb{E}|D\xi(t; x)h_1|^2 |D\xi(t; x)h_2|^2 \right)^{1/2}$$

$$+2 [\varphi]_2 \sum_{\substack{i,j=1 \\ i \neq j}}^{2} \left(\mathbb{E}|D\xi(t; x)h_i - D\xi(t; y)h_i|^2 \right)^{1/2} \mathbb{E}|D\xi(t; x)h_j|^2.$$

Now, due to the Theorem 1.3.6, for any $x, y, h_1, \ldots, h_j \in \mathbb{R}^d$ and $j = 0, 1, 2$

$$(D^j\xi(t; x) - D^j\xi(t; y))(h_1, \ldots, h_j)$$

$$= \int_0^1 D^{j+1}\xi(t; \theta x + (1 - \theta)y)(h_1, \ldots, h_j, x - y) \, d\theta,$$

so that, thanks to (1.3.17) we have

$$\mathbb{E}|D^j\xi(t; x) - D^j\xi(t; y))|^p \leq f_{j+1,p}(t)|x - y|^p. \tag{1.7.9}$$

This implies that

$$|I_1(t; x, y)| \leq c(t) \|\varphi\|_{C_b^{2+\theta}(\mathbb{R}^d)} \left(|x - y|^\theta + |x - y| \right) |h_1||h_2|.$$

Similarly, we have

$$|I_2| \leq [\varphi]_2 \left(\mathbb{E}|\xi(t; x) - \xi(t; y)|^2 \right)^{1/2} \left(\mathbb{E}|D^2\xi(t; x)(h_1, h_2)|^2 \right)^{1/2}$$

$$+[\varphi]_1 \, \mathbb{E}|D^2\xi(t; x)(h_1, h_2) - D^2\xi(t; y)(h_1, h_2)|,$$

so that

$$|I_2(t; x, y)| \leq c \|\varphi\|_{C_b^{2+\theta}(\mathbb{R}^d)}|x - y||h_1||h_2|.$$

This yields that

$$|D^2(P_t\varphi)(x) - D^2(P_t\varphi)(y)| \leq c \|\varphi\|_{C_b^{2+\theta}(\mathbb{R}^d)}|x - y|^\theta,$$

and (1.7.8) follows. □

Chapter 2

Asymptotic behaviour of solutions

We are here concerned with the study of the asymptotic behaviour of the solution of the stochastic equation

$$d\xi(t) = b(\xi(t))\,dt + \sigma(\xi(t))\,dw(t), \qquad \xi(0) = x \in \mathbb{R}^d. \tag{2.0.1}$$

If the coefficients b and σ are Lipschitz-continuous, then the problem is well studied (for a bibliography see for example [48]). But here, as in the previous chapter, we are considering coefficients b and σ which are only locally Lipschitz. In fact, we assume the following conditions for b and σ.

Hypothesis 2.1. *1. The function $\sigma : \mathbb{R}^d \to \mathcal{L}(\mathbb{R}^d)$ is continuous and there exists $k \geq 0$ such that*

$$\sup_{x \in \mathbb{R}^d} \frac{\|\sigma(x)\|}{1 + |x|^k} < +\infty.$$

2. The function $b : \mathbb{R}^d \to \mathbb{R}^d$ is locally Lipschitz-continuous and there exists $m \geq k$ such that

$$\sup_{x \in \mathbb{R}^d} \frac{|b(x)|}{1 + |x|^{2m+1}} < +\infty.$$

3. For any $p \geq 1$ there exists $c_p \in \mathbb{R}$ such that for each $x, y \in \mathbb{R}^d$

$$\langle b(x) - b(y), x - y \rangle + p \|\sigma(x) - \sigma(y)\|_2^2 \leq c_p |x - y|^2.$$

4. There exist $a, \gamma > 0$ and $c \in \mathbb{R}$ such that for any $x, h \in \mathbb{R}^d$

$$\langle b(x + h) - b(x), h \rangle \leq -a|h|^{2m+2} + c\left(|x|^\gamma + 1\right).$$

Thus, as in the previous chapter we are dealing with coefficients b and σ which have polynomial growth and enjoy suitable dissipativity conditions. Unlike in the previous chapter, here we do not assume the coefficients to be differentiable.

Our aim is to show that there exists a unique invariant measure for the equation (2.0.1) and to prove some further regularity properties of such invariant measure, as the *strongly mixing* property, the *exponential convergence* to equilibrium, the absolute continuity with respect to the Lebesgue measure.

In the section 2.2 we prove the existence of an invariant measure μ. Since we are working in finite dimension, it simply follows by proving an estimate for the second moment of the solution $\xi(t; x)$.

In the section 2.3 we prove the uniqueness and other ergodic properties of the invariant measure μ. The main tools are given by the Khas'minskii and the Doob theorems. This is why we have to prove that the semigroup P_t is *strongly Feller* and *irreducible*. Notice that in the previous chapter we have proved that if the coefficients b and σ are of class C^h, then the semigroup P_t maps Borel functions into C^h functions. Here we prove that P_t is strongly Feller, without assuming b and σ to be differentiable. In order to prove the irreducibility of P_t, as in [42] we consider the restrictions of the semigroup to bounded sets and then we use general results of irreducibility for elliptic operators with continuous and bounded coefficients. We give also an alternative proof of irreducibility, which works for bounded diffusion terms and which is based on the *approximate controllability* of the deterministic controlled system associated with the stochastic equation (2.0.1).

In the section 2.4 we show that the law of the process $\xi(t; x)$ is absolutely continuous with respect to the Lebesgue measure, for any $t > 0$ and $x \in \mathbb{R}^d$. In particular this implies that the invariant measure μ is absolutely continuous with respect to the Lebesgue measure.

Finally, in the section 2.5, by assuming a stronger dissipativity condition for b, we estimate the rate of convergence of $P_t\varphi$ to equilibrium. Namely we show that

$$\|P_t\varphi - \langle \varphi, \mu \rangle\|_0 \leq c\, (t \vee 1)^{-\frac{1}{\rho}} \|\varphi\|_0,$$

for some $\rho > 0$. Moreover, we show that in some cases the rate of convergence to equilibrium is of exponential type

$$\|P_t\varphi - \langle \varphi, \mu \rangle\|_0 \leq c\, e^{-ct} \|\varphi\|_0,$$

for some $c > 0$.

2.1 Notations and preliminary results

We introduce here some notations and some classical results on the existence and uniqueness of invariant measures corresponding to stochastic problems with values in general separable Hilbert spaces H.

Let H be a separable Hilbert space. We consider the following stochastic problem

$$d\xi(t) = f(\xi(t))\, dt + g(\xi(t))\, dw(t), \quad \xi(0) = x \in H, \tag{2.1.1}$$

where $f : H \to H$ and $g : H \to \mathcal{L}(H)$ are functions satisfying sufficient assumptions ensuring the existence and uniqueness of the solution $\xi(t; x)$ and the continuity of the trajectories.

For any $\varphi \in B_b(H)$, the *Markov transition semigroup* P_t corresponding to the problem (2.1.1) is defined for $\varphi \in B_b(H)$ and $t \geq 0$ by

$$P_t\varphi(x) = \mathbb{E}\varphi(\xi(t; x)), \quad x \in H.$$

It is easy to show that $P_t\varphi \in B_b(H)$.

We say that a semigroup P_t acting on $B_b(H)$ enjoys the *Feller* property if

$$\varphi \in C_b(H) \Rightarrow P_t\varphi \in C_b(H), \quad t \geq 0.$$

Instead, we say that P_t enjoys the *strong Feller* property (or that is strongly Feller) if

$$\varphi \in B_b(H) \Rightarrow P_t\varphi \in C_b(H), \quad t > 0.$$

Moreover, we say that P_t is *irreducible* at time t_0 if for every non empty open set $A \subseteq H$ and $x \in H$ we have that $P_{t_0}(x, A) > 0$, or equivalently

$$\mathbb{P}(|\xi(t_0; x) - z| \leq \epsilon) > 0 \tag{2.1.2}$$

for any $z \in H$ and $\epsilon > 0$. Clearly, the semigroup P_t is said to be *irreducible* if it is irreducible at any time.

In what follows we shall denote by $M_1(H)$ the space of all probability measures defined on $(H, \mathcal{B}(H))$, where $\mathcal{B}(H)$ is the σ-algebra of Borelian sets. A measure $\mu \in M_1(H)$ is *invariant* for the semigroup P_t if for any $t \geq 0$ it holds

$$P_t^\star \mu = \mu$$

(here $P_t^\star$ denotes the transpose of P_t), or equivalently

$$\int_H P_t\varphi(y)\, \mu(dy) = \int_H \varphi(y)\, \mu(dy).$$

In connection with the equation (2.1.1) we can introduce the *Markov transition probabilities* defined for $x \in H$ and $t \geq 0$ by

$$P_t(x, \Gamma) = P_t\chi_\Gamma(x), \quad \Gamma \in \mathcal{B}(H).$$

By definition $P_t(x, \cdot)$ is the law of $\xi(t; x)$, so that for any $\varphi \in B_b(H)$ we have

$$P_t\varphi(x) = \int_H \varphi(y)\, P_t(x, dy), \quad t \geq 0 \ x \in H.$$

Thus, for each $t \geq 0$, $x \in H$ and $\Gamma \in \mathcal{B}(H)$ we have

1. $P_t(x, \cdot)$ is probability measure on $(H, \mathcal{B}(H))$,

2. $P_t(\cdot, \Gamma)$ is a $\mathcal{B}(H)$-measurable function,

3. $P_{t+s}(x, \Gamma) = \int_H P_t(y, \Gamma) P_s(x, dy)$ (*Chapman-Kolmogorov law*),

4. $P_0(x, \Gamma) = \chi_\Gamma(x)$.

We also recall that an invariant measure μ for the semigroup P_t is *ergodic* if and only if one of the following equivalent conditions holds.

1. If $\varphi \in L^2(H; \mu)$ and $P_t \varphi = \varphi$, μ-almost surely for any $t > 0$, then φ is constant μ-almost surely.

2. If $\Gamma \in \mathcal{B}(H)$ and $P_t \chi_\Gamma = \chi_\Gamma$, μ-almost surely for any $t > 0$, then either $\mu(\Gamma) = 1$ or $\mu(\Gamma) = 0$.

3. For any $\varphi \in L^2(H; \mu)$ it holds

$$\lim_{T \to +\infty} \frac{1}{T} \int_0^T P_t \varphi \, dt = \int_H \varphi \, d\mu, \quad \text{in } L^2(H; \mu).$$

It is possible to show (see for example [48]) that if $\mu \in M_1(H)$ is the unique invariant measure for the semigroup P_t, then μ is ergodic.

An invariant measure μ for the semigroup P_t is said to be *strongly mixing* if for arbitrary $\varphi \in L^2(H; \mu)$ it holds

$$\lim_{t \to +\infty} P_t \varphi = \int_H \varphi \, d\mu, \quad \text{weakly in } L^2(H; \mu).$$

Clearly if for any $x \in H$

$$P_t(x, \cdot) \rightharpoonup \mu, \quad \text{weakly},$$

then μ is strongly mixing. Obviously the strong mixing property implies the ergodicity of the system. Finally, if for any $\varphi \in C_b(H)$ and $t \geq 0$

$$\left| P_t \varphi(x) - \int_H \varphi \, d\mu \right| \leq c(x) e^{-\alpha t}, \quad x \in H,$$

for some $\alpha > 0$ and some function $c(x)$, we say that μ is *exponentially mixing*. It will be important to verify when the exponential convergence of $P_t \varphi$ to equilibrium is uniform with respect to $x \in H$.

Once introduced the main notations, we describe some fundamental results which enable us to prove existence and uniqueness of invariant measures for the equation (2.1.1). The proofs of all quoted results can be found for instance in the book by Da Prato and Zabczyk [48].

We recall that a set of probability measures $\Lambda \subset M_1(H)$ is said to be *tight* if for any $\epsilon > 0$ there exists a compact set $K_\epsilon \subset H$ such that for any $\mu \in \Lambda$

$$\mu(K_\epsilon) \geq 1 - \epsilon.$$

The *Prohorov theorem* gives a characterization of tight subsets $\Lambda \subset M_1(H)$. Actually it states that $\Lambda \subset M_1(H)$ is tight if and only if from any sequence $\{\mu_n\} \subset M_1(H)$ it is possible to extract a subsequence $\{\mu_{n_k}\}$ which is *weakly convergent* to some $\mu \in M_1(H)$, that is

$$\lim_{k \to +\infty} \int_H \varphi(x) \mu_{n_k}(dx) = \int_H \varphi(x) \mu(dx),$$

for any $\varphi \in C_b(H)$.

For any $\nu \in M_1(H)$ and $t > 0$ we define

$$R_t^\star \nu = \frac{1}{t} \int_0^t P_s^\star \nu \, ds.$$

This means that for any $\varphi \in B_b(H)$

$$\int_H \varphi(y) \, R_t^\star \nu(dy) = \frac{1}{t} \int_0^t \int_H P_s \varphi(y) \nu(dy) \, ds.$$

The following theorem describes a method of constructing an invariant measure for the system (2.1.1), by using the measures $R_t^\star \nu$.

Theorem 2.1.1 (Krylov-Bogoliubov). *Assume that P_t is a Feller semigroup. If for some $\nu \in M_1(H)$ and some sequence $\{t_n\}$ increasing to infinity we have that $R_{t_n}^\star \nu$ converges weakly to some $\mu \in M_1(H)$, as n goes to infinity, then μ is an invariant measure for P_t.*

Due to the previous theorem and the Prohorov theorem, the notion of tightness plays an important rôle in order to prove the existence of an invariant measure. Actually, the following result holds true.

Proposition 2.1.2. *If for some $\nu \in M_1(H)$ and $a \geq 0$ the family of probability measures $\{R_t^\star \nu \, ; \, t \geq a\}$ is tight, then there exists an invariant measure for P_t.*

A further problem is to state the uniqueness of the invariant measure. The *Doob theorem* provides an important tool.

Theorem 2.1.3 (Doob). *Let P_t, $t \geq 0$ be the Markov semigroup associated with the stochastic system (2.1.1) and let μ be an invariant measure for P_t. Assume that there exists $t_0 > 0$ such that all probability measures $P_t(x, \cdot)$, $x \in H$ and $t > t_0$, are mutually equivalent. Then the following statements hold*

1. μ is the unique invariant measure for the semigroup P_t and in particular is ergodic;

2. μ is equivalent to all the probability measures $P_t(x, \cdot)$, for any $x \in H$ and $t \geq t_0$;

3. μ is strongly mixing.

We recall that Seidler in [107] and Stettner in [110] have proved that under the hypothesis of the Doob theorem something more than the strongly mixing property holds. Namely they have shown that

$$\lim_{\substack{t \to +\infty}} \sup_{\substack{\varphi \in C_b(H) \\ \|\varphi\|_0 \leq 1}} \left| P_t \varphi(x) - \int_H \varphi(y) \, \mu(dy) \right| = 0, \qquad x \in H,$$

that is $P_t(x, \cdot)$ converges to μ, as t goes to infinity, in the *total variation norm*.

The next theorem is due to Khas'minskii and describes a situation in which the assumptions of the Doob theorem 2.1.3 are satisfied.

Theorem 2.1.4 (Khas'minskii). *If the semigroup P_t is strongly Feller and irreducible, then all probability measures $P_t(x, \cdot)$, with $x \in H$ and $t > 0$, are equivalent.*

2.2 Existence

Our first step is proving the existence of an invariant measures for the stochastic problem

$$d\xi(t) = b(\xi(t)) \, dt + \sigma(\xi(t)) \, dw(t), \qquad \xi(0) = x \in \mathbb{R}^d. \tag{2.2.1}$$

For $r \geq 0$ and $x \in \mathbb{R}^d$ we define the stopping time τ_r^x by setting

$$\tau_r^x = \inf \left\{ t \geq 0 \, : \, |\xi(t;x)| \geq r \right\}.$$

By proceeding as in the proof of the Lemma 1.2.3, due to the Hypothesis 2.1-4 we have

$$\mathbb{E}|\xi(t \wedge \tau_r^x)|^2 \leq |x|^2 + c\,\mathbb{E}\int_0^t \chi_{\tau_r^x \geq s} \left(-a\,|\xi(s)|^{2+2m} + c\left(|\xi(s)|^{2k} + 1 \right) \right) \, ds$$

$$= |x|^2 + c\,\mathbb{E}\int_0^t \left(-a\,|\xi(s \wedge \tau_r^x)|^{2+2m} - a\,r^{2+2m}\chi_{\tau_r^x < s} + c\left(|\xi(s \wedge \tau_r^x)|^{2k} + 1 \right) \chi_{\tau_r^x \geq s} \right) \, ds.$$

Thanks to the Young inequality and the Hölder inequality we obtain

$$\mathbb{E}|\xi(t \wedge \tau_r^x)|^2 \leq |x|^2 + c\int_0^t \left(1 - a\,\mathbb{E}|\xi(s \wedge \tau_r^x)|^{2+2m} - a\,r^{2+2m}\mathbb{P}(\tau_r^x < s) \right) \, ds$$

$$\leq |x|^2 + c\int_0^t \left(1 - a\left(\mathbb{E}|\xi(s \wedge \tau_r^x)|^2 \right)^{1+m} - a\,r^{2+2m}\mathbb{P}(\tau_r^x < s) \right) \, ds.$$

By comparison, this easily implies that

$$\int_0^t \mathbb{E}\,|\xi(s \wedge \tau_r^x)|^2\,ds \le \frac{1}{a}|x|^2 + \int_0^t e^{-a(t-s)}\left(s - a\,r^{2+2m}\int_0^s \mathbb{P}(\tau_r^x < \rho)\,d\rho\right)ds$$

Now, thanks to the Chebyschev inequality and to (1.2.13), we have

$$r^{2+2m}\int_0^s \mathbb{P}(\tau_r^x < \rho)\,d\rho \le r^{2+2m}\int_0^s \mathbb{P}\left(\sup_{\sigma \in [0,\rho]}|\xi(\sigma)|^{2m+3} \ge r^{2m+3}\right)d\rho$$

$$\le \frac{1}{r}\int_0^s \mathbb{E}\,\sup_{\sigma \in [0,\rho]}|\xi(\sigma)|^{2m+3}\,d\rho \le \frac{1}{r}\int_0^s c_{2m+3}(\rho)\,d\rho\left(|x|^{2m+3} + 1\right).$$

Therefore, by taking the limit as r goes to infinity, due to the Fatou lemma we obtain that

$$\sup_{t \ge 0}\mathbb{E}\,|\xi(t;x)|^2 \le c\left(|x|^2 + 1\right). \tag{2.2.2}$$

This means that for any fixed $x_0 \in \mathbb{R}^d$ the family of measures $\{P_t(x_0, \cdot)\}_{t \ge 0}$ is tight. According to the Proposition 2.1.2 we can conclude that the following existence result holds true.

Theorem 2.2.1. *Assume that the Hypothesis 2.1 holds. Then there exists at least one invariant measure μ for the equation (2.2.1).*

2.3 Uniqueness

In the Theorem 2.2.1 we have shown that if b and σ satisfy the Hypothesis 2.1 then the Markov semigroup P_t associated with the equation (2.2.1) admits an invariant measure μ. The aim of this section is to show that such an invariant measure is unique. To this purpose we need to require that σ does not degenerate.

Hypothesis 2.2. *The matrix $\sigma(x)$ is invertible for any $x \in \mathbb{R}^d$ and*

$$\sup_{x \in \mathbb{R}^d}\|\sigma^{-1}(x)\| < \infty.$$

Our strategy is to prove that the semigroup P_t is strongly Feller and irreducible. In this way, by applying the Khas'minskii theorem 2.1.4, we have that the probability measures $P_t(x, \cdot)$, $t > 0$ and $x \in \mathbb{R}^d$, are all mutually equivalent, and thanks to the Doob theorem we obtain the following theorem.

Theorem 2.3.1. *Under the Hypotheses 2.1 and 2.2*

1. *the system (2.2.1) admits a unique invariant measure μ which is equivalent to all probability measures $P_t(x, \cdot)$, $t > 0$ and $x \in \mathbb{R}^d$;*

2. μ is strongly mixing and for any $x \in \mathbb{R}^d$ and $\varphi \in B_b(\mathbb{R}^d)$

$$\lim_{t \to +\infty} \frac{1}{t} \int_0^t \varphi(\xi(s;x)) \, ds = \int_{\mathbb{R}^d} \varphi(y) \, \mu(dy), \quad \mathbb{P} - a.s.$$

3. For any $\nu \in M_1(\mathbb{R}^d)$ and $B \in \mathcal{B}(\mathbb{R}^d)$ it holds

$$\lim_{t \to +\infty} P_t^\star \nu(B) = \nu(B).$$

2.3.1 Strong Feller property

In the previous chapter we have proved that if the coefficients b and σ are smooth, then the semigroup P_t has a regularizing effect. In this section we want to show that, even if b and σ are only continuous, the semigroup P_t is *strongly Feller*, as well.

Proposition 2.3.2. *Assume the Hypotheses 2.1 and 2.2. Then the semigroup P_t is strongly Feller.*

Proof. We approximate b and σ by two sequences $\{b_n\} \subset C^\infty(\mathbb{R}^d; \mathbb{R}^d)$ and $\{\sigma_n\} \subset C^\infty(\mathbb{R}^d; \mathcal{L}(\mathbb{R}^d))$, which are defined respectively by

$$b_n(x) = \int_{\mathbb{R}^d} \rho_n(x - y) b(y) \, dy$$

$$\sigma_n(x) = \int_{\mathbb{R}^d} \rho_n(x - y) \sigma(y) \, dy,$$

where ρ_n are usual mollifiers in $\mathbb{R}^d$. It is not difficult to show that for each $n \in \mathbb{N}$ the coefficients b_n and σ_n satisfy the Hypotheses 2.1 and 2.2. In particular, for each $n \in \mathbb{N}$ the problem

$$d\xi(t) = b_n(\xi(t)) \, dt + \sigma_n(\xi(t)) \, dw(t), \quad \xi(0) = x,$$

has a unique solution $\xi_n(t;x)$. This allows us to introduce the transition semigroup $P_t^n \varphi(x)$ corresponding to the problem above. Due to the Proposition 1.5.1, P_t^n enjoys the strong Feller property and for any $t > 0$ and $\varphi \in B_b(\mathbb{R}^d)$

$$|P_t^n \varphi(x) - P_t^n \varphi(y)| \leq c \, (t \wedge 1)^{-\frac{1}{2}} \|\varphi\|_0 |x - y|, \quad x, y \in \mathbb{R}^d. \tag{2.3.1}$$

Moreover, it can be easily checked that all constants appearing in the Hypotheses 2.1 and 2.2 for b_n and σ_n are independent of $n \in \mathbb{N}$, and then the constant c in (2.3.1) does not depend on $n \in \mathbb{N}$. Thus, if we show that for any fixed $t \geq 0$ and $x \in \mathbb{R}^d$ it holds

$$\lim_{n \to +\infty} \mathbb{E} |\xi_n(t;x) - \xi(t;x)|^2 = 0, \tag{2.3.2}$$

by taking the limit as $n \to +\infty$ in (2.3.1), thanks to the dominated convergence theorem if $\varphi \in C_b(\mathbb{R}^d)$ we have

$$|P_t\varphi(x) - P_t\varphi(y)| \leq c\,(t \wedge 1)^{-\frac{1}{2}} \|\varphi\|_0 |x - y|, \quad x, y \in \mathbb{R}^d.$$

This means that

$$\mathrm{Var}\,(P_t(x, \cdot) - P_t(y, \cdot)) \leq c\,(t \wedge 1)^{-\frac{1}{2}} |x - y|, \quad x, y \in \mathbb{R}^d,$$

and then $P_t\varphi$ is Lipschitz-continuous for any $\varphi \in B_b(\mathbb{R}^d)$.

By using Itô's formula and taking first the expectation and then the derivative with respect to t, we have

$$\frac{d}{dt} \mathbb{E}\,|\xi_n(t; x) - \xi(t; x)|^2 = \mathbb{E}\,\|\sigma_n\,(\xi_n(t; x)) - \sigma(\xi(t; x))\|_2^2$$

$$+ 2\,\mathbb{E}\,\langle b_n(\xi_n(t; x)) - b(\xi(t; x)), \xi_n(t; x) - \xi(t; x) \rangle$$

$$\leq 2\,\mathbb{E}\,\|\sigma_n\,(\xi_n(t; x)) - \sigma_n(\xi(t; x))\|_2^2 + 2\,\mathbb{E}\,\|\sigma_n\,(\xi(t; x)) - \sigma(\xi(t; x))\|_2^2$$

$$+ 2\,\mathbb{E}\,\langle b_n(\xi_n(t; x)) - b_n(\xi(t; x)), \xi_n(t; x) - \xi(t; x) \rangle$$

$$+ 2\,\mathbb{E}\,\langle b_n(\xi(t; x)) - b(\xi(t; x)), \xi_n(t; x) - \xi(t; x) \rangle.$$

Therefore, by using the Hypothesis 2.1 for b_n and σ_n we easily have

$$\frac{d}{dt} \mathbb{E}\,|\xi_n(t; x) - \xi(t; x)|^2 \leq c\,\mathbb{E}\,|\xi_n(t; x) - \xi(t; x)|^2$$

$$+ c\,\mathbb{E}\,\|\sigma_n(\xi(t; x)) - \sigma(\xi(t; x))\|_2^2 + c\,\mathbb{E}\,|b_n\,(\xi(t; x)) - b(\xi(t; x))|^2$$

and thanks to the Gronwall lemma, this implies that

$$\mathbb{E}\,|\xi_n(t; x) - \xi(t; x)|^2$$

$$\leq c \int_0^t e^{c\,(t-s)} \mathbb{E}\,\left(\|\sigma_n(\xi(s; x)) - \sigma(\xi(s; x))\|^2 + |b_n(\xi(s; x)) - b(\xi(s; x))|^2 \right)\,ds.$$

Now, for any $x \in \mathbb{R}^d$ we have

$$\|\sigma_n(\xi(s; x)) - \sigma(\xi(s; x))\|^2 + |b_n(\xi(s; x)) - b(\xi(s; x))|^2 \leq c\,\left(1 + |\xi(s; x)|^{4m+2}\right)$$

and $b_n(z)$ and $\sigma_n(z)$ converge respectively to $b(z)$ and $\sigma(z)$, as n goes to infinity, for any $z \in \mathbb{R}^d$. Therefore, thanks to the estimate (1.2.13), we can apply the dominated convergence theorem and (2.3.2) follows. □

2.3.2   Irreducibility

In the next proposition we prove that the semigroup P_t is *irreducible*. Notice that such a proof adapts to the case of coefficients having any growth.

Proposition 2.3.3. *If the Hypotheses 2.1 and 2.2 hold, then the semigroup P_t is irreducible.*

Proof. For any $n \in \mathbb{N}$ and $x \in \mathbb{R}^d$, we introduce the stopping time

$$\tau_n^x = \inf \{ t \geq 0 : |\xi(t; x)| \geq n \}$$

and for any $\varphi \in B_b(\mathbb{R}^d)$ and $t \geq 0$ we define

$$P_t^n \varphi(x) = \mathbb{E}\varphi(t \wedge \tau_n^x; x), \qquad x \in \mathbb{R}^d.$$

It can be checked that the generator L_n of the semigroup P_t^n is a second order elliptic operator with continuous and bounded coefficients in the ball $B_n(0)$ equipped with Dirichlet boundary conditions. Then, as proved for example in the book by Davies [50] (see the Theorem 3.3.5), the semigroup P_t^n is irreducible.

Now, for $\epsilon > 0$ and $z \in \mathbb{R}^d$ fix the ball $B_\epsilon(z) \subset \mathbb{R}^d$ of center z and radius ϵ. For each $n \in \mathbb{N}$ such that $\overline{B}_\epsilon(z) \subset B_n(0)$ we have

$$\mathbb{P}(\xi(t; x) \in B_\epsilon(z)) = \mathbb{P}(\xi(t; x) \in B_\epsilon(z), \ \tau_n^x > t) + \mathbb{P}(\xi(t; x) \in B_\epsilon(z), \ \tau_n^x \leq t)$$

$$\geq \mathbb{P}(\xi(t; x) \in B_\epsilon(z), \ \tau_n^x > t) = \mathbb{P}((\xi(t \wedge \tau_n^x; x) \in B_\epsilon(z), \ \tau_n^x > t)$$

$$= \mathbb{P}(\xi(t \wedge \tau_n^x; x) \in B_\epsilon(z)) - \mathbb{P}(\xi(\tau_n^x; x) \in B_\epsilon(z), \ \tau_n^x \leq t).$$

Now, we remark that $|\xi(\tau_n^x; x)| = n$ and then $\mathbb{P}(\xi(\tau_n^x; x) \in B_\epsilon(z)) = 0$. This implies that

$$\mathbb{P}(\xi(t; x) \in B_\epsilon(z)) \geq P_t^n I_{B_\epsilon(z)}(x)$$

and since P_t^n is irreducible we have $\mathbb{P}(\xi(t; x) \in B_\epsilon(z)) > 0$. $\qquad \square$

Remark 2.3.4. The method used in the proof of the proposition above is strictly related to the finite dimensional situation and it is not clear how to extend it to the infinite dimensional case. We give here a different proof of the irreducibility of P_t, which has the advantage that it can be extended to the infinite dimensional situation, and has the disadvantage of requiring bounded diffusion terms.

Let us fix $t_0, \epsilon > 0$ and $z \in \mathbb{R}^d$. We want to prove that

$$\mathbb{P}(\xi(t_0; x) \in B_\epsilon(z)) > 0. \tag{2.3.3}$$

We consider the deterministic controlled problem associated with (2.2.1)

$$\frac{dy}{dt}(t) = b(y(t)) + \sigma(y(t))u(t), \qquad y(0) = x \in \mathbb{R}^d, \tag{2.3.4}$$

with $u \in L^2([0, t_0]; \mathbb{R}^d)$. As for the stochastic problem, even in this case it is possible to prove that for any u there exists a unique solution $y(t; x, u)$.

Since for any $y \in \mathbb{R}^d$ there exists $\sigma^{-1}(y)$ bounded, the problem (2.3.4) is *controllable* at time t_0, that is for any $z \in \mathbb{R}^d$ there exists $u_0 \in L^2(0, t_0; \mathbb{R}^d)$ such that

$$y(t_0; x, u_0) = z. \qquad (2.3.5)$$

Actually, if we define

$$u_0(t) = \sigma^{-1}\left(x + \frac{t}{t_0}(z - x)\right)\left(\frac{1}{t_0}(z - x) - b\left(x + \frac{t}{t_0}(z - x)\right)\right),$$

then $y(t; x, u_0) = x + t(z - x)/t_0$ fulfills (2.3.5) and is a solution of (2.3.4). Therefore, in order to prove (2.3.3) it is enough to show that

$$\mathbb{P}(|\xi(t_0; x) - y(t_0; x, u_0)| < \epsilon) > 0. \qquad (2.3.6)$$

We define

$$f(t) = \int_0^t \sigma(y(s; x, u_0)) u_0(s) \, ds,$$

and we fix $R > 2\bar{R} + \epsilon$, where

$$\bar{R} = \sup_{t \le t_0} |y(t; x, u_0)|.$$

Moreover we define

$$w^\sigma(t; x) = \int_0^t \sigma(\xi(s; x)) \, dw(s).$$

Since b is locally Lipschitz-continuous, then there exists a constant $c_R > 0$ such that

$$|b(x) - b(y)| \le c_R |x - y|, \quad x, y \in B_R(0).$$

Hence for any $t \le t_0$ and $\omega \in \Omega$ we have

$$\sup_{s \le t} |\xi(s; x)(\omega)| \le R \Rightarrow |\xi(t; , x)(\omega) - y(t; x, u_0)|$$

$$\le \int_0^t |b(\xi(s; x)(\omega)) - b(y(s; x, u_0))| \, ds + |w^\sigma(t; x)(\omega) - f(t)|$$

$$\le c_R \int_0^t |\xi(s; x)(\omega) - y(s; x, u_0)| \, ds + |w^\sigma(t; x)(\omega) - f(t)|,$$

so that we obtain

$$\sup_{s \le t} |\xi(s; x)(\omega)| \le R \Rightarrow$$

$$\qquad (2.3.7)$$

$$|\xi(t; x)(\omega) - y(t; x, u_0)| \le \int_0^t e^{c_R(t - s)} |w^\sigma(s; x)(\omega) - f(s)| \, ds.$$

Now, for every $\epsilon > 0$ we define

$$\Omega_\epsilon = \left\{ \omega \in \Omega \; : \; \int_0^{t_0} e^{c_R(t_0-s)} \, |w^\sigma(s;x)(\omega) - f(s)| \; ds < \epsilon \right\}.$$

If σ is bounded, by general support theorems (see [114] for example), we have that $\mathbb{P}(\Omega_\epsilon) > 0$ and this allows us to conclude the proof. Actually, for any $\omega \in \Omega$ we introduce the set

$$E(\omega) = \left\{ t \leq t_0 \; : \; \sup_{s \leq t} |\xi(s;x)(\omega) - y(s;x,u_0)| \right.$$

$$\left. \leq \int_0^t e^{c_R(t-s)} \, |w^\sigma(s;x)(\omega) - f(s)| \; ds \right\}.$$

Since $0 \in E(\omega)$, we have that $E(\omega) \neq \emptyset$ for any $\omega \in \Omega_\epsilon$. Then we can define

$$T(\omega) = \sup E(\omega), \quad \omega \in \Omega.$$

As $\xi(t;x)$ has continuous trajectories and $y(t;x,u_0)$ is continuous, it is immediate to check that $E(\omega)$ is closed and $T(\omega) \in E(\omega)$. Hence if we show that $T(\omega) = t_0$, for any $\omega \in \Omega_\epsilon$, it follows that

$$|\xi(t_0;x)(\omega) - y(t_0;x,u_0)| \leq \int_0^{t_0} e^{c_R(t_0-s)} \, |w^\sigma(s;x)(\omega) - f(s)| \; ds < \epsilon$$

and since $\mathbb{P}(\Omega_\epsilon) > 0$, (2.3.6) turns out to be true and the proposition is completely proved.

Fix $\omega \in \Omega_\epsilon$ and assume that $T(\omega) < t_0$. For any $t \leq T(\omega)$ we have

$$|\xi(t;x)(\omega)| \leq |\xi(t;x)(\omega) - y(t;x,u_0)| + |y(t;x,u_0)|$$

$$\leq \int_0^t e^{c_R(t-s)} \, |w^\sigma(s;x)(\omega) - f(s)| \; ds + \bar{R} < \epsilon + \bar{R} < R - \bar{R}.$$

Then, as $\xi(t;x)$ has continuous trajectories, there exists $T'(\omega) > T(\omega)$ such that

$$\sup_{t \leq T'(\omega)} |\xi(t;x)(\omega)| \leq R,$$

and according to (2.3.7) it is easy to show that

$$\sup_{t \leq T'(\omega)} |\xi(t;x)(\omega) - y(t;x,u_0)| \leq \int_0^{T'(\omega)} e^{c_R(T'(\omega)-s)} \, |w^\sigma(s;x)(\omega) - f(s)| \; ds,$$

so that $T'(\omega) \in E(\omega)$, which is a contradiction. Therefore $T(\omega) = t_0$ for any $\omega \in \Omega_\epsilon$.

2.4   Absolute continuity

Our aim here is proving the existence of a density with respect to the Lebesgue measure for the law of the process $\xi(t; x)$.

Proposition 2.4.1. *Under the Hypotheses 2.1 and 2.2, all the transition probabilities $P_t(x, \cdot)$, $t > 0$ and $x \in \mathbb{R}^d$, associated with the problem (2.2.1) are absolutely continuous with respect to the Lebesgue measure.*

Proof. For any $n \in \mathbb{N}$, the semigroup P_t^n introduced in the proof of the Proposition 2.3.3 admits a kernel for $t > 0$ and then the measures $P_t^n(x, \cdot)$ are absolutely continuous with respect to the Lebesgue measure λ (for a proof see for example the book by Davies [50]). Now, let $\Gamma \in \mathcal{B}(\mathbb{R}^d)$ and assume that $\lambda(\Gamma) = 0$. For any $n \in \mathbb{N}$ we have

$$P_t(x, \Gamma) = \mathbb{P}(\xi(t; x) \in \Gamma)$$

$$= \mathbb{P}(\xi(t; x) \in \Gamma, \ \tau_n^x < t) + \mathbb{P}(\xi(t; x) \in \Gamma, \ \tau_n^x \geq t)$$

$$\leq \mathbb{P}(\xi(t; x) \in \Gamma, \ \tau_n^x < t) + \mathbb{P}(\xi(t \wedge \tau_n^x; x) \in \Gamma) \qquad (2.4.1)$$

$$= \mathbb{P}(\xi(t; x) \in \Gamma, \ \tau_n^x < t) + P_t^n(x, \Gamma) = \mathbb{P}(\xi(t; x) \in \Gamma, \ \tau_n^x < t),$$

last equality following from the absolute continuity of $P_t^n(x, \cdot)$ with respect to the Lebesgue measure, as $\lambda(\Gamma) = 0$. Now, according to the estimate (2.2.2), it is easy to verify that

$$\lim_{n \to +\infty} \mathbb{P}(\tau_n^x < t) = 0,$$

and hence, from arbitrariness of n, it follows that $P_t(x, \Gamma) = 0$. $\square$

2.5   The strongly dissipative case

In this section we show that when b enjoys a stronger dissipativity condition, then it is possible to describe the rate of convergence of $P_t\varphi$ to the equilibrium, as t goes to infinity.

To this purpose we have to assume

Hypothesis 2.3. *There exist $\rho, c_\rho > 0$ such that for all $x, y \in \mathbb{R}^d$*

$$2 \langle b(x) - b(y), x - y \rangle + \|\sigma(x) - \sigma(y)\|_2^2 \leq -c_\rho |x - y|^{2+\rho}.$$

Proposition 2.5.1. *Assume that the Hypotheses 2.1, 2.2 and 2.3 hold. Then for any $\varphi \in B_b(\mathbb{R}^d)$ we have*

$$\|P_t\varphi - \langle \varphi, \mu \rangle\|_0 \leq c \, (t \vee 1)^{-1/\rho} \|\varphi\|_0.$$

Proof. If $v(t)$, $t \geq 0$, is a Wiener process independent of $w(t)$, we define for any $t \in \mathbb{R}$

$$\bar{w}(t) = \begin{cases} w(t) & \text{if } t \geq 0 \\ \\ v(-t) & \text{if } t \leq 0, \end{cases}$$

with the filtration $\bar{\mathcal{F}}_t = \sigma(\bar{w}(s), \, s \leq t)$. For any $\lambda \in \mathbb{R}$, we consider the problem

$$d\xi(t) = b(\xi(t)) \, dt + \sigma(\xi(t)) \, d\bar{w}(t), \quad t \geq -\lambda, \quad \xi(-\lambda) = x. \tag{2.5.1}$$

By proceeding as in the chapter 1, it is possible to prove that there exists a unique solution $\xi_\lambda(t; x)$ for the problem (2.5.1).

Now, for any $x \in \mathbb{R}^d$ and $\lambda > 0$ we set

$$\zeta_\lambda(t; x) = \xi_\lambda(t; x) - \xi_\lambda(t; 0). \tag{2.5.2}$$

Due to Itô's formula, we have

$$|\zeta_\lambda(t; x)|^2 = |x|^2 + 2 \int_{-\lambda}^t \langle b(\xi_\lambda(s; x)) - b(\xi_\lambda(s; 0)), \zeta_\lambda(s; x) \rangle \, ds$$

$$+ \int_{-\lambda}^t \text{Tr}\left[a_\lambda(s; x) \right] ds + \int_{-\lambda}^t \langle \sigma(\xi_\lambda(s; x)) - \sigma(\xi_\lambda(s; 0)), \zeta_\lambda(s; x) dw(s) \rangle,$$

where for any $\lambda > 0$ and $x \in \mathbb{R}^d$

$$a_\lambda(t; x) = \left[\sigma(\xi_\lambda(t; x)) - \sigma(\xi_\lambda(t; 0)) \right] \left[\sigma(\xi_\lambda(t; x)) - \sigma(\xi_\lambda(t; 0)) \right]^*.$$

By taking the expectation and by differentiating with respect to t, we get

$$\frac{d}{dt} \mathbb{E} |\zeta_\lambda(t; x)|^2 = 2 \mathbb{E} \, \langle b(\xi_\lambda(t; x)) - b(\xi_\lambda(t; 0)), \zeta_\lambda(t; x) \rangle$$

$$+ \mathbb{E} \|\sigma(\xi_\lambda(t; 0)) - \sigma(\xi_\lambda(t; x))\|_2^2 \leq -c_\rho \mathbb{E} |\zeta_\lambda(t; x)|^{2+\rho} \leq -c_\rho \left(\mathbb{E} |\zeta_\lambda(t; x)|^2 \right)^{1 + \frac{\rho}{2}}.$$

Then, by comparison we get

$$\mathbb{E} |\zeta_\lambda(t; x)|^2 \leq |x|^2 \left(1 + |x|^\rho c_\rho (t + \lambda) \right)^{-\frac{2}{\rho}}.$$

This implies that

$$\lim_{\lambda \to +\infty} \mathbb{E} |\xi_\lambda(0; x) - \xi_\lambda(0; 0)|^2 = \lim_{\lambda \to +\infty} \mathbb{E} |\zeta_\lambda(0; x)|^2 = 0. \tag{2.5.3}$$

Now we want to show that the sequence $\{\xi_\lambda(0; x); \, \lambda \geq 0\}$ is a Cauchy sequence in $L^2(\Omega; \mathbb{R}^d)$, for every $x \in \mathbb{R}^d$. This should imply that

$$\exists \lim_{\lambda \to +\infty} \xi_\lambda(0; x) = \eta, \quad \text{in } L^2(\Omega; \mathbb{R}^d),$$

for a suitable random variable η which is independent of x because of (2.5.3). Then $\mathcal{L}(\eta) = \mu$ is the unique invariant measure (for a proof see [47, chapter 11] or [48, chapter 6]). For any $0 < \gamma < \delta$ and $t \geq -\gamma$, we define

$$\zeta_{\gamma,\delta}(t;x) = \xi_\gamma(t;x) - \xi_\delta(t;x), \quad x \in \mathbb{R}^d.$$

We have

$$\mathbb{E}|\zeta_{\gamma,\delta}(t;x)|^2 = \mathbb{E}|x - \xi_\delta(-\gamma;x)|^2$$

$$+ \mathbb{E}\left(\int_{-\gamma}^t 2\langle b(\xi_\gamma(s;x)) - b(\xi_\delta(s;x)), \zeta_{\gamma,\delta}(s;x)\rangle + \mathrm{Tr}\,[a_{\gamma,\delta}(s;x)]\,ds\right),$$

where

$$a_{\gamma,\delta}(t;x) = [\sigma(\xi_\gamma(t;x)) - \sigma(\xi_\delta(t;,x))][\sigma(\xi_\gamma(t;x)) - \sigma(\xi_\delta(t;x))]^*.$$

By using again the Hypothesis 2.3 we get

$$\mathbb{E}|\xi_\gamma(0;x) - \xi_\delta(0;x)|^2 \leq \frac{\mathbb{E}|x - \xi_\delta(-\gamma;x)|^2}{\left(1 + c_\rho\gamma\left(\mathbb{E}|x - \xi_\delta(-\gamma;x)|^2\right)^{\frac{1}{2}}\right)^{\frac{2}{\rho}}},$$

and from this it follows that for any $0 < \gamma < \delta$

$$\mathbb{E}|\xi_\gamma(0;x) - \xi_\delta(0;x)|^2 \leq (c_\rho\gamma)^{-\frac{2}{\rho}}, \quad x \in \mathbb{R}^d.$$

Next, we fix $\varphi \in B_b(\mathbb{R}^d)$ and $1 < t \leq s$. As proved in the Proposition 2.3.2 the semigroup P_t is *strongly Feller* and it holds $\|P_1\varphi\|_{0,1} \leq c\|\varphi\|_0$. Then, since $\mathcal{L}(\xi_{t-1}(0;x)) = \mathcal{L}(\xi(t-1;x))$, we have

$$|P_t\varphi(x) - P_s\varphi(x)| = |\mathbb{E}P_1\varphi(\xi_{t-1}(0;,x)) - \mathbb{E}P_1\varphi(\xi_{s-1}(0;x))|$$

$$\leq \|P_1\varphi\|_{0,1}\left(\mathbb{E}|\xi_{t-1}(0;x) - \xi_{s-1}(0;x)|^2\right)^{1/2} \leq c\left(c_\rho(t-1)\right)^{-\frac{1}{\rho}}\|\varphi\|_0.$$

(2.5.4)

Therefore, since

$$\|P_t\varphi - <\varphi,\mu>\|_0 \leq 2\|\varphi\|_0, \quad t \geq 0$$

by taking the limit as $s \to +\infty$ in (2.5.4), we get

$$\|P_t\varphi - <\varphi,\mu>\|_0 \leq c\,(t \vee 1)^{-\frac{1}{\rho}}\|\varphi\|_0, \quad t \geq 0.$$ (2.5.5)

$\square$

If we assume that b satisfies an even stronger dissipativity condition then we have exponential convergence to equilibrium.

Proposition 2.5.2. *Assume that the Hypotheses 2.1, 2.2 hold. Moreover assume that there exist $\rho > 0$ and c_ρ, $d_\rho > 0$ such that*

$$\langle b(x) - b(y), x - y \rangle + \|\sigma(x) - \sigma(y)\|^2 \leq -c_\rho |x - y|^{2+\rho} - d_\rho |x - y|^2. \tag{2.5.6}$$

Then for any $\varphi \in B_b(\mathbb{R}^d)$ we have

$$\|P_t \varphi - \langle \varphi, \mu \rangle \|_0 \leq c e^{-\frac{d_\rho}{2} t} \|\varphi\|_0, \quad t \geq 0. \tag{2.5.7}$$

Proof. Proceeding as in the proof of the previous proposition, for any $0 < \gamma < \delta$ we have

$$\frac{d}{dt} \mathbb{E}|\xi_\gamma(t; x) - \xi_\delta(t; x)|^2$$

$$\leq -2c_\rho \, \mathbb{E}|\xi_\gamma(t; x) - \xi_\delta(t; x)|^{2+\rho} - 2d_\rho \, \mathbb{E}|\xi_\gamma(t; x) - \xi_\delta(t; x)|^2.$$

and after some computations this yields

$$\mathbb{E}|\xi_\gamma(0; x) - \xi_\delta(0; x)|^2 \leq c e^{-\delta_\rho \gamma} \left(1 - e^{-\frac{\delta_\rho \gamma}{2}} \right)^{-\frac{2}{\delta}}.$$

Now, recalling once more that for any $t \geq 0$

$$|P_t \varphi(x) - \langle \varphi, \mu \rangle| \leq 2 \|\varphi\|_0, \quad x \in \mathbb{R}^d$$

by applying the same arguments used in the previous proposition, for any $t > 0$ we have

$$\|P_t \varphi - \langle \varphi, \mu \rangle \|_0 \leq c e^{-\frac{\delta_\rho}{2} t} \|\varphi\|_0.$$

$\square$

Chapter 3

Analyticity of the semigroup in a degenerate case

We consider the following class of second order differential operators

$$\mathcal{L}_0(x, D) = \sum_{i,j=1}^{d} a_{ij}(x) D_{ij} + \sum_{i=1}^{d} b_i(x) D_i, \quad x \in \mathbb{R}^d,$$

where $a(x)$ is a positive semi-definite symmetric matrix which has quadratic growth and $b(x)$ is a vector field of class C^2 which has linear growth. We assume that the mapping $a : \mathbb{R}^d \to \mathcal{L}(\mathbb{R}^d)$ is also of class C^2 with bounded second derivatives, so that it can be written as

$$a(x) = \frac{1}{2} \sigma(x) \sigma^\star(x), \quad x \in \mathbb{R}^d,$$

for some matrix valued function $\sigma : \mathbb{R}^d \to \mathcal{L}(\mathbb{R}^d)$ which is Lipschitz-continuous (for a proof of this fact see [66] and also [114]). Here we assume further regularity for a; namely we assume that a can be factorized by some σ which is twice differentiable with bounded derivatives. Moreover a compatibility condition is given among σ and b. More precisely we assume that there exists a bounded vector field $\beta : \mathbb{R}^d \to \mathbb{R}^d$ of class C^2 such that

$$b(x) = \sigma(x)\beta(x), \quad x \in \mathbb{R}^d, \quad \text{and} \quad \sup_{x \in \mathbb{R}^d} |D^j \beta(x)| < \infty, \quad j = 0, 1, 2.$$

Notice that an analogous condition is also considered in many papers dealing with the generation of analytic semigroups by second order elliptic operators and it seems to be quite natural. In the case of the Ornstein-Uhlenbeck operator (where $a(x) = A > 0$ does not depend on $x \in \mathbb{R}^d$ and $b(x) = Bx$, for a non-zero matrix B) the boundedness of $\beta(x) = A^{-1}Bx$ fails to be true and in fact, as proved in [44], the semigroup is not analytic.

Our goal is proving that the operator $\mathcal{L}_0$ generates an analytic semigroup in the space of bounded uniformly continuous functions. Since $\mathcal{L}_0$ is the diffusion operator corresponding to the stochastic differential problem

$$d\xi(t) = b(\xi(t))\, dt + \sigma(\xi(t))\, dw(t), \quad \xi(0) = x, \tag{3.0.1}$$

by using the usual representation formula for the corresponding transition semigroup

$$P_t\varphi(x) = \mathbb{E}\varphi(\xi(t; x)), \quad x \in \mathbb{R}^d,$$

we show that for any $t > 0$ and $\varphi \in C_b^2(\mathbb{R}^d)$ it is possible to express explicitly $\mathcal{L}_0(P_t\varphi)(x)$ in terms of φ and not of its derivatives. This allows us to get the fundamental estimate

$$\sup_{x \in \mathbb{R}^d} |\mathcal{L}_0(P_t\varphi)(x)| \leq c\,(t \wedge 1)^{-1} \sup_{x \in \mathbb{R}^d} |\varphi(x)|, \tag{3.0.2}$$

from which the analyticity of the semigroup follows, by general functional analysis arguments.

In the section 3.2 we prove some crucial identities for the solution $\xi(t; x)$ of the stochastic problem (3.0.1). As we are assuming b and σ to be twice differentiable with bounded derivatives, $\xi(t; x)$ is twice mean-square differentiable with respect to the initial datum $x \in \mathbb{R}^d$. We show that, under suitable assumptions for σ and b, for any $x, h \in \mathbb{R}^d$ there exists a d-dimensional continuous process $v^h(t; x)$ such that

$$D\xi(t; x)\sigma(x)h = \sigma(\xi(t; x))v^h(t; x) \tag{3.0.3}$$

and

$$\sup_{x \in \mathbb{R}^d} \mathbb{E}|v^h(t; x)|^2 \leq c(t)|h|^2,$$

for some continuous increasing function $c(t)$. Similarly we prove that there exists a d-dimensional continuous process $u^h(t; x)$ such that

$$D^2\xi(t; x)(\sigma(x)h, \sigma(x)h) = \sigma(\xi(t; x))u^h(t; x)$$

and

$$\sup_{x \in \mathbb{R}^d} \mathbb{E}|u^h(t; x)|^2 \leq c(t)|h|^4.$$

As proved in the first chapter, when $\mathcal{L}_0$ is strictly elliptic, that is when $\sigma(x)$ is invertible for any $x \in \mathbb{R}^d$ and its inverse is bounded, the semigroup P_t has a smoothing effect, that is P_t maps bounded Borel functions into twice differentiable functions (here we are assuming σ and b of class C^2). Moreover, the classical Bismut-Elworthy formula holds for any $\varphi \in C_b(\mathbb{R}^d)$

$$\langle D(P_t\varphi)(x), h \rangle = \frac{1}{t}\mathbb{E}\varphi(\xi(t; x)) \int_0^t \langle \sigma^{-1}(\xi(s; x))D\xi(s; x)h, dw(s) \rangle,$$

where $D\xi(s;x)h$ is the first mean-square derivative of $\xi(s;x)$ with respect to $x \in \mathbb{R}^d$ along the direction $h \in \mathbb{R}^d$. But in our case, $\mathcal{L}_0$ is possibly degenerate and we cannot prove that the semigroup P_t is regularizing.

Nevertheless, in the section 3.3. we prove a generalization of the Bismut-Elworthy formula. We apply such a general formula to our situation and in the section 3.4 we show that for any twice differentiable function φ it holds

$$\langle D(P_t\varphi)(x), \sigma(x)h \rangle = \frac{1}{t}\mathbb{E}\,\varphi(\xi(t;x)) \int_0^t \left\langle v^h(s;x), dw(s) \right\rangle,$$

where $v^h(s;x)$ is the process introduced in (3.0.3). A similar formula is proved for the second derivative $\left\langle D^2(P_t\varphi)(x)\sigma(x)h, \sigma(x)h \right\rangle$, so that we give an expression for the derivatives of $P_t\varphi$ involving φ and not its derivatives, even if only along the directions $\sigma(x)h$, for $x, h \in \mathbb{R}^d$. This allows us to prove the estimate (3.0.2).

In the section 3.5 we show that P_t is a weakly continuous semigroup and we introduce its weak generator $\mathcal{L}$. We show that $P_t\varphi \in D(\mathcal{L})$, for any $t > 0$ and $\varphi \in C_b(\mathbb{R}^d)$ and

$$\|\mathcal{L}(P_t\varphi)\|_0 \le c\,(t \wedge 1)^{-1}\|\varphi\|_0, \quad t > 0.$$

By standard functional analysis arguments we conclude that P_t is analytic in $C_b(\mathbb{R}^d)$.

3.1 Assumptions

We recall that we are assuming that the symmetric matrix $a(x)$ is positive semi-definite and of class C^2 with bounded derivatives. Thus, as proved for example in [66, Theorem 3.2.1], we have the following result

Theorem 3.1.1. *There exists $\sigma : \mathbb{R}^d \to \mathcal{L}(\mathbb{R}^d)$ Lipschitz-continuous such that*

$$a(x) = \frac{1}{2}\sigma(x)\sigma^*(x), \qquad x \in \mathbb{R}^d. \tag{3.1.1}$$

Here σ has to fulfill stronger regularity assumptions than Lipschitz-continuity. But at present it is not clear what sort of stronger regularity assumptions has to be satisfied by a in order to have stronger regularity for σ. In fact, if we consider

$$a(x) = |x|^2 I, \qquad x \in \mathbb{R}^d,$$

we have that $a : \mathbb{R}^d \to \mathcal{L}(\mathbb{R}^d)$ is of class C^∞ with bounded derivatives beginning from order 2, but its square root $|x|I$ is not even once differentiable.

Hypothesis 3.1.   1. *The functions $\sigma : \mathbb{R}^d \to \mathcal{L}(\mathbb{R}^d)$ and $b : \mathbb{R}^d \to \mathbb{R}^d$ are of class C^2 and have bounded first and second derivatives.*

2. *There exists a bounded vector field $\beta : \mathbb{R}^d \to \mathbb{R}^d$ of class C^2 having bounded derivatives such that*

$$b(x) = \sigma(x)\beta(x), \qquad x \in \mathbb{R}^d. \tag{3.1.2}$$

3. *There exist two bounded maps* $\gamma_1 : \mathbb{R}^d \to \mathcal{L}(\mathbb{R}^d; \mathcal{L}(\mathbb{R}^d))$ *and* $\gamma_2 : \mathbb{R}^d \to \mathcal{L}(\mathbb{R}^d \times \mathbb{R}^d; \mathcal{L}(\mathbb{R}^d))$ *such that for any* $x, h, k \in \mathbb{R}^d$ *it holds*

$$D\sigma(x)\sigma(x)h = \sigma(x)\gamma_1(x)h,$$

$$D^2\sigma(x)\,(\sigma(x)h, \sigma(x)k) = \sigma(x)\gamma_2(x)(h, k). \tag{3.1.3}$$

Notice that from (3.1.2) and (3.1.3) it immediately follows that there exist two bounded maps $\rho_1 : \mathbb{R}^d \to \mathcal{L}(\mathbb{R}^d)$ and $\rho_2 : \mathbb{R}^d \to \mathcal{L}(\mathbb{R}^d \times \mathbb{R}^d; \mathbb{R}^d)$ such that for any $x, h, k \in \mathbb{R}^d$ it holds

$$Db(x)\sigma(x)h = \sigma(x)\rho_1(x)h,$$

$$D^2b(x)(\sigma(x)h, \sigma(x)k) = \sigma(x)\rho_2(x)(h, k). \tag{3.1.4}$$

Remark 3.1.2.   1. If the operator $\mathcal{L}_0$ is assumed to be *strongly* elliptic, that is there exists $\nu > 0$ such that

$$\inf_{x \in \mathbb{R}^d} \sum_{i,j=1}^{d} a_{ij}(x)h_i h_j \geq \nu |h|^2, \quad h \in \mathbb{R}^d,$$

and if $a : \mathbb{R}^d \to \mathcal{L}(\mathbb{R}^d)$ is of class C^2, then it is possible to show that there exists $\sqrt{a} : \mathbb{R}^d \to \mathcal{L}(\mathbb{R}^d)$ of class C^2. The matrix $\sqrt{a(x)}$ is clearly invertible for any $x \in \mathbb{R}^d$ and then, if b is any vector field of class C^2, having bounded derivatives, the Hypotheses 3.1 is satisfied.

2. The boundedness assumption for the function γ_2 implies that the second derivatives of σ vanish at infinity as $1/|x|$. In particular, by some calculations we easily have that the hypothesis of boundedness for the second derivatives of a is necessary in order to have (3.1.3).

3. In order to use deterministic techniques it is often assumed that the coefficients a and b do not vanish outside a negligible set (see e.g. [73], [59], [10] and [9]). Here no conditions are given on the set where a degenerates. Actually, a can also be taken identically zero; in this case b is zero and the semigroup generated by $\mathcal{L}_0$ is the semigroup identically equal to identity, which is analytic.

4. Let $\psi : \mathbb{R}^d \to \mathcal{L}(\mathbb{R}^r)$ be a mapping of class C^2 with bounded first and second derivatives and assume that for any $x \in \mathbb{R}^d$ the matrix $\psi(x)$ is invertible and the mapping $\psi^{-1} : \mathbb{R}^d \to \mathcal{L}(\mathbb{R}^r)$ is bounded. Then, if a is any matrix valued function which satisfies the Hypothesis 3.1, the matrix $\sqrt{\psi}a\sqrt{\psi}$ can be factorized as in (3.1.1) and $\sqrt{\psi}\sigma$ satisfies the Hypothesis 3.1, as well.

5. The conditions described in the Hypotheses 3.1-3 are satisfied if we take

$$a(x) = g(x)a, \quad x \in \mathbb{R}^d,$$

for any function $g : \mathbb{R}^d \to \mathbb{R}$ of class C^2 and any matrix $a \in \mathcal{L}(\mathbb{R}^d)$ which does not depend on $x \in \mathbb{R}^d$.

6. If we assume that for any $i, j = 1, \ldots, d$

$$a_{ij}(x) = a_i(x)\delta_{ij}, \quad x \in \mathbb{R}^d,$$

then the conditions of the Hypothesis 3.1-3 can be written as

$$\frac{\partial \sigma_h}{\partial x_i}(x)\sigma_i = \sigma_h(\gamma_1 e_i)_h,$$

$$\frac{\partial^2 \sigma_h}{\partial x_i \partial x_j}\sigma_i \sigma_j = \sigma_h(\gamma_2(e_i, e_j))_h,$$

for any $i, j, h = 1, \ldots, d$. A simple case in which they are satisfied is when for any $h = 1, \ldots, d$

$$a_h(x) = f_h^2(x_h), \quad x \in \mathbb{R}^d,$$

and $f_h : \mathbb{R} \to \mathbb{R}$ is any function of class C^2 with bounded derivatives. Another situation in which the conditions above are both satisfied is when for any $h = 1, \ldots, d$

$$a_h(x) = f_h^6(x), \quad x \in \mathbb{R}^d,$$

for some function $f_h : \mathbb{R}^d \to \mathbb{R}$ of class C^2 such that for any $x \in \mathbb{R}^d$

$$\left| f_i^3(x) \frac{\partial f_h}{\partial x_i}(x) \right| \leq c|f_h(x)|$$

and

$$\left| f_i^3(x) f_j^3(x) \left(2\frac{\partial f_h}{\partial x_i}(x)\frac{\partial f_h}{\partial x_j}(x) + f_h(x)\frac{\partial^2 f_h}{\partial x_i \partial x_j}(x) \right) \right| \leq c|f_h^2(x)|.$$

7. The class of operators considered in the chapters 1 and 2 is not covered by the present situation. In fact, the coefficients b and σ are not necessarily Lipschitz-continuous and the compatibility condition described in the Hypothesis 3.1-2 does not hold.

3.2 Some properties of the solution of the SDE

The operator $\mathcal{L}_0$ can be rewritten in the following form

$$\mathcal{L}_0(x, D) = \frac{1}{2}\text{Tr}\left[\sigma(x)\sigma^*(x)D^2\right] + \langle b(x), D \rangle, \quad x \in \mathbb{R}^d.$$

Since we are assuming that both b and σ are Lipschitz-continuous, then for any $x \in \mathbb{R}^d$ the problem (3.0.1) admits a unique solution having continuous trajectories, such that for any $x \in \mathbb{R}^d$

$$\mathbb{E} \, |\xi(t;x)|^2 \le c(t) \, \left(|x|^2 + 1\right), \quad t \ge 0,$$

for a suitable continuous increasing function $c(t)$.

Moreover, since b and σ are twice differentiable with bounded derivatives, it can be proved that $\xi(t;x)$ is twice mean-square differentiable with respect to $x \in \mathbb{R}^d$. As well known (see also chapter 1) the derivatives of $\xi(t;x)$ are solutions of suitable stochastic differential equations that one obtains from (3.0.1) by differentiating the coefficients. Namely, if $\eta^h(t;x) = D\xi(t;x)h$ denotes the first mean-square derivative of $\xi(t;x)$ with respect to $x \in \mathbb{R}^d$ along the direction $h \in \mathbb{R}^d$, then we have

$$d\eta^h(t) = Db(\xi(t;x))\eta^h(t) \, dt + [D\sigma(\xi(t;x))\eta^h(t)] \, dw(t), \quad \eta^h(0) = h. \qquad (3.2.1)$$

Similarly, if $\zeta^{hk}(t;x) = D^2\xi(t;x)(h,k)$ denotes the second mean-square derivative of $\xi(t;x)$ with respect to $x \in \mathbb{R}^d$ along the directions $h, k \in \mathbb{R}^d$, we have

$$d\zeta^{hk}(t) = Db(\xi(t;x))\zeta^{hk}(t) \, dt + [D\sigma(\xi(t;x))\zeta^{hk}(t)] \, dw(t) + d\theta^{hk}(t), \quad \zeta^{hk}(0) = 0,$$
$$\qquad (3.2.2)$$

where

$$d\theta^{hk}(t) = D^2 b(\xi(t;x))(\eta^h(t;x), \eta^k(t;x)) \, dt + D^2\sigma(\xi(t;x))(\eta^h(t;x), \eta^k(t;x)) \, dw(t).$$

In what follows we shall denote by $\{g_i\}_{i=1}^{2d}$ and $\{e_i\}_{i=1}^{d}$ the standard orthonormal basis in $\mathbb{R}^{2d}$ and $\mathbb{R}^d$, respectively.

Lemma 3.2.1. *Assume that the Hypothesis 3.1 holds and let $y(t)$ be defined by*

$$y(t) = y(0) + \int_0^t y_1(s) \, ds + \int_0^t y_2(s) \, dw(s),$$

for some $y_1 : [0, +\infty) \to \mathbb{R}^d$ and $y_2 : [0, +\infty) \to \mathcal{L}(\mathbb{R}^d)$. Then, if the function $\varphi : \mathbb{R}^d \times \mathbb{R}^d \to \mathbb{R}^d$ is defined by $\varphi(x,y) = \sigma(x)y$ and if $z(t) = (\xi(t), y(t))^t$, where $\xi(t) = \xi(t;x)$ is the solution of the problem (3.0.1), we have

$$d\varphi(z(t)) = \sigma(\xi(t)) \left([\gamma_1(\xi(t))\beta(t)]y(t) + y_1(t)\right) dt$$

$$+ \sigma(\xi(t)) \sum_{i=1}^{d} \left([\gamma_1(\xi(t))y_2^*(t)e_i]e_i + \frac{1}{2}[\gamma_2(\xi(t))(e_i, e_i)]y(t) \right) dt \qquad (3.2.3)$$

$$+ \sigma(\xi(t)) \left([\gamma_1(\xi(t))dw(t)]y(t) + y_2(t)dw(t)\right).$$

Proof. The process $z(t) = (\xi(t; x), y(t))^t$ is the solution of the problem

$$dz(t) = B(z(t))dt + \Sigma(z(t))dw(t), \quad z(0) = (x, y(0))^t,$$

where, for any $t \geq 0$

$$B(z(t)) = \begin{pmatrix} b(\xi(t)) \\ y_1(t) \end{pmatrix}, \quad \Sigma(z(t)) = \begin{pmatrix} \sigma(\xi(t)) \\ y_2(t) \end{pmatrix}.$$

Now, since σ is twice continuously differentiable, the function φ is twice differentiable with continuous derivatives and it holds

$$D\varphi(x, y)(h, k) = [D\sigma(x)h]y + \sigma(x)k$$

$$D^2\varphi(x, y)((h_1, k_1), (h_2, k_2)) = [D^2\sigma(x)(h_1, h_2)]y + [D\sigma(x)h_1]k_2 + [D\sigma(x)h_2]k_1.$$

Then we can apply Itô's formula to the function φ and to the process $z(t)$ and we get

$$d\varphi(z(t)) = D\varphi(z(t))\,dz(t) + \frac{1}{2}\mathrm{Tr}\left[D^2\varphi(z(t))(\Sigma\Sigma^*)(z(t))\right]\,dt$$

$$= D\varphi(z(t))B(z(t))\,dt + D\varphi(z(t))\Sigma(z(t))dw(t) + \frac{1}{2}\mathrm{Tr}\left[D^2\varphi(z(t))(\Sigma\Sigma^*)(z(t))\right]\,dt.$$

We have

$$(\Sigma\Sigma^*)(z(t)) = \begin{pmatrix} (\sigma\sigma^*)(\xi(t)) & \sigma(\xi(t))y_2^*(t) \\ y_2(t)\sigma^*(\xi(t)) & (y_2(t)y_2^*(t)) \end{pmatrix}$$

and then, if $i = 1, \ldots, d$

$$D^2\varphi(z(t))((\Sigma\Sigma^*)(z(t))g_i, g_i)$$

$$= [D^2\sigma(\xi(t))((\sigma\sigma^*)(\xi(t))e_i, e_i)]y(t) + [D\sigma(\xi(t))e_i]y_2(t)\sigma^*(\xi(t))e_i$$

$$\tag{3.2.4}$$

and if $i = d+1, \ldots, 2d$

$$D^2\varphi(z(t))((\Sigma\Sigma^*)(z(t))g_i, g_i) = [D\sigma(\xi(t))\sigma(\xi(t))y_2^*(t)e_i]e_i. \tag{3.2.5}$$

It is easy to check that, if $A \in \mathcal{L}(\mathbb{R}^d)$ and $x \in \mathbb{R}^d$, it holds

$$\sum_{i=1}^d [D\sigma(x)e_i]Ae_i = \sum_{i=1}^d [D\sigma(x)A^*e_i]e_i.$$

In the same way, if $B \in \mathcal{L}(\mathbb{R}^d)$ and $x \in \mathbb{R}^d$

$$\sum_{i=1}^d D^2\sigma(x)(BB^*e_i, e_i) = \sum_{i=1}^d D^2\sigma(x)(Be_i, Be_i).$$

Due to (3.2.4) and (3.2.5) this implies that

$$\frac{1}{2}\mathrm{Tr}\left[D^2\varphi(z(t))(\Sigma\Sigma^\star)(z(t))\right]$$

$$=\sum_{i=1}^{d}\left(\frac{1}{2}[D^2\sigma(\xi(t))(\sigma(\xi(t))e_i,\sigma(\xi(t))e_i)]y(t)+[D\sigma(\xi(t))\sigma(\xi(t))y_2^\star(t)e_i]e_i\right).$$

Hence we can conclude that

$$d\varphi(z(t))=([D\sigma(\xi(t))b(\xi(t))]y(t)+\sigma(\xi(t))y_1(t))\ dt$$

$$+\sum_{i=1}^{d}\left(\frac{1}{2}[D^2\sigma(\xi(t))(\sigma(\xi(t))e_i,\sigma(\xi(t))e_i)]y(t)+[D\sigma(\xi(t))\sigma(\xi(t))y_2^\star(t)e_i]e_i\right)\ dt$$

$$+[D\sigma(\xi(t))\sigma(\xi(t))dw(t)]y(t)+\sigma(\xi(t))y_2(t)dw(t).$$

From the Hypothesis 3.1-2 we have

$$[D\sigma(\xi(t))b(\xi(t))]y(t)=[D\sigma(\xi(t))(\sigma\beta)(\xi(t))]y(t)$$

and then from the Hypothesis 3.1-3 it follows

$$[D\sigma(\xi(t))b(\xi(t))]y(t)=\sigma(\xi(t))[\gamma_1(\xi(t))\beta(\xi(t))]y(t).$$

Again, due to the Hypothesis 3.1-3 we have

$$\left[D^2\sigma(\xi(t))(\sigma(\xi(t))e_i,\sigma(\xi(t))e_i)\right]y(t)=\sigma(\xi(t))[\gamma_2(\xi(t))(e_i,e_i)]y(t)$$

and

$$[D\sigma(\xi(t))\sigma(\xi(t))dw(t)]y(t)=\sigma(\xi(t))[\gamma_1(\xi(t))dw(t)]y(t).$$

Finally, by using once more the Hypothesis 3.1-3

$$[D\sigma(\xi(t))\sigma(\xi(t))y_2^\star(t)e_i]e_i=\sigma(\xi(t))[\gamma_1(\xi(t))y_2^\star(\xi(t))e_i]e_i.$$

Therefore, collecting all terms, we get (3.2.3).   □

Proposition 3.2.2. *Under the Hypothesis 3.1, for any* $x,h\in\mathbb{R}^d$ *there exists a* d-*dimensional continuous process* $v^h(t;x)$ *such that*

$$D\xi(t;x)\sigma(x)h=\sigma(\xi(t;x))v^h(t;x),\quad\mathbb{P}-a.s.\qquad(3.2.6)$$

and

$$\sup_{x\in\mathbb{R}^d}\mathbb{E}|v^h(t;x)|^2\le c(t)|h|^2,\qquad(3.2.7)$$

for some continuous increasing function $c(t)$

Proof. Since $D\xi(t;x)\sigma(x)h$ is the unique solution of the problem (3.2.1) with initial value equal to $\sigma(x)h$, in order to prove (3.2.6) it is sufficient to show that there exist suitable mappings $v_1^h(\cdot;x) : [0,+\infty) \to \mathbb{R}^d$ and $v_2^h(\cdot;x) : [0,+\infty) \to \mathcal{L}(\mathbb{R}^d)$ such that

$$v^h(t;x) = h + \int_0^t v_1^h(s;x)\,ds + \int_0^t v_2^h(s;x)\,dw(s)$$

and

$$d\varphi(z(t)) = Db(\xi(t))\sigma(\xi(t))v^h(t)\,dt + [D\sigma(\xi(t))\sigma(\xi(t))v^h(t)]\,dw(t),$$

where $z(t) = (\xi(t;x),v^h(t;x))^t$ and $\varphi(x,y) = \sigma(x)y$ is the function introduced in the previous lemma.

By using (3.1.4) we have

$$Db(\xi(t))\sigma(\xi(t))v^h(t) = \sigma(\xi(t))\rho_1(\xi(t))v^h(t)$$

and from the Hypothesis 3.1-3 we have

$$[D\sigma(\xi(t))\sigma(\xi(t))v^h(t)]dw(t) = \sigma(\xi(t))[\gamma_1(\xi(t))v^h(t)]dw(t).$$

Therefore, if we assume that

$$\sigma(\xi(t))v_1^h(t) = \sigma(\xi(t))\left(\rho_1(\xi(t))v^h(t) - [\gamma_1(\xi(t))\beta(\xi(t))]v^h(t)\right)$$

$$-\sigma(\xi(t))\sum_{i=1}^d \left(\frac{1}{2}[\gamma_2(\xi(t))(e_i,e_i)]v^h(t) + [\gamma_1(\xi(t))(v_2^h(t))^*e_i]e_i\right)$$

and for any $k \in \mathbb{R}^d$

$$\sigma(\xi(t))v_2^h(t)k = \sigma(\xi(t))\left([\gamma_1(\xi(t))v^h(t)]k - [\gamma_1(\xi(t))k]v^h(t)\right),$$

due to (3.2.3), we are done. A possible choice for $v_2^h(t;x)k$ is

$$v_2^h(t;x)k = A_2^h(t;x)v^h(t;x) = [\gamma_1(\xi(t))v^h(t)]k - [\gamma_1(\xi(t))k]v^h(t),$$

so that for $v_1^h(t;x)$ we can take

$$v_1^h(t;x) = A_1^h(t;x)v^h(t;x) = \rho_1(\xi(t))v^h(t) - [\gamma_1(\xi(t))\beta(\xi(t))]v^h(t)$$

$$-\sum_{i=1}^d \left(\frac{1}{2}[\gamma_2(\xi(t))(e_i,e_i)]v^h(t) + \left[\gamma_1(\xi(t))\left(A_2^h(t;x)v^h(t;x)\right)^*e_i\right]e_i\right).$$

This means that the process $v^h(t;x)$ is the solution of the following linear stochastic equation with random coefficients

$$dv^h(t) = A_1^h(t;x)v^h(t)\,dt + A_2^h(t;x)v^h(t)\,dw(t), \quad v^h(0) = h. \tag{3.2.8}$$

Thanks to the boundedness of the functions γ_i and ρ_i, $i = 1, 2$, we have that there exists $c > 0$ such that

$$\sup_{t \geq 0,\, x \in \mathbb{R}^d} |A_1^h(t; x)|_{\mathcal{L}(\mathbb{R}^d)} + |A_2^h(t; x)|_{\mathcal{L}(\mathbb{R}^d; \mathcal{L}(\mathbb{R}^d))} \leq c, \quad \mathbb{P} - a.s.,$$

so that the equation (3.2.8) admits a unique strong solution, which is $v^h(t; x)$. Moreover, thanks to Itô's formula we have

$$d\,|v^h(t)|^2 = \left(2 \left\langle A_1^h(t; x)v^h(t), v^h(t) \right\rangle + \mathrm{Tr}\,[A_2^h(t; x)(A_2^h(t; x))^\star] \right) dt$$

$$+ 2 \left\langle A_2^h(t; x)v^h(t), v^h(t) \right\rangle dw(t),$$

so that, by using the boundedness of $A_1^h(t; x)$ and $A_2^h(t; x)$, from standard calculations we get (3.2.7). Finally, we remark that, since $\xi(t; x)$ has continuous trajectories and the mappings γ_i and ρ_i, $i = 1, 2$, are continuous, the process $v^h(t; x)$ is continuous. $\qquad \square$

Now we prove a similar result concerning the second derivative of $\xi(t; x)$.

Proposition 3.2.3. *Under the Hypothesis 3.1, for any $x, h \in \mathbb{R}^d$ there exists a d-dimensional continuous process $u^h(t; x)$ such that*

$$D^2 \xi(t; x)(\sigma(x)h, \sigma(x)h) = \sigma(\xi(t; x))u^h(t; x), \quad \mathbb{P} - a.s. \tag{3.2.9}$$

and

$$\sup_{x \in \mathbb{R}^d} \mathbb{E}\,|u^h(t; x)|^2 \leq c(t)|h|^4, \tag{3.2.10}$$

for some continuous increasing function $c(t)$.

Proof. The proof is analogous to that of the previous proposition. Actually, the process $D^2 \xi(t; x)(\sigma(x)h, \sigma(x)h)$ is the unique solution of the problem (3.2.2), with

$$d\theta(t) = D^2 b(\xi(t; x))(D\xi(t; x)\sigma(x)h, D\xi(t; x)\sigma(x)h)\, dt$$

$$+ D^2 \sigma(\xi(t; x))(D\xi(t; x)\sigma(x)h, D\xi(t; x)\sigma(x)h)\, dw(t).$$

Then, if $z(t) = (\xi(t; x), u^h(t; x))^t$ and $u^h(t; x)$ is a semi-martingale having drift term $u_1^h(t; x)$ and covariance term $u_2^h(t; x)$ and such that $u^h(0; x) = 0$, we have to impose the condition

$$d\varphi(z(t)) = Db(\xi(t))\sigma(\xi(t; x))u^h(t)\, dt + [D\sigma(\xi(t))\sigma(\xi(t; x))u^h(t)]\, dw(t) + d\theta(t).$$

By using the conditions described in (3.1.3) and (3.1.4), as in the proof of the previous proposition, we get an explicit expression for $u_1^h(t; x)$ and $u_2^h(t; x)$ and we prove the estimate (3.2.10), due to the boundedness of the coefficients. $\qquad \square$

3.3   A generalization of the Bismut-Elworthy formula

When the solution $\xi(t; x)$ of a general stochastic problem in a general separable Hilbert space H

$$d\xi(t) = b(\xi(t)) \, dt + \sigma(\xi(t)) \, dw(t), \quad \xi(0) = x \in H, \qquad (3.3.1)$$

is mean-square differentiable with respect to $x \in H$ and σ is strictly non-degenerate, that is there exists $\sigma^{-1}(x)$ for any $x \in H$ and $\sup_{x \in H} |\sigma^{-1}(x)| < \infty$, it is possible to prove the following Bismut-Elworthy formula for the first derivative of $P_t \varphi$

$$\langle (D(P_t\varphi))(x), h \rangle_H = \frac{1}{t} \mathbb{E} \varphi(\xi(t; x)) \int_0^t \left\langle \sigma^{-1}(\xi(s; x)) D\xi(s; x) h, dw(s) \right\rangle_H \qquad (3.3.2)$$

(see [7] and [57] for a proof in finite dimension and [40] and [104] for a proof in infinite dimension; see also the chapter 1 and the chapters 4,6 and 7). Such a formula is first proved for functions $\varphi \in C_b^2(H)$ and hence is extended to all functions $\varphi \in C_b(H)$. By using the semigroup law, the differentiability of $P_t \varphi$ follows for any $\varphi \in B_b(H)$.

In the next proposition we show that we can enlarge the use of such a formula to a wider class of situations.

Proposition 3.3.1. *Let P_t be the transition semigroup corresponding to (3.3.1). Assume that for any $\varphi \in C_b^1(H)$ and $t \geq 0$ the function $P_t \varphi$ is differentiable and for $t > 0$ and some fixed $x, h \in H$ it holds*

$$\langle (D(P_t\varphi))(x), h \rangle_H = \mathbb{E} \left\langle D\varphi(\xi(t; x)), \sigma(\xi(t; x)) v^h(t; x) \right\rangle_H, \qquad (3.3.3)$$

for some predictable process $v^h(t; x)$ such that

$$\mathbb{E} \int_0^t |v^h(s; x)|_H^2 \, ds < \infty. \qquad (3.3.4)$$

Then for the same x, h we have

$$\langle (D(P_t\varphi))(x), h \rangle_H = \frac{1}{t} \mathbb{E} \varphi(\xi(t; x)) \int_0^t \left\langle v^h(s; x), dw(s) \right\rangle_H. \qquad (3.3.5)$$

Proof. Let us fix $\varphi \in C_b^2(H)$, such that for any $x \in H$

$$\text{Tr} \, [D^2\varphi(x)] < \infty.$$

According to Itô's formula it is possible to prove that

$$\varphi(\xi(t; x)) = P_t\varphi(x) + \int_0^t \langle D(P_{t-s}\varphi)(\xi(s; x)), \sigma(\xi(s; x)) dw(s) \rangle_H.$$

Thanks to (3.3.4) we can multiply each side by $\int_0^t \left\langle v^h(s;x), dw(s) \right\rangle_H$ and, by taking the expectation, we get

$$\mathbb{E}\varphi(\xi(t;x)) \int_0^t \langle v(s;x,h), dw(s) \rangle_H$$

$$= \mathbb{E} \int_0^t \left\langle D(P_{t-s}\varphi)(\xi(s;x)), \sigma(\xi(s;x))v^h(s;x) \right\rangle_H ds.$$

Due to (3.3.3), this implies that

$$\mathbb{E}\varphi(\xi(t;x)) \int_0^t \left\langle v^h(s;x), dw(s) \right\rangle_H = \int_0^t \langle D(P_t\varphi)(x), h \rangle_H \, ds,$$

so that (3.3.5) follows.

Now, if $\varphi \in C_b^1(H)$, there exists a sequence $\{\varphi_n\} \subset C_b^2(H)$ such that

$$\text{Tr}\,[D^2\varphi_n(x)] < \infty \qquad \text{and} \qquad \sup_{n \in \mathbb{N}} \|\varphi_n\|_1 < \infty$$

and such that, for any $x, h \in H$, it holds

$$\begin{cases} \lim_{n \to +\infty} \varphi_n(x) = \varphi(x), \\[2mm] \lim_{n \to +\infty} \langle D\varphi_n(x), h \rangle_H = \langle D\varphi(x), h \rangle_H. \end{cases} \qquad (3.3.6)$$

Actually we can define for each $n \in \mathbb{N}$

$$\varphi_n(x) = \int_{\mathbb{R}^n} \varphi(T_n\xi)\rho_n(\Pi_n x - \xi)\, d\xi, \qquad x \in H,$$

where $\{\rho_n\}$ is a sequence in $C^\infty(\mathbb{R}^n)$ such that

$$\text{supp}\,\rho_n \subset \{\xi \in \mathbb{R}^n \,;\, |\xi| \leq 1/n\} \qquad \text{and} \qquad \int_{\mathbb{R}^n} \rho_n(\xi)\, d\xi = 1$$

and $T_n : \mathbb{R}^n \to \mathbb{R}$ and $\Pi_n : H \to \mathbb{R}^n$ are defined respectively by

$$T_n(\xi) = \sum_{k=1}^n \xi_k e_k,$$

$$\Pi_n(x) = (\langle x, e_1 \rangle_H, \ldots, \langle x, e_n \rangle_H)$$

for a complete orthonormal basis $\{e_k\}$ in H.

Now, as proved before, for each $n \in \mathbb{N}$ we have

$$\langle D(P_t\varphi_n)(x), h \rangle_H = \mathbb{E}\left\langle D\varphi_n(\xi(t;x)), \sigma(\xi(t;x))v^h(t;x) \right\rangle_H$$

and then, due to (3.3.4) and (3.3.6), by the dominated convergence theorem we have

$$\lim_{n \to +\infty} \langle D(P_t \varphi_n)(x), h \rangle_H$$

$$= \mathbb{E} \left\langle D\varphi(\xi(t;x)), \sigma(\xi(t;x))v^h(t;x) \right\rangle_H = \langle D(P_t \varphi)(x), h \rangle_H .$$

On the other hand, for each $n \in \mathbb{N}$ we have

$$\langle D(P_t \varphi_n)(x), h \rangle_H = \frac{1}{t} \mathbb{E} \varphi_n(\xi(t;x)) \int_0^t \left\langle v^h(s;x), dw(s) \right\rangle_H .$$

Thus, by using once more (3.3.4) and (3.3.6), from the dominated convergence theorem we have

$$\lim_{n \to +\infty} \langle D(P_t \varphi_n)(x), h \rangle_H = \frac{1}{t} \mathbb{E} \varphi(\xi(t;x)) \int_0^t \left\langle v^h(s;x), dw(s) \right\rangle_H .$$

This implies that (3.3.5) holds, for any $\varphi \in C_b^1(H)$. $\qquad \square$

Remark 3.3.2. 1. The formula (3.3.5) coincides with the formula (3.3.2), when σ is non-degenerate and the solution $\xi(t;x)$ is mean-square differentiable. As a matter of fact, in this case the process $v^h(t;x)$ introduced in (3.3.3) is given by

$$v^h(t;x) = \sigma^{-1}(\xi(t;x))D\xi(t;x)h.$$

2. In the book by Krylov [83, chapter 5], it is presented a finite dimensional case in which $\xi(t;x)$ is not mean-square differentiable, but for any $\varphi \in C_b^1(\mathbb{R}^d)$ the function $P_t \varphi$ is differentiable and for any $x, h \in \mathbb{R}^d$ it holds

$$\langle D(P_t \varphi)(x), h \rangle = \mathbb{E} \left\langle D\varphi(\xi(t;x)), \eta^h(t;x) \right\rangle,$$

where $\eta^h(t;x)$ is the solution of the first variation equation corresponding to the problem (3.3.1) (see also section 1.3). Then, if σ is non-degenerate, the formula (3.3.5) yields

$$\langle D(P_t \varphi)(x), h \rangle = \frac{1}{t} \mathbb{E} \varphi(\xi(t;x)) \int_0^t \left\langle \sigma^{-1}(\xi(s;x))\eta^h(s;x), dw(s) \right\rangle.$$

Actually, in this case the process $v^h(t;x)$ is given by

$$v^h(t;x) = \sigma^{-1}(\xi(t;x))\eta^h(t;x).$$

3. The formula (3.3.5) will be also applied in the second part to some systems of stochastic partial differential equations whose solutions are not mean-square differentiable with respect to initial datum, in general, and/or whose covariance operator has no bounded inverse.

3.4   The transition semigroup

For any $\varphi \in B_b(\mathbb{R}^d)$ and $t \geq 0$ we set

$$P_t\varphi(x) = \mathbb{E}\varphi(\xi(t;x)) = \int_{\mathbb{R}^d} \varphi(y) \, P_t(x, dy),$$

where $P_t(x, dy)$, $t \geq 0$ and $x \in \mathbb{R}^d$, is the transition probabilities family corresponding to the equation (3.0.1). Clearly, P_t defines a semigroup of contractions from $B_b(\mathbb{R}^d)$ into itself. Moreover, by proceeding as in the first chapter, it is possible to prove that for any $t \geq 0$ and $x, y \in \mathbb{R}^d$

$$\mathbb{E}|\xi(t;x) - \xi(t;y)|^2 \leq c(t) \, |x - y|^2.$$

Then, we easily have that P_t is a Feller semigroup, that is maps $C_b(\mathbb{R}^d)$ into itself.

As we recalled in the previous section, since we are assuming b and σ to be twice differentiable with bounded derivatives, the solution $\xi(t;x)$ of the problem (3.0.1) is twice mean-square differentiable with respect to $x \in \mathbb{R}^d$. Then, if $\varphi \in C_b^2(\mathbb{R}^d)$, by differentiating under the integration sign, we get that $P_t\varphi \in C_b^2(\mathbb{R}^d)$ and for any $x, h, k \in \mathbb{R}^d$ it holds

$$\langle D(P_t\varphi)(x), h \rangle = \mathbb{E}\langle D\varphi(\xi(t;x)), D\xi(t;x)h \rangle,$$

$$\langle D^2(P_t\varphi)(x)h, k \rangle = \mathbb{E}\langle D^2\varphi(\xi(t;x))D\xi(t;x)h, D\xi(t;x)k \rangle$$

$$+\mathbb{E}\langle D\varphi(\xi(t;x)), D^2\xi(t;x)(h, k) \rangle.$$

In the first chapter we have seen that if $\sigma(x)$ is invertible with bounded inverse, P_t maps Borel bounded functions into differentiable functions. In the present case, in which σ is possibly degenerate, the semigroup P_t has no regularizing effect. Nevertheless, by using the results proved in the previous section, we can give an expression for $\langle D(P_t\varphi)(x), \sigma(x)h \rangle$ and $\langle D^2(P_t\varphi)(x)\sigma(x)h, \sigma(x)h \rangle$, which does not involve the derivatives of φ.

Proposition 3.4.1. *Under the Hypothesis 3.1, for any $\varphi \in C_b^2(\mathbb{R}^d)$, $t > 0$ and $x, k \in \mathbb{R}^d$, we have*

$$\langle D(P_t\varphi)(x), \sigma(x)k \rangle = \frac{1}{t}\mathbb{E}\varphi(\xi(t;x)) \int_0^t \langle v^k(s;x), dw(s) \rangle, \qquad (3.4.1)$$

where $v^k(t;x)$ is the process introduced in the Proposition 3.2.2. Moreover it holds

$$\sup_{x \in \mathbb{R}^d} |\langle D(P_t\varphi)(x), \sigma(x)k \rangle| \leq c \, (t \wedge 1)^{-\frac{1}{2}} \|\varphi\|_0 |k|. \qquad (3.4.2)$$

Proof. If $v^k(t;x)$ is the process introduced in the Proposition 3.2.2, for any $\psi \in C_b^1(\mathbb{R}^d)$ we have

$$\mathbb{E}\Big\langle D\psi(\xi(t;x)), \sigma(\xi(t;x))v^k(t;x)\Big\rangle$$

$$= \mathbb{E}\langle D\psi(\xi(t;x)), D\xi(t;x)\sigma(x)k\rangle = \langle D(P_t\psi)(x), \sigma(x)k\rangle.$$

Then, by applying the Proposition 3.3.1 with $h = \sigma(x)k$ we get (3.4.1).

The estimate (3.4.2) easily follows from (3.2.7) and the semigroup law. Indeed, if $0 < t \leq 1$ we have

$$\frac{1}{t}\left|\mathbb{E}\varphi(\xi(t;x))\int_0^t \Big\langle v^k(s;x), dw(s)\Big\rangle\right| \leq \frac{1}{t}\|\varphi\|_0 \left(\mathbb{E}\left|\int_0^t \Big\langle v^k(s;x), dw(s)\Big\rangle\right|^2\right)^{\frac{1}{2}}$$

$$= \frac{1}{t}\|\varphi\|_0 \left(\mathbb{E}\int_0^t \Big|v^k(s;x)\Big|^2 ds\right)^{\frac{1}{2}} \leq ct^{-\frac{1}{2}}\|\varphi\|_0|k|.$$

If $t > 1$ we have

$$\langle D(P_t\varphi)(x), h\rangle = \langle D(P_1(P_{t-1}\varphi))(x), k\rangle$$

and then

$$\sup_{x\in\mathbb{R}^d} |\langle D(P_t\varphi)(x), k\rangle| \leq c\|P_{t-1}\varphi\|_0|k| \leq c\|\varphi\|_0|k|.$$

This implies (3.4.2).   □

Concerning the second derivative of $P_t\varphi$ we have the following result.

Proposition 3.4.2. *Assume that the Hypothesis 3.1 holds. Then, if $\varphi \in C_b^2(\mathbb{R}^d)$, for any $x, k \in \mathbb{R}^d$ and $t > 0$ we have*

$$\langle D^2(P_t\varphi)(x)\sigma(x)k, \sigma(x)k\rangle$$

$$= \frac{2}{t}\mathbb{E}\Big\langle D(P_{t/2}\varphi)(\xi(t/2;x)), D\xi(t/2;x)\sigma(x)k\Big\rangle \int_0^{t/2} \Big\langle v^k(s;x), dw(s)\Big\rangle$$

$$-\frac{2}{t}\mathbb{E}\int_0^{t/2} \Big\langle D(P_{t-s}\varphi)(\xi(s;x)), [D\sigma(\xi(s;x))D\xi(s;x)\sigma(x)k]\,v^k(s;x)\Big\rangle ds \qquad (3.4.3)$$

$$+\frac{2}{t}\mathbb{E}\int_0^{t/2} \Big\langle D(P_{t-s}\varphi)(\xi(s;x)), D^2\xi(s;x)(\sigma(x)k, \sigma(x)k)\Big\rangle ds$$

and the following estimate holds

$$\sup_{x\in\mathbb{R}^d} |\langle D^2(P_t\varphi)(x)\sigma(x)k, \sigma(x)k\rangle| \leq c\,(t \wedge 1)^{-1}\|\varphi\|_0|k|^2. \qquad (3.4.4)$$

Proof. Let us fix $t > 0$ and $0 \leq s \leq t$. Since $\varphi \in C_b^2(\mathbb{R}^d)$, due to Itô's formula we have

$$P_{t-s}\varphi(\xi(t;x)) = P_t\varphi(x) + \int_0^s \langle D(P_{t-r}\varphi)(\xi(r;x)), \sigma(\xi(r;x))dw(r) \rangle .$$

Then, by taking for each side the derivative with respect to $x \in \mathbb{R}^d$ along the direction $\sigma(x)k$, we get

$$\langle D(P_{t-s}\varphi)(\xi(s;x)), D\xi(s;x)\sigma(x)k \rangle$$

$$= \langle D(P_t\varphi)(x), \sigma(x)k \rangle + \int_0^s \langle D^2(P_{t-r}\varphi)(\xi(r;x))D\xi(r;x)\sigma(x)k, \sigma(\xi(r;x))dw(r) \rangle$$

$$+ \int_0^s \langle D(P_{t-r}\varphi)(\xi(r;x)), [D\sigma(\xi(r;x))D\xi(r;x)\sigma(x)k] \, dw(r) \rangle .$$

Thus, if we set $s = t$ we have

$$\langle D\varphi(\xi(t;x)), D\xi(t;x)\sigma(x)k \rangle = \langle D(P_t\varphi)(x), \sigma(x)k \rangle$$

$$+ \int_0^t \langle D^2(P_{t-r}\varphi)(\xi(r;x))D\xi(r;x)\sigma(x)k, \sigma(\xi(r;x))dw(r) \rangle$$

$$+ \int_0^t \langle D(P_{t-r}\varphi)(\xi(r;x)), [D\sigma(\xi(r;x))D\xi(r;x)\sigma(x)k] \, dw(r) \rangle .$$

Next, we multiply each side by $\int_0^t \langle v^k(r;x), dw(r) \rangle$ and by taking the expectation and recalling (3.2.6) it follows

$$\mathbb{E}\langle D\varphi(\xi(t;x)), D\xi(t;x)\sigma(x)k \rangle \int_0^t \langle v^k(r;x), dw(r) \rangle$$

$$= \mathbb{E}\int_0^t \langle D^2(P_{t-r}\varphi)(\xi(r;x))D\xi(r;x)\sigma(x)k, D\xi(r;x)\sigma(x)k \rangle \, dr$$

$$+ \mathbb{E}\int_0^t \langle D(P_{t-r}\varphi)(\xi(r;x)), [D\sigma(\xi(r;x))D\xi(r;x)\sigma(x)k] v^k(r;x) \rangle \, dr.$$

We recall that for any $0 \leq r \leq s$ it holds $\mathbb{E}P_{s-r}\varphi(\xi(r;x)) = P_s\varphi(x)$. Then, if we differentiate twice with respect to $x \in \mathbb{R}^d$ each side (with $s = t$) along the directions $\sigma(x)k$, we get

$$\langle D^2(P_t\varphi)(x)\sigma(x)k, \sigma(x)k \rangle$$

$$= \mathbb{E}\langle D^2(P_{t-r}\varphi)(\xi(r;x))D\xi(r;x)\sigma(x)k, D\xi(r;x)\sigma(x)k \rangle$$

$$+ \mathbb{E}\langle D(P_{t-r}\varphi)(\xi(r;x)), D^2\xi(r;x)(\sigma(x)k, \sigma(x)k) \rangle .$$

Therefore we can conclude that

$$t \left\langle D^2(P_t\varphi)(x)\sigma(x)k, \sigma(x)k \right\rangle$$

$$= \mathbb{E}\langle D\varphi(\xi(t;x)), D\xi(t;x)\sigma(x)k \rangle \int_0^t \left\langle v^k(r;x), dw(r) \right\rangle$$

$$+\mathbb{E} \int_0^t \left\langle D(P_{t-r}\varphi)(\xi(r;x)), D^2\xi(r;x)(\sigma(x)k, \sigma(x)k)) \right\rangle \, dr$$

$$-\mathbb{E} \int_0^t \left\langle D(P_{t-r}\varphi)(\xi(r;x)), [D\sigma(\xi(r;x))D\xi(r;x)\sigma(x)k] \, v^k(r;x) \right\rangle \, dr.$$

If we replace t with $s/2$ and φ with $P_{s/2}\varphi$, then (3.4.3) follows.

Finally, we prove the estimate (3.4.4). We can assume $0 < t \leq 1$. Actually, the general case $t > 0$ follows from the semigroup law as in the proof of the previous proposition. Due to (3.2.6) we have

$$\frac{2}{t}\mathbb{E}\langle D(P_{t/2}\varphi)(\xi(t/2;x)), D\xi(t/2;x)\sigma(x)k \rangle \int_0^{t/2} \left\langle v^k(s;x), dw(s) \right\rangle$$

$$= \frac{2}{t}\mathbb{E}\left\langle D(P_{t/2}\varphi)(\xi(t/2;x)), \sigma(\xi(t/2;x))v^k(t/2;x) \right\rangle \int_0^{t/2} \left\langle v^k(s;x), dw(s) \right\rangle,$$

then, since (3.4.2) holds, we have

$$\frac{2}{t}\left| \mathbb{E}\langle D(P_{t/2}\varphi)(\xi(t/2;x)), D\xi(t/2;x)\sigma(x)k \rangle \int_0^{t/2} \left\langle v^k(s;x), dw(s) \right\rangle \right|$$

$$\leq ct^{-3/2}\|\varphi\|_0 \left(\mathbb{E}|v^k(t/2;x)|^2\right)^{\frac{1}{2}} \left(\mathbb{E}\int_0^{t/2} |v^k(s;x)|^2 \, ds\right)^{\frac{1}{2}}.$$

Thanks to (3.2.7), it follows that

$$\frac{2}{t}\left| \mathbb{E}\langle D(P_{t/2}\varphi)(\xi(t/2;x)), D\xi(t/2;x)\sigma(x)k \rangle \int_0^{t/2} \left\langle v^k(s;x), dw(s) \right\rangle \right|$$

$$\leq ct^{-1}\|\varphi\|_0|k|^2.$$

By using (3.2.9) we have

$$\frac{2}{t}\mathbb{E}\int_0^{t/2} \left\langle D(P_{t-s}\varphi)(\xi(s;x)), D^2\xi(s;x)(\sigma(x)k, \sigma(x)k) \right\rangle \, ds$$

$$= \frac{2}{t}\mathbb{E}\int_0^{t/2} \left\langle D(P_{t-s}\varphi)(\xi(s;x)), \sigma(\xi(s;x))u^k(s;x) \right\rangle \, ds,$$

and then, due to (3.2.10) and (3.4.2), we get

$$\frac{2}{t}\left|\mathbb{E}\int_0^{t/2}\left\langle D(P_{t-s}\varphi)(\xi(s;x)), D^2\xi(s;x)(\sigma(x)k,\sigma(x)h)\right\rangle\, ds\right|$$

$$\leq c\,t^{-1}\|\varphi\|_0\int_0^{t/2}(t-s)^{-1/2}\mathbb{E}|u^k(s;x)|\,ds \leq c\,t^{-1}\|\varphi\|_0|k|^2.$$

Moreover, due to the Hypothesis 3.1-3 and due to (3.2.6), we have

$$[D\sigma(\xi(s;x))D\xi(s;x)\sigma(x)k]\,v^k(s;x) = \left[D\sigma(\xi(s;x))\sigma(\xi(s;x))v^k(s;x)\right]v^k(s;x)$$

$$= \sigma(\xi(s;x))\left[\gamma_1(\xi(s;x))v^k(s;x)\right]v^k(s;x).$$

Therefore, by using (3.4.2), we obtain

$$\frac{2}{t}\left|\mathbb{E}\int_0^{t/2}\left\langle D(P_{t-s}\varphi)(\xi(s;x)), [D\sigma(\xi(s;x))D\xi(s;x)\sigma(x)k]\,v^k(s;x)\right\rangle\, ds\right|$$

$$\leq c\,t^{-1}\|\varphi\|_0\int_0^{t/2}(t-s)^{-1/2}\mathbb{E}|v^k(s;x)|^2\,ds \leq c\,t^{-1}\|\varphi\|_0|k|^2.$$

Hence, collecting all terms, the estimate (3.4.4) follows. $\square$

As an immediate consequence of Itô's formula and of the previous two propositions we have

Corollary 3.4.3. *Assume the Hypothesis 3.1. Then for any* $\varphi \in C_b^2(\mathbb{R}^d)$ *and* $x \in \mathbb{R}^d$, *the mapping* $(0,+\infty) \to \mathbb{R},\ t \mapsto P_t\varphi(x)$, *is differentiable and it holds*

$$\frac{d}{dt}P_t\varphi(x) = \mathcal{L}_0(P_t\varphi)(x).$$

Moreover we have

$$\sup_{x\in\mathbb{R}^d}|\frac{d}{dt}P_t\varphi(x)| = \|\mathcal{L}_0(P_t\varphi)\|_0 \leq c\,(t\wedge 1)^{-1}\,\|\varphi\|_0. \qquad (3.4.5)$$

Proof. Since φ is assumed to be twice differentiable with bounded derivative, we can apply Itô's formula and we have that

$$\frac{d}{dt}P_t\varphi(x) = \frac{1}{2}\sum_{i=1}^d\langle D^2(P_t\varphi)(x)\sigma(x)e_i, \sigma(x)e_i\rangle + \langle D(P_t\varphi)(x), b(x)\rangle = \mathcal{L}_0(P_t\varphi)(x).$$

Now, since $b(x) = \sigma(x)\beta(x)$, we have that

$$\langle D(P_t\varphi)(x), b(x)\rangle = \langle D(P_t\varphi)(x), \sigma(x)\beta(x)\rangle$$

and recalling that β is assumed to be bounded, by applying the Propositions 3.4.1 and 3.4.2, we get (3.4.5). $\square$

3.5   The generation result

In the present section we want to use the previous results to show that the semigroup P_t is analytic in $C_b(\mathbb{R}^d)$. To this purpose we use the notion of *weak generator* introduced in [17](see the appendix B for definitions).

Actually, as proved in the Proposition 1.4.1, the semigroup P_t is weakly continuous. Thus, we can define its generator as the unique closed linear operator $\mathcal{L} : D(\mathcal{L}) \subseteq C_b(H) \to C_b(H)$ such that

$$R(\lambda, \mathcal{L})\varphi(x) = \int_0^{+\infty} e^{-\lambda t} P_t\varphi(x)\, dt, \qquad \operatorname{Re}\lambda > 0,$$

for any $\varphi \in C_b(H)$ and $x \in \mathbb{R}^d$.

As proved in the Lemma B.2.1, $P_t D(\mathcal{L}) \subseteq D(\mathcal{L})$, for any $t \geq 0$, and if $\varphi \in D(\mathcal{L})$ then $P_t(\mathcal{L}\varphi) = \mathcal{L}(P_t\varphi)$. Moreover, if $\varphi \in D(\mathcal{L})$, for any fixed $x \in \mathbb{R}^d$ the mapping $[0, +\infty) \to \mathbb{R}, \ t \mapsto P_t\varphi(x)$, is differentiable and

$$\frac{d}{dt} P_t\varphi(x) = \mathcal{L}(P_t\varphi)(x) = P_t(\mathcal{L}\varphi)(x)$$

(see the Proposition B.2.2 for a proof). In particular, we have

Lemma 3.5.1. *If $\varphi \in C_b^2(\mathbb{R}^d)$, we have that $P_t\varphi \in D(\mathcal{L})$, for any $t \geq 0$, and*

$$\mathcal{L}(P_t\varphi) = \mathcal{L}_0(P_t\varphi).$$

Proof. We first remark that if $\varphi \in C_b^2(\mathbb{R}^d)$ then

$$\mathcal{L}_0(P_t\varphi) = P_t(\mathcal{L}_0\varphi). \tag{3.5.1}$$

Indeed, from Itô's formula we have

$$\frac{P_{t+h}\varphi(x) - P_t\varphi(x)}{h} = \frac{1}{h}\mathbb{E}\left(\varphi(\xi(t+h; x)) - \varphi(\xi(t; x))\right)$$

$$= \frac{1}{h}\mathbb{E} \int_t^{t+h} \left[\operatorname{Tr}\left[a(\xi(s; x))D^2\varphi(\xi(s; x))\right] + \langle D\varphi(\xi(s; x)), b(\xi(s; x))\rangle \right] ds.$$

Then, by using the dominated convergence theorem, it follows

$$\lim_{h \to 0+} \frac{P_{t+h}\varphi(x) - P_t\varphi(x)}{h}$$

$$= \mathbb{E}\left(\operatorname{Tr}\left[a(\xi(t; x))D^2\varphi(\xi(t; x))\right] + \langle D\varphi(\xi(t; x)), b(\xi(t; x))\rangle\right) = P_t(\mathcal{L}_0\varphi)(x).$$

Recalling that

$$\frac{d}{dt} P_t\varphi(x) = \mathcal{L}_0(P_t\varphi)(x),$$

this yields (3.5.1).

Now, for any $t \geq 0$ we define

$$\psi_t = \lambda P_t \varphi - \mathcal{L}_0(P_t \varphi).$$

By using (3.5.1) and the definition of $\mathcal{L}$, we have

$$R(\lambda, \mathcal{L})\psi_t(x) = \int_0^{+\infty} e^{-\lambda s} P_s \left(\lambda P_t \varphi - \mathcal{L}_0(P_t \varphi)\right)(x)\, ds$$

$$= \int_0^{+\infty} e^{-\lambda s} \left(\lambda P_{t+s}\varphi(x) - \mathcal{L}_0(P_{t+s}\varphi)(x)\right) ds,$$

so that

$$R(\lambda, \mathcal{L})\psi_t(x) = -\int_0^{+\infty} \frac{d}{ds}\left(e^{-\lambda s} P_{t+s}\varphi(x)\right) ds = P_t\varphi(x).$$

This means that $P_t \varphi \in D(\mathcal{L})$ and $\mathcal{L}(P_t \varphi) = \mathcal{L}_0(P_t \varphi)$. $\square$

Theorem 3.5.2. *Assume the Hypothesis 3.1. Then, for any $\varphi \in C_b(\mathbb{R}^d)$ and $t > 0$ we have that $P_t \varphi \in D(\mathcal{L})$. In particular, for any fixed $x \in \mathbb{R}^d$, the mapping $(0, +\infty) \to \mathbb{R}$, $t \mapsto P_t\varphi(x)$, is differentiable and*

$$\sup_{x \in \mathbb{R}^d} |\frac{d}{dt}(P_t\varphi)(x)| = \|\mathcal{L}(P_t\varphi)\|_0 \leq c\,(t \wedge 1)^{-1}\|\varphi\|_0, \qquad (3.5.2)$$

so that P_t is analytic in $C_b(\mathbb{R}^d)$.

Proof. We have seen above that if $\varphi \in C_b^2(\mathbb{R}^d)$, then $P_t\varphi \in D(\mathcal{L})$, for any $t > 0$, and (3.5.2) holds. Now, for any $\varphi \in C_b(\mathbb{R}^d)$ let us fix a sequence $\{\varphi_n\} \subset C_b^2(\mathbb{R}^d)$ converging to φ in $C_b(\mathbb{R})$. We have

$$\lim_{n \to +\infty} P_t\varphi_n = P_t\varphi \qquad \text{in } C_b(\mathbb{R}^d)$$

and, by using (3.4.5) we have

$$\|\mathcal{L}(P_t\varphi_n) - \mathcal{L}(P_t\varphi_m)\|_0 = \|\mathcal{L}_0(P_t(\varphi_n - \varphi_m))\|_0 \leq c\,(t \wedge 1)^{-1}\|\varphi_n - \varphi_m\|_0,$$

so that $\{\mathcal{L}(P_t\varphi_n)\}$ is a Cauchy sequence in $C_b(\mathbb{R}^d)$. Then, as $\mathcal{L}$ is a closed operator, we have that $P_t\varphi \in D(\mathcal{L})$ and

$$\mathcal{L}(P_t\varphi) = \lim_{n \to +\infty} \mathcal{L}(P_t\varphi_n).$$

In particular, as (3.5.2) holds for φ_n, (3.5.2) follows for any $\varphi \in C_b(\mathbb{R}^d)$.

Now, by proceeding as in the standard case of analytic semigroups, it easily follows that, for any $\varphi \in C_b(\mathbb{R}^d)$ and $x \in \mathbb{R}^d$, the mapping $(0, +\infty) \to \mathbb{R}$, $t \mapsto P_t\varphi(x)$ is infinitely differentiable and

$$\frac{d^n}{dt^n}(P_t\varphi(x)) = (\mathcal{L}(P_{t/n}))^n \varphi(x), \quad t > 0,$$

so that from (3.5.2) we have

$$\sup_{x \in \mathbb{R}^d} \left| \frac{d^n}{dt^n}(P_t\varphi(x)) \right| \leq c^n \left(\frac{t}{n} \wedge 1 \right)^{-n} \|\varphi\|_0.$$

As well known, this implies that the mapping $(0, +\infty) \to \mathbb{R}$, $t \mapsto P_t\varphi(x)$, admits an analytic extension to a sector around the positive real axis, which is independent of $x \in \mathbb{R}^d$. Therefore, as proved in [119, Theorem IX-10], we have that there exists a constant $c > 0$ such that for any $\varphi \in C_b(\mathbb{R}^d)$

$$\|\lambda R(\lambda, \mathcal{L})\varphi\|_0 \leq c\|\varphi\|_0, \quad \operatorname{Re}\lambda > 0.$$

This means that $\mathcal{L}$ generates an analytic semigroup (see [91, Proposition 2.1.11]) and such a semigroup coincides clearly with P_t.   □

Part II

Infinite dimension

Chapter 4

Smooth dependence on data for the SPDE: the Lipschitz case

In this second part we are dealing with a certain class of reaction-diffusion systems in bounded domains of $\mathbb{R}^d$, perturbed by a gaussian random field.

In the present chapter and in the following one we consider the case of Lipschitz-continuous reaction terms. We first prove the differentiable dependence on initial datum for the solution of the system and then we study the regularizing properties of the associated transition semigroup. In the next chapter we apply these results to the study of the corresponding Kolmogorov equations.

In the chapters 6 and 7 we consider the case of reaction terms having polynomial growth. We study the smoothing effect of the transition semigroup both in the space of square integrable functions and in the space of continuous functions. In the last three chapters we study the asymptotic behaviour of the system and some stochastic optimal control problems.

Let $\mathcal{O}$ be a bounded domain in $\mathbb{R}^d$, with $d \leq 3$, having a regular boundary $\partial\mathcal{O}$. We consider the following reaction-diffusion system perturbed by a stochastic term

$$
\begin{cases}
\dfrac{\partial u_k}{\partial t}(t,\xi) = \mathcal{A}_k(\xi, D)u_k(t,\xi) + f_k(\xi, u_1(t,\xi), \ldots, u_n(t,\xi)) + Q_k\dfrac{\partial^2 w_k}{\partial t \partial \xi}(t,\xi) \\[2mm]
u_k(0,\xi) = x_k(\xi), \quad t > 0, \ \xi \in \overline{\mathcal{O}} \\[2mm]
\mathcal{B}_k(\xi, D)u_k(t,\xi) = 0, \quad t > 0, \ \xi \in \partial\mathcal{O}, \qquad k = 1, \ldots, n.
\end{cases}
$$

$$(4.0.1)$$

For each k, $\mathcal{A}_k(\xi, D)$ is a uniformly elliptic operator with regular real coefficients given by

$$
\mathcal{A}_k(\xi, D) = \sum_{i,j=1}^{d} a_{ij}^k(\xi)\frac{\partial^2}{\partial\xi_i\,\partial\xi_j} + \sum_{i=1}^{d} b_i^k(\xi)\frac{\partial}{\partial\xi_i}, \qquad \xi \in \overline{\mathcal{O}}
$$

and $\mathcal{B}_k(\xi, D)$ is a linear operator acting on the boundary of $\mathcal{O}$, which may be taken either equal to the identity operator (Dirichlet boundary condition) or equal to a first order operator satisfying a non-tangentiality condition (oblique boundary condition). The function $f = (f_1, \ldots, f_n) : \overline{\mathcal{O}} \times \mathbb{R}^n \to \mathbb{R}^n$ fulfills a Charatheodory condition and for any fixed $\xi \in \overline{\mathcal{O}}$ the function $f(\xi, \cdot) : \mathbb{R}^n \to \mathbb{R}^n$ is of class C^r, for some $r \geq 2$. Q_k are non-negative bounded linear operators from $L^2(\mathcal{O})$ into itself and $\partial^2 w_k / \partial t \, \partial \xi$ are independent space-time white noises defined on the stochastic basis $(\Omega, \mathcal{F}, \mathcal{F}_t, \mathbb{P})$. Throughout this second part the reference Hilbert space H will be $L^2(\mathcal{O}; \mathbb{R}^n)$. For all notations see the introduction.

In the present chapter we assume that $f(\xi, \cdot)$ is Lipschitz-continuous, uniformly with respect to $\xi \in \overline{\mathcal{O}}$, so that the system (4.0.1) admits an unique mild solution $u(\cdot; x)$, for any initial datum $x \in H$. Our aim is to study the regularizing properties of the transition semigroup P_t associated with (4.0.1). P_t is a Feller semigroup, that is it maps $C_b(H)$ into itself, but in general is not strongly continuous in $C_b(H)$. Nevertheless, we show that it is a weakly continuous semigroup (see the appendix B for the definition and main properties). However, the main effort of this chapter is to show that P_t has a smoothing effect. Namely, we prove that for some k depending on the dimension d and on the order of differentiability of f it holds

$$\varphi \in B_b(H) \Rightarrow P_t \varphi \in C_b^k(H), \quad t > 0.$$

Moreover we prove that for any $j = 1, \ldots, k$ the following estimate holds

$$\sup_{x \in H} |D^j (P_t \varphi)(x)| \leq c \, (t \wedge 1)^{-\frac{j(1+\epsilon)}{2}} \sup_{x \in H} |\varphi(x)|,$$

where ϵ is a constant which depends on Q_k and $\mathcal{A}_k$.

In the section 4.2 we study the differential dependence on initial datum for the solution $u(\cdot; x)$ of the system (4.0.1). The problem of differential dependence on initial datum for diffusion processes is well studied if the coefficients are assumed to be Fréchet differentiable (see for example [47] or [12], [14]). But here $f(\xi, \cdot)$ is not assumed to be necessarily linear for any $\xi \in \overline{\mathcal{O}}$, so that the corresponding Nemytskii operator F is not Fréchet differentiable in H with bounded derivatives and we cannot proceed in the usual way. However, concerning the first derivative, as F is Gâteaux differentiable at any point $x \in H$, by using a modification of the principle of contractions depending on parameters, we can proceed as in the standard case. More serious problems arise for higher order derivatives. Indeed, F is not twice differentiable along any directions of H, but only along directions in more regular spaces, like for example $L^4(\mathcal{O}; \mathbb{R}^n)$, and then, in order to prove the existence of the second order derivative, we have to prove as a preliminary step that $D_x u(t; x) h$ is a $L^4(\mathcal{O}; \mathbb{R}^n)$-valued process, for any $h \in H$ and $t > 0$. We show that this is true since the semigroup e^{tA} generated by the operator $(\mathcal{A}_1, \ldots, \mathcal{A}_n)$ satisfies a *ultracontractivity* property, that is maps H into $L^\infty(\mathcal{O}; \mathbb{R}^n)$, for any $t > 0$ and

$$|e^{tA} x|_\infty \leq c \, t^{-\frac{d}{4}} |x|_H, \quad x \in H.$$

As far as the higher order derivative are concerned, we use similar arguments, with $L^4(\mathcal{O}; \mathbb{R}^n)$ replaced by $L^{2j}(\mathcal{O}; \mathbb{R}^n)$.

It is worth-while to notice that we are not able to use these arguments for any order derivatives, because after a certain number of derivatives the singularities arising from ultracontractivity property are no more integrable. This difficulty seems to be not only technical, so that we can conjecture that even if $f \in C^\infty(\mathcal{O} \times \mathbb{R}^n; \mathbb{R}^n)$ nevertheless $u(\cdot; x)$ is in general only k times mean-square differentiable, with

$$k < \frac{4}{d} + 2.$$

In the section 4.3 we introduce the transition semigroup P_t associated with the system (4.0.1) and we study its properties in $C_b(H)$. Namely, we show that it is a Feller semigroup and is weakly continuous.

In the section 4.4, by using the differentiability properties of the solution $u(t; x)$, we show that P_t maps $B_b(H)$ into $C_b^k(H)$. The main tool is the following generalization of the Bismut-Elworthy formula

$$\langle D(P_t\varphi)(x), h\rangle_H = \frac{1}{t}\, \mathbb{E}\,\varphi(u(t; x)) \int_0^t \langle Q^{-1} D_x u(s; x)h, dw(s)\rangle_H, \qquad (4.0.2)$$

where $D_x u(s; x)h$ is the first mean-square derivative of $u(t; x)$ along the direction $h \in H$ and Q^{-1} is the *pseudo-inverse* of $Q = (Q_1, \ldots, Q_n)$.

Notice that the present situation is not covered by the papers by Da Prato, Elworthy and Zabczyk [40] and by Peaszat and Zabczyk [104], where they extend to the infinite dimensional case (respectively in the case of additive and multiplicative noise) the classical Bismut-Elworthy formula which is given in a finite dimensional setting (see [7] and [57]). Actually, here Q is not assumed to have a bounded inverse. Thus, in order to give a meaning to the Itô integral appearing in the right side of (4.0.2), we have first to prove that $D_x u(s; x)h \in D(Q^{-1})$ for any $x, h \in H$ and $s > 0$ and then that

$$\mathbb{E} \int_0^t |Q^{-1} D_x u(s; x)h|_H^2 \, ds < +\infty.$$

As usual, higher order differentiability of P_t, up to the k-th order, follows from the formula (4.0.2), Itô's formula and the semigroup law.

4.1 Notations and assumptions

4.1.1 The operator A

All the results concerning the theory of elliptic operators and analytic semigroups that will be used in what follows can be found in Agmon [1], Davies [50] and Lunardi [91]. For the proofs we refer to those monographs.

We shall denote by $\mathcal{A}$ the second order differential operator defined by

$$\mathcal{A}x = (\mathcal{A}_1 x_1, \dots, \mathcal{A}_n x_n), \quad x = (x_1, \dots, x_n) \in H,$$

where for any $k = 1, \dots, n$

$$\mathcal{A}_k(\xi, D) = \sum_{i,j=1}^{d} a_{ij}^k(\xi) \frac{\partial^2}{\partial \xi_i \partial \xi_j} + \sum_{i=1}^{d} b_i^k(\xi) \frac{\partial}{\partial \xi_i}, \quad \xi \in \overline{\mathcal{O}}.$$

The coefficients a_{ij}^k and b_i^k are assumed to be in $C^1(\overline{\mathcal{O}})$ and for any $\xi \in \overline{\mathcal{O}}$ and $k \le n$ the matrix $[a_{ij}^k(\xi)]$ is non-negative and symmetric and satisfies the uniform ellipticity condition

$$\inf_{\xi \in \overline{\mathcal{O}}} \sum_{i,j=1}^{d} a_{ij}^k(\xi) h_i h_j \ge \nu |h|^2, \quad h \in \mathbb{R}^d,$$

for some $\nu > 0$. The boundary operator $\mathcal{B}$ is defined by $\mathcal{B}x = (\mathcal{B}_1 x_1, \dots, \mathcal{B}_n x_n)$ and for each $k \le n$

$$\mathcal{B}_k(\xi, D) = \sum_{i=1}^{d} \beta_i^k(\xi) \frac{\partial}{\partial \xi_i} + \gamma^k(\xi) I, \quad \xi \in \mathcal{O},$$

for some β_i^k and γ^k which belong to $C^1(\overline{\mathcal{O}})$. We assume that either $\beta_i^k \equiv 0$ and $\gamma^k \equiv 1$ on $\partial\mathcal{O}$ (Dirichlet boundary condition), or

$$\inf_{\xi \in \partial\mathcal{O}} \left| \sum_{i=1}^{d} \beta_i^k(\xi) \nu_i(\xi) \right| > 0,$$

where $\nu(\xi) = (\nu_1(\xi), \dots, \nu_d(\xi))$ is the normal vector to the boundary of $\overline{\mathcal{O}}$ (oblique boundary condition).

In what follows we denote by A' the realization in H of the elliptic operator $\mathcal{A}$, with the boundary conditions given by $\mathcal{B}$

$$D(A') = \left\{ x \in H : \mathcal{A}x \in H, \ \mathcal{B}x_{|\partial\mathcal{O}} = 0 \right\}, \quad A'x = \mathcal{A}x, \ x \in D(A').$$

The operator A' generates an analytic semigroup $e^{tA'}$. Moreover, if we denote by $e^{tA'_k}$ the analytic semigroup generated by the realization A'_k in $L^2(\mathcal{O})$ of the elliptic operator $\mathcal{A}_k$ with the boundary condition given by $\mathcal{B}_k$, it is trivial to check that for any $x = (x_1, \dots, x_n) \in H$ it holds

$$e^{tA'} x = (e^{tA'_1} x_1, \dots, e^{tA'_n} x_n).$$

It is useful to remark that all properties satisfied by A' are derived from analogous properties fulfilled by each operator A'_k.

In a similar way, for any $p \in (1, \infty]$ the realization in $L^p(\mathcal{O}; \mathbb{R}^n)$ of the operator $\mathcal{A}$, with the boundary conditions given by $\mathcal{B}$, generates an analytic semigroup $e^{t(A^p)'}$. All these semigroups $e^{t(A^p)'}$, for $p \in (1, \infty]$, are consistent, in the sense that $e^{t(A^p)'}x = e^{t(A^q)'}x$, for any $x \in L^p(\mathcal{O}; \mathbb{R}^n) \cap L^q(\mathcal{O}; \mathbb{R}^n)$. This is why we shall omit the superscript p and we shall denote them by $e^{tA'}$. As the semigroups $e^{tA'_k}$ are all analytic in $L^p(\mathcal{O})$, for each $p \in (1, \infty]$, the semigroup $e^{tA'}$ is analytic in $L^p(\mathcal{O}; \mathbb{R}^n)$, for each $p \in (1, +\infty]$, so that there exist two constants $M_p > 0$ and $\rho_p \in \mathbb{R}$ such that

$$|e^{tA'}x|_p \le M_p e^{\rho_p t}|x|_p.$$

Notice that, due to the Riesz-Thorin theorem, it is possible to fix two constants $M > 0$ and $\rho \in \mathbb{R}$ independent of p such that

$$|e^{tA'}x|_p \le M e^{\rho t}|x|_p, \tag{4.1.1}$$

for any $p \in (1, \infty]$. In what follows it will be convenient to define

$$A = A' - (\rho + 1)I.$$

The operator A generates the semigroup $e^{tA} = e^{-(\rho+1)t}e^{tA'}$, which enjoys the same properties as A'. Moreover, according to (4.1.1), for any $p \in (1, \infty]$ we have

$$|e^{tA}x|_p \le M e^{-(\rho+1)t}e^{\rho t}|x|_p = M e^{-t}|x|_p. \tag{4.1.2}$$

This means that e^{tA} is of *negative type* For any $p \in (1, \infty)$, the domain of A in $L^p(\mathcal{O}; \mathbb{R}^n)$ coincides with the functions φ in $W_0^{2,p}(\mathcal{O}; \mathbb{R}^n)$, for the Dirichlet boundary conditions, and with the functions φ in $W^{2,p}(\mathcal{O}; \mathbb{R}^n)$ such that $\mathcal{B}\varphi = 0$ on ∂O, for the oblique boundary conditions. Moreover, for any $t > 0$ the semigroup e^{tA} maps $L^p(\mathcal{O}; \mathbb{R}^n)$ into $D(A)$ and it holds

$$|e^{tA}x|_{2,p} \le c \left(|e^{tA}x|_p + |Ae^{tA}x|_p \right) \le c e^{-t}(t \wedge 1)^{-1}|x|_p,$$

for some $c > 0$ depending on p. Now, recalling that for any $s \in (0, 2)$

$$W^{s,p}(\mathcal{O}; \mathbb{R}^n) = \left(L^p(\mathcal{O}; \mathbb{R}^n), W^{2,p}(\mathcal{O}; \mathbb{R}^n) \right)_{\frac{s}{2}, \infty},$$

by interpolation we get that e^{tA} maps $L^p(\mathcal{O}; \mathbb{R}^n)$ into $W^{s,p}(\mathcal{O}; \mathbb{R}^n)$, for any $t > 0$ and

$$|e^{tA}x|_{s,p} \le c e^{-\frac{s}{2}t}(t \wedge 1)^{-\frac{s}{2}}|x|_p. \tag{4.1.3}$$

From the Sobolev embedding theorem we know that $W^{s,2}(\mathcal{O}; \mathbb{R}^n)$ is continuously embedded into $L^\infty(\mathcal{O}; \mathbb{R}^n)$, for any $s > d/2$. Then the semigroup e^{tA} is *ultracontractive*, that is it maps H into $L^\infty(\mathcal{O}; \mathbb{R}^n)$ and thanks to (4.1.3) we have

$$|e^{tA}x|_\infty \le c e^{-\frac{d}{4}t}(t \wedge 1)^{-\frac{d}{4}}|x|_H. \tag{4.1.4}$$

In what follows it will be essential that the ultracontractivity exponent is less than 1. This is the reason why we assume $d \leq 3$. Now, we recall that $W^{\frac{d(2-p)}{2p}, p}(\mathcal{O}; \mathbb{R}^n)$ is continuously embedded into H, for any $p \in (1, 2)$. Therefore, from (4.1.3) we get

$$|e^{tA}x|_H \leq c_p \, t^{-\frac{(2-p)d}{4p}} |x|_p, \qquad x \in L^p(\mathcal{O}; \mathbb{R}^n).$$

By using the Riesz-Thorin interpolation theorem, due to (4.1.4) this implies that for any $1 < p \leq q \leq \infty$ and $t > 0$

$$|e^{tA}x|_p \leq c_p \, t^{-\frac{d(p-q)}{2pq}} |x|_q, \qquad x \in L^q(\mathcal{O}; \mathbb{R}^n). \tag{4.1.5}$$

Moreover, we recall that with the boundary conditions given by B the operator A is of negative type, that is

$$\langle Ax, x \rangle_H \leq 0, \qquad x \in H. \tag{4.1.6}$$

Since the operator A generates an analytic semigroup e^{tA} of negative type, for any $\delta \geq 0$ we can introduce the *fractional power* $(-A)^\delta$ of $-A$, defined by

$$(-A)^\delta = \frac{1}{\Gamma(\delta)} \int_0^{+\infty} t^{\delta-1} e^{tA} \, dt,$$

where Γ is the Euler function. It is possible to show that $\mathrm{Range}(e^{tA}) \subset D((-A)^\delta)$, for any $t > 0$, and

$$\left| (-A)^\delta e^{tA} x \right|_H \leq M_\delta \, t^{-\delta} |x|_H, \qquad x \in H, \tag{4.1.7}$$

for a suitable positive constant M_δ. Moreover, it is easy to verify that if $\delta_1 \leq \delta \leq \delta_2$ and $x \in D((-A)^{\delta_2})$ the following interpolatory inequality holds

$$\left| (-A)^\delta x \right|_H \leq c \left| (-A)^{\delta_1} x \right|_H^{1-\theta} \left| (-A)^{\delta_2} x \right|_H^{\theta}, \qquad \theta = \frac{\delta - \delta_1}{\delta_2 - \delta_1}. \tag{4.1.8}$$

4.1.2   The operator Q and the stochastic convolution $w^A(t)$

Let $\{w_i(t)\}$ be a sequence of mutually independent real Brownian motions defined on a stochastic basis $(\Omega, \mathcal{F}, \mathcal{F}_t, \mathbb{P})$ and adapted to the non-anticipative filtration $\mathcal{F}_t$, $t \geq 0$. We define the cylindrical Wiener process $w(t)$ as

$$w(t) = \sum_{i=1}^{\infty} e_i w_i(t), \tag{4.1.9}$$

where $\{e_i\}$ is a complete orthonormal system of H. The series (4.1.9) defining $w(t)$ does not converge in H, but it is convergent in any Hilbert space U such that the embedding $H \subset U$ is Hilbert-Schmidt (see [47, chapter 4]).

Now, for any $x \in H$ we define

$$Qx = (Q_1 x_1, \ldots, Q_n x_n).$$

We assume that Q fulfills the following conditions.

Hypothesis 4.1.   *1. The bounded linear operator $Q : H \to H$ is non-negative and it holds*

$$\int_0^t \mathrm{Tr}\left[e^{sA} QQ^\star e^{sA^\star}\right] ds < \infty, \qquad t \geq 0. \tag{4.1.10}$$

2. There exists $\epsilon < 1$ such that

$$\mathrm{Range}\left((-A)^{-\frac{\epsilon}{2}}\right) \subset \mathrm{Range}\,(Q). \tag{4.1.11}$$

Remark 4.1.1. Due to the Hypothesis 4.1-1 the Ornstein-Uhlenbeck system associated with the equation (4.0.1)

$$dz(t) = Az(t)\, dt + Q dw(t), \qquad z(s) = 0 \tag{4.1.12}$$

admits a unique solution $w^A(s,t)$, which is called *stochastic convolution* and which we denote by $w^A(t)$ when $s = 0$. Actually, as well known (see [47, Theorems 5.2 and 5.4] for a proof), if (4.1.10) holds, then there exists a unique solution $w^A(s,t)$ for the problem (4.1.12), which is the mean-square continuous gaussian process with values in H given by

$$w^A(s,t) = \int_s^t e^{(t-r)A} Q\, dw(r).$$

It is possible to check that for any $T > 0$

$$\mathbb{E}|w^A(s,t)|_H^2 = \int_s^t \mathrm{Tr}\left[e^{\sigma A} QQ^\star e^{\sigma A^\star}\right] d\sigma,$$

so that, due to the assumption (4.1.10),

$$\sup_{t \in [s,T]} \mathbb{E}|w^A(s,t)|_H^2 = c_T < \infty.$$

In particular, due to the gaussianity of the process, for any $p \geq 1$ this implies

$$\sup_{t \in [s,T]} \mathbb{E}|w^A(s,t)|_H^p = c_{T,p} < \infty. \tag{4.1.13}$$

As far as the Hypothesis 4.1-2 is concerned, it is a sort of non-degeneracy condition for the operator Q and it is assumed in order to have the regularizing effect of the semigroup P_t. Notice that due to the closed graph theorem, the operator $\Gamma_\epsilon = Q^{-1}(-A)^{-\epsilon/2}$ is bounded and by (4.1.7) and (4.1.11), we have that $\mathrm{Range}(e^{tA}) \subset \mathrm{Range}(Q)$, for any $t > 0$. This implies that if we set $\Gamma(t) = Q^{-1} e^{tA}$ then

$$|\Gamma(t)x|_H = \left|\Gamma_\epsilon (-A)^{\epsilon/2} e^{tA} x\right|_H \leq c\|\Gamma_\epsilon\| t^{-\epsilon/2} |x|_H, \qquad x \in H.$$

4.1.3 The Nemytskii operator

We assume here that the reaction term $f = (f_1, \ldots, f_n)$ enjoys the following conditions.

Hypothesis 4.2. *The function $f : \overline{\mathcal{O}} \times \mathbb{R}^n \to \mathbb{R}^n$ is measurable in $\xi \in \overline{\mathcal{O}}$, for all $\sigma \in \mathbb{R}^n$. Moreover, for almost all $\xi \in \overline{\mathcal{O}}$ the function $f(\xi, \cdot)$ is of class C^r, for some $r \geq 2$ and it holds*

$$\sup_{\xi \in \overline{\mathcal{O}}} \sup_{\sigma \in \mathbb{R}^n} |D_\sigma^j f(\xi, \sigma)| < \infty, \qquad j = 1, \ldots, h.$$

Let ρ be the constant introduced in (4.1.1). The Nemytskii operator F associated with the function $(\xi, \sigma) \mapsto f(\xi, \sigma) + (\rho + 1)\sigma$ is defined for any $x : \mathcal{O} \to \mathbb{R}^n$ as

$$F(x)(\xi) = f(\xi, x(\xi)) + (1 + \rho)x(\xi), \qquad \xi \in \mathcal{O}.$$

As f is Lipschitz-continuous with respect to $\sigma \in \mathbb{R}^n$, uniformly with respect to $\xi \in \overline{\mathcal{O}}$, we have that F is well defined and Lipschitz-continuous from $L^p(\mathcal{O}; \mathbb{R}^n)$ into itself, for any $p \geq 1$. Concerning the differentiability of F, it is not difficult to prove that for any $p \neq 2$ it is Fréchet differentiable from $L^p(\mathcal{O}; \mathbb{R}^n)$ into $L^q(\mathcal{O}; \mathbb{R}^n)$ and the derivative is given by

$$(DF(x)y)(\xi) = D_\sigma f(\xi, x(\xi))y(\xi) + (\rho + 1)y(\xi), \qquad \xi \in \mathcal{O}. \tag{4.1.14}$$

The case $p = 2$ is more delicate. In fact, even if f is assumed to be smooth, F is not even once Fréchet differentiable in H. Actually, it is possible to prove that F is Fréchet differentiable in H if and only if $f(\xi, \cdot)$ is linear (for a proof see [2]). Nevertheless, $F : H \to H$ is Gâteaux differentiable at any point $x \in H$ along any direction $y \in H$ and the Gâteaux derivative is given by (4.1.14).

For higher order derivatives there are further problems, because the higher order directional derivatives of F do not exist along any directions, but only along more regular directions. Actually, if we fix $y \in L^{p_1}(\mathcal{O}; \mathbb{R}^n)$, with $p_1 \geq p$, then the mapping $DF(\cdot)y : L^p(\mathcal{O}; \mathbb{R}^n) \to L^p(\mathcal{O}; \mathbb{R}^n)$ is differentiable at any point $x \in L^p(\mathcal{O}; \mathbb{R}^n)$ along any direction $z \in L^{p_2}(\mathcal{O}; \mathbb{R}^n)$, with $p_2 = pp_1/(p_1 - p)$, and it holds

$$D^2 F(x)(y, z)(\xi) = D_\sigma^2 f(\xi, x(\xi))(yz)(\xi), \qquad \xi \in \mathcal{O}.$$

In a similar way, for any $j \geq 2$ and $y_i \in L^{p_i}(\mathcal{O}; \mathbb{R}^n)$, with $i \leq j$ and $p_1^{-1} + \cdots + p_j^{-1} \leq 1/p$, we have that the mapping

$$D^j F(\cdot)(y_1, \ldots, y_j) : L^p(\mathcal{O}; \mathbb{R}^n) \to L^p(\mathcal{O}; \mathbb{R}^n)$$

is differentiable at any point $x \in H$ along any direction $y_{j+1} \in L^{p_{j+1}}(\mathcal{O}; \mathbb{R}^n)$, with $p_{j+1}^{-1} = 1/p - (p_1^{-1} + \cdots + p_j^{-1})$, and

$$D^{j+1} F(x)(y_1, \ldots, y_j, y_{j+1})(\xi) = D_\sigma^j f(\xi, x(\xi))(y_1 \cdots y_{j+1})(\xi), \qquad \xi \in \mathcal{O}.$$

4.2 Differential dependence on initial data

With the notations introduced in the previous section, the system (4.0.1) can be rewritten as the following stochastic evolution equation with values in H

$$du(t) = [Au(t) + F(u(t))]\, dt + Q\, dw(t), \quad u(0) = x. \qquad (4.2.1)$$

In what follows we shall denote its solution by $u(t;x)$.

Let X be a Banach space. For any $T > 0$ and $p \geq 1$ fixed, we denote by $\mathcal{H}_p(T, X)$ the set of adapted processes $u \in C([0,T]; L^p(\Omega; X))$. $\mathcal{H}_p(T, X)$ is a Banach space, endowed with norm

$$\|u\|_{\mathcal{H}_p(T,X)}^p = \sup_{t \in [0,T]} e^{-\rho t} \mathbb{E}\, |u(t)|_X^p, \qquad (4.2.2)$$

where ρ is a suitable constant to be chosen later. Moreover, we denote by $C_p(T, X)$ the subspace of processes in $L^p(\Omega; C([0,T]; X))$, endowed with the norm

$$\|u\|_{C_p(T,X)}^p = \mathbb{E} \sup_{t \in [0,T]} |u(t)|_X^p.$$

Now, for any $x \in H$ and $u \in \mathcal{H}_p(T, H)$ we set

$$\mathcal{F}(x, u)(t) = e^{tA} x + \int_0^t e^{(t-s)A} F(u(s))\, dr + w^A(t), \quad t \geq 0, \qquad (4.2.3)$$

where $w^A(t)$ is the unique solution of (4.1.12). Due to (4.1.13), $w^A(t) \in \mathcal{H}_p(T, H)$ and then, since F has linear growth, the mapping

$$\mathcal{F} : H \times \mathcal{H}_p(T, H) \to \mathcal{H}_p(T, H)$$

is well defined, for any $T > 0$ and $p \geq 1$. Moreover, as F is Lipschitz-continuous, there exists some $\rho_0 > 0$ such that $\mathcal{F}(x, \cdot)$ is a contraction in $\mathcal{H}_p(T, H)$, endowed with the norm corresponding to ρ_0.

Actually, if we denote by L the Lipschitz constant of F and by M the constant introduced in (4.1.2), for any $\rho > 0$ and $u, v \in \mathcal{H}_p(T, H)$ we have

$$e^{-\rho t} \mathbb{E} \left| \int_0^t e^{(t-s)A} [F(u(s)) - F(v(s))]\, ds \right|_H^p \leq (ML)^p e^{-\rho t} \mathbb{E} \left(\int_0^t |u(s) - v(s)|_H\, ds \right)^p$$

$$\leq (ML)^p \int_0^t e^{-\rho s}\, \mathbb{E} |u(s) - v(s)|_H^p\, ds \left(\int_0^t e^{-\frac{\rho(t-s)}{p-1}}\, ds \right)^{p-1}$$

$$\leq (ML)^p\, t \left(\frac{p-1}{\rho} \right)^{p-1} \left(1 - e^{-\frac{\rho t}{p-1}} \right)^{p-1} \|u - v\|_{\mathcal{H}_p(T,H)}^p.$$

Therefore

$$\|\mathcal{F}(x, u) - \mathcal{F}(x, v)\|_{\mathcal{H}_p(T,H)} \leq T^{1/p} ML \left(\frac{p-1}{\rho} \right)^{\frac{p-1}{p}} \left(1 - e^{-\frac{\rho T}{p-1}} \right)^{\frac{p-1}{p}} \|u - v\|_{\mathcal{H}_p(T,H)}$$

and then, if we fix $\rho > 0$ such that

$$T^{1/p} ML \left(\frac{p-1}{\rho}\right)^{\frac{p-1}{p}} \left(1 - e^{-\frac{\rho T}{p-1}}\right)^{\frac{p-1}{p}} < 1,$$

we have that $\mathcal{F}(x, \cdot)$ is a contraction in the space $\mathcal{H}_p(T, H)$ endowed with the norm (4.2.2). Thanks to the theorem of contractions depending on parameters (see appendix C for a generalization of it), this means that $\mathcal{F}(x, \cdot)$ admits a unique fixed point in $\mathcal{H}_p(T, H)$, for any fixed $x \in H$. That is, for any $x \in H$ there exists a unique solution $u(x) \in \mathcal{H}_p(T, H)$ of the problem

$$\mathcal{F}(x, u(x)) = u(x).$$

4.2.1   First derivative

If F is Lipschitz-continuous and Fréchet differentiable, due to the classical theorem of contractions depending on parameters, it is possible to prove that for any $p \geq 1$ the solution $u(\cdot; x)$ is *differentiable in the mean of order p* with respect to $x \in H$, that is the mapping

$$H \to \mathcal{H}_p(T, H), \qquad x \mapsto u(\cdot; x),$$

is differentiable. When $p = 2$ we say that $u(\cdot; x)$ is *mean-square differentiable*.

Next proposition shows that the same result holds, even if F is not Fréchet differentiable as in the present case of Nemytskii operators.

Proposition 4.2.1. *Under the Hypotheses 4.1 and 4.2, for any $p \geq 1$ the solution $u(\cdot; x)$ of the problem (4.2.1) is differentiable in the mean of order p with respect to initial data and the derivative $D_x u(\cdot; x)h$ at $x \in H$ along any direction $h \in H$ is the solution of the first variation equation*

$$\frac{dz}{dt}(t) = Az(t) + DF(u(t; x))z(t), \qquad z(0) = h. \tag{4.2.4}$$

Moreover $D_x u(t; x)h \in L^q(\mathcal{O}; \mathbb{R}^n)$, $\mathbb{P}$-a.s. for any $t \geq 0$ and $h \in L^q(\mathcal{O}; \mathbb{R}^n)$, with $q \geq 1$, and there exists an increasing continuous function $\nu_q(t)$ such that

$$\sup_{x \in H} |D_x u(t; x)h|_q \leq \nu_q(t)|h|_q, \qquad \mathbb{P} - a.s. \tag{4.2.5}$$

In particular, $D_x u(\cdot; x)h \in C_p(T, L^q(\mathcal{O}; \mathbb{R}^n))$, for any $x \in H$, $h \in L^q(\mathcal{O}; \mathbb{R}^n)$ and $p \geq 1$.

Proof. All norms

$$u \mapsto \sup_{t \in [0,T]} e^{-\rho t} \mathbb{E}|u(t)|_H^p, \qquad \rho \in \mathbb{R},$$

are equivalent in $\mathcal{H}_p(T, H)$. Actually, if $\rho_1 < \rho_2$

$$e^{-\rho_2 t} \mathbb{E}|u(t)|_H^p \leq e^{-\rho_1 t} \mathbb{E}|u(t)|_H^p \leq e^{(\rho_2 - \rho_1)T} e^{-\rho_2 t} \mathbb{E}|u(t)|_H^p$$

and then the equivalence follows by taking the supremum over $t \in [0, T]$. Thus, for the sake of simplicity, we can study the differentiability properties of the operator $\mathcal{F}$ defined in (4.2.3) in the space $\mathcal{H}_p(T, H)$, endowed with the norm corresponding to $\rho = 0$.

Step 1: We have seen that $\mathcal{F}(x, \cdot) : \mathcal{H}_p(T, H) \to \mathcal{H}_p(T, H)$ is a contraction, for any fixed $x \in H$. Thus, if we show that $\mathcal{F}$ enjoys the Hypothesis C.1, with $\Lambda = H$ and $E = \mathcal{H}_p(T, H)$, we can apply the Proposition C.0.3 and we obtain the existence of the first derivative of $u(\cdot; x)$.

For any fixed $u \in \mathcal{H}_p(T, H)$, the mapping

$$\mathcal{F}(\cdot, u) : H \to \mathcal{H}_p(T, H), \quad x \mapsto \mathcal{F}(x, u),$$

is differentiable and its derivative along any direction $h \in H$ is given by

$$\left(\frac{\partial \mathcal{F}}{\partial x}(x, u)h \right)(t) = e^{(t-s)A} h. \tag{4.2.6}$$

It is clearly continuous from $H \times \mathcal{H}_p(T, H)$ into $\mathcal{L}(H; \mathcal{H}_p(T, H))$.

Next, for any fixed $x \in H$ and $u, v \in \mathcal{H}_p(T, H)$, the mapping

$$\mathbb{R} \to \mathcal{H}_p(T, H), \quad \lambda \mapsto \mathcal{F}(x, u + \lambda v)$$

is differentiable and if we define

$$\frac{\partial \mathcal{F}}{\partial u}(x, u)v = \frac{d}{d\lambda} \mathcal{F}(x, u + \lambda v) \Big|_{\lambda = 0},$$

we have

$$\left(\frac{\partial \mathcal{F}}{\partial u}(x, u)v \right)(t) = \int_0^t e^{(t-s)A} DF(u(s))v(s)\, ds. \tag{4.2.7}$$

Indeed, we have

$$\frac{1}{\lambda} \left(\mathcal{F}(x, u + \lambda v) - \mathcal{F}(x, u) \right)(t) - \int_0^t e^{(t-s)A} DF(u(s))v(s)\, ds$$

$$= \int_0^t e^{(t-s)A} \left(\frac{1}{\lambda} [F(u(s) + \lambda v(s)) - F(u(s))] - DF(u(s))v(s) \right) ds$$

and then

$$\sup_{t \in [0,T]} \left| \frac{1}{\lambda} \left(\mathcal{F}(x, u + \lambda v) - \mathcal{F}(x, u) \right)(t) - \int_0^t e^{(t-s)A} DF(u(s))v(s)\, ds \right|_H$$

$$\tag{4.2.8}$$

$$\leq M \int_0^T \left| \frac{1}{\lambda} [F(u(s) + \lambda v(s)) - F(u(s))] - DF(u(s))v(s) \right|_H ds.$$

Now, since F is Gâteaux differentiable in H, for any fixed $s \in [0,t]$ it holds

$$\lim_{\lambda \to 0} \left| \frac{1}{\lambda} [F(u(s) + \lambda v(s)) - F(u(s))] - DF(u(s))v(s) \right|_H = 0, \qquad \mathbb{P} - \text{a.s.} \qquad (4.2.9)$$

Moreover, for any $\xi \in \mathcal{O}$

$$\left(\frac{1}{\lambda} [F(u(s) + \lambda v(s)) - F(u(s))] - DF(u(s))v(s) \right)(\xi)$$

$$= \int_0^1 [D_\sigma f\left(\xi, u(s)(\xi) + \theta \lambda v(s)(\xi)\right) - D_\sigma f(\xi, u(s)(\xi))]\, v(s)(\xi)\, d\theta,$$

so that, due to the boundedness of $D_\sigma f$

$$\left| \frac{1}{\lambda} [F(u(s) + \lambda v(s)) - F(u(s))] - DF(u(s))v(s) \right|_H \leq c\, |v(s)|_H.$$

This implies

$$\left(\int_0^T \left| \frac{1}{\lambda} (F(u(s) + \lambda v(s)) - F(u(s))) - DF(u(s))v(s) \right|_H ds \right)^p$$

$$\leq c\, T^{p-1} \int_0^T |v(s)|_H^p\, ds, \qquad \mathbb{P} - \text{a.s.}$$

and then, by using the dominated convergence theorem, thanks to (4.2.8) and (4.2.9) we can conclude that

$$\lim_{\lambda \to 0} \sup_{t \in [0,T]} \mathbb{E} \left| \frac{1}{\lambda} (\mathcal{F}(x, u + \lambda v) - \mathcal{F}(x, u))(t) - \left(\frac{\partial \mathcal{F}}{\partial u}(x, u)v \right)(t) \right|_H^p = 0.$$

Clearly, $\partial \mathcal{F}/\partial u(x, u)$ is a bounded linear operator on $\mathcal{H}_p(T, H)$, for any $x \in H$ and $u \in \mathcal{H}_p(T, H)$, thus in order to apply the Proposition C.0.4 it remains to show that the mapping

$$\frac{\partial \mathcal{F}}{\partial u}(\cdot, \cdot)v : H \times \mathcal{H}_p(T, H) \to \mathcal{H}_p(T, H), \qquad (x, u) \mapsto \frac{\partial \mathcal{F}}{\partial u}(x, u)v,$$

is continuous, for any fixed $v \in \mathcal{H}_p(T, H)$. If $x \in H$ and $u \in \mathcal{H}_p(T, H)$, we fix any two sequences $\{x_n\} \subset H$ and $\{u_n\} \subset \mathcal{H}_p(T, H)$, converging respectively to x and u. We have

$$\frac{\partial \mathcal{F}}{\partial u}(x, u)v(t) - \frac{\partial \mathcal{F}}{\partial u}(x_n, u_n)v(t) = \int_0^t e^{(t-s)A} \left(DF(u(s)) - DF(u_n(s)) \right) v(s)\, ds,$$

and then

$$\left\| \frac{\partial \mathcal{F}}{\partial u}(x, u)v - \frac{\partial \mathcal{F}}{\partial u}(x_n, u_n)v \right\|^p_{\mathcal{H}_p(T,H)}$$

$$\leq M^p T^{p-1} \int_0^T \mathbb{E} \left\| [DF(u(t)) - DF(u_n(t))] v(t) \right\|^p_H dt.$$

Now, as u_n converge to u in $\mathcal{H}_p(T, H)$, there exists a subsequence $\{u_{n_k}\}$ which converges to u almost surely in $[0, T] \times \mathcal{O} \times \Omega$, with respect to product measure $\lambda_{[0,T]} \otimes \lambda_d \otimes \mathbb{P}$, where $\lambda_{[0,T]}$ and λ_d are the Lebesgue measure respectively in $[0, T]$ and $\mathbb{R}^d$. Thus, since

$$\lim_{n \to +\infty} \left(D_\sigma f(\xi, u(s)(\xi)) - D_\sigma f(\xi, u_{n_k}(s)(\xi)) \right) v(s)(\xi) = 0,$$

and

$$\left| (D_\sigma f(\xi, u(s)(\xi)) - D_\sigma f(\xi, u_{n_k}(s)(\xi))) v(s)(\xi) \right| \leq c \left| v(s)(\xi) \right|, \qquad \mathbb{P} - \text{a.s.}$$

from the dominated convergence theorem we can conclude that

$$\lim_{n \to +\infty} \left\| \frac{\partial \mathcal{F}}{\partial u}(x, u)v - \frac{\partial \mathcal{F}}{\partial u}(x_n, u_n)v \right\|_{\mathcal{H}_p(T,H)} = 0.$$

According to the Proposition C.0.3 this implies that $u(\cdot; x)$ is differentiable in the mean of order p at any point $x \in H$ along any direction $h \in H$ and the derivative $D_x u(\cdot; x)h$ is the mild solution of the problem (4.2.4).

Step 2: Thanks to the Hypothesis 4.2, if $z^h(\cdot; x)$ is the unique mild solution of the problem (4.2.4) then for any $q \geq 1$

$$DF(u(\cdot; x))z^h(\cdot; x) \in L^q(0, T; H), \qquad \mathbb{P} - \text{a.s.}$$

so that for any initial datum $h \in D(A)$ we have

$$z^h(\cdot; x) \in L^q(0, T; D(A)), \qquad \mathbb{P}\text{-a.s.}$$

Thus, multiplying each side of (4.2.4) by $|z(t)|^{q-2} z(t)$ and integrating over $\mathcal{O}$, we get

$$\frac{1}{q} \frac{d}{dt} |z(t)|^q_q = \int_{\mathcal{O}} \langle Az(t), z(t) \rangle |z(t)|^{q-2} \, d\xi + \int_{\mathcal{O}} |z(t)|^{q-2} \langle DF(u(t; x))z(t), z(t) \rangle \, d\xi.$$

Since e^{tA} is a contraction semigroup in $L^q(\mathcal{O}; \mathbb{R}^n)$, we have

$$\int_{\mathcal{O}} |z(t)|^{q-2} \langle Az(t), z(t) \rangle \, d\xi = \left\langle Az(t), |z(t)|^{q-2} z(t) \right\rangle_{q,q'} \leq 0$$

and then

$$\frac{1}{q}\frac{d}{dt}\,|z(t)|_q^q \leq \int_{\mathcal{O}} |z(t)|^{q-2}\,\langle D_\sigma f(\xi, u(t;x))z(t), z(t)\rangle \; d\xi \leq c\,|z(t)|_q^q.$$

Thanks to the Gronwall lemma, this implies that for any $h \in D(A)$

$$\sup_{x\in H}\left|z^h(t;x)\right|_q \leq \nu_q(t)|h|_q, \quad \mathbb{P}\text{-a.s.}$$

with $\nu_q(t) = e^{ct}$. Finally, if $h \in L^q(\mathcal{O}; \mathbb{R}^n)$, there exists a sequence $\{h_n\} \subset D(A)$ converging to h in $L^q(\mathcal{O}; \mathbb{R}^n)$ such that $|h_n|_q \leq c\,|h|_q$, for some constant c. Hence, it easily follows

$$\lim_{n\to+\infty} z^{h_n}(t;x) = z^h(t;x), \quad \text{in } L^q(\mathcal{O}; \mathbb{R}^n), \qquad \mathbb{P}-\text{a.s.}$$

Since

$$\sup_{x\in H}\left|z^{h_n}(t;x)\right|_q \leq \nu_q(t)|h_n|_q, \quad \mathbb{P}\text{-a.s.},$$

by taking the limit as $n \to +\infty$, for any $h \in L^q(\mathcal{O}; \mathbb{R}^n)$ we get

$$\sup_{x\in H}\left|z^h(t;x)\right|_q \leq \nu_q(t)|h|_q, \quad \mathbb{P}\text{-a.s.}$$

which implies (4.2.5). Moreover,

$$\sup_{t\in [0,T]} |D_x u(t;x)h|_q^p \leq \nu_q^p(T)|h|_q^p, \qquad \mathbb{P}-\text{a.s.}$$

and then $D_x u(\cdot;x)h \in \mathcal{C}_p(T, L^q(\mathcal{O}; \mathbb{R}^n))$, for any $p \geq 1$. $\qquad\square$

4.2.2   Higher order derivatives

As for the first order derivative, for any $j \geq 1$ we say that the process $u(\cdot;x)$ is *j-times differentiable in the mean of order p* with respect to $x \in H$ if the mapping

$$H \to \mathcal{L}^{j-1}(H; \mathcal{H}_p(T, H)), \qquad x \mapsto D_x^{j-1} u(\cdot;x),$$

is differentiable.

Before proving the existence of higher order derivatives, we need to introduce some notations. For any $T > 0$ and $r, p, q \geq 1$, we define $\mathcal{K}_q^{r,p}(T, H)$ as the space of processes $u \in \mathcal{H}_q(T, H)$ such that $u(t) \in L^p(\mathcal{O}; \mathbb{R}^n)$, for any $t > 0$, $\mathbb{P}$-a.s., and

$$\langle u \rangle_{r,p} = \sup_{t\in (0,T]} t^{\frac{d(p-r)}{2rp}}\,|u(t)|_p \in L^q(\Omega).$$

$\mathcal{K}_q^{r,p}(T, H)$ is a Banach space, endowed with the norm

$$\|u\|_{\mathcal{K}_q^{r,p}(T,H)} = \|u\|_{\mathcal{H}_q(T,H)} + \left(\mathbb{E}\,\langle u \rangle_{r,p}^q\right)^{1/q}.$$

If $r = 2$ we set $\langle u \rangle_{r,p} = \langle u \rangle_p$ and $\mathcal{K}_q^{2,p}(T, H) = \mathcal{K}_q^p(T, H)$. Notice that if $p = 2$, then $\mathcal{K}_q^p(T, H) = \mathcal{C}_q(T, H)$, for any $q \geq 1$.

Lemma 4.2.2. *Let us fix $1 \leq r \leq p \leq \infty$. Then under the Hypotheses 4.1 and 4.2, for any $x \in H$ and $h \in L^r(\mathcal{O}; \mathbb{R}^n)$ we have that $D_x u(t; x)h \in L^p(\mathcal{O}; \mathbb{R}^n)$, $\mathbb{P}$-a.s. for $t > 0$ and*

$$\sup_{x \in H} |D_x u(t; x)h|_p \leq \mu_{r,p}(t) t^{-\frac{d(p-r)}{2rp}} |h|_r, \qquad \mathbb{P} - a.s. \qquad (4.2.10)$$

for an increasing continuous function $\mu_{r,p}(t)$. In particular

$$\sup_{x \in H} \langle D_x u(\cdot; x)h \rangle_{r,p} \leq \mu_{r,p}(T)|h|_r, \qquad \mathbb{P} - a.s. \qquad (4.2.11)$$

and then $D_x u(\cdot; x)h \in \mathcal{K}_q^{r,p}(T, H)$, for any $q \geq 1$.

Proof. In the previous proposition we have seen that $D_x u(\cdot; x)h$ is the unique mild solution of the problem

$$D_x u(t; x)h = e^{tA}h + \int_0^t e^{(t-s)A} DF(u(s; x)) D_x u(s; x)h \, ds.$$

We have to show that $D_x u(\cdot; x)h \in L^p(\mathcal{O}; \mathbb{R}^n)$, for any $t > 0$, $\mathbb{P}$-a.s., and the estimate (4.2.10) holds true.

Due to (4.1.5), if $h \in L^r(\mathcal{O}; \mathbb{R}^n)$ we have

$$\sup_{t \in (0,T]} t^{\frac{d(p-r)}{2rp}} |e^{tA}h|_p \leq c_{r,p} |h|_r.$$

Moreover, from the boundedness of DF we have

$$\left| \int_0^t e^{(t-s)A} DF(u(s; x)) D_x u(s; x)h \, ds \right|_p \leq c_{r,p} \int_0^t (t - s)^{-\frac{d(p-r)}{2rp}} |D_x u(s; x)h|_r \, ds$$

and then from (4.2.5)

$$t^{\frac{d(p-r)}{2rp}} \left| \int_0^t e^{(t-s)A} DF(u(s; x)) D_x u(s; x)h \, ds \right|_p \leq c_{r,p} \, t \, \nu_r(t)|h|_r.$$

Thus we can conclude that (4.2.10) and (4.2.11) hold true with

$$\mu_{r,p}(t) = c_{r,p} \left(1 + t \, \nu_r(t) \right).$$

$\square$

Remark 4.2.3. Let us denote by $z^h(t; x)$ the unique mild solution of the problem

$$\frac{dz}{dt}(t) = Az(t) + DF(u(t; x))z(t) + \theta(t), \quad t \geq 0, \quad z(0) = h,$$

where θ is any process in $\mathcal{H}_p(T, H)$ with $p \geq 2$. Then $z^h(t; x)$ is given by

$$z^h(t; x) = U_x(t, 0)h + \int_0^t U_x(t, s)\theta(s) \, ds,$$

where $U_x(t, s)h$ is the unique solution of the linear problem

$$\frac{dy}{dt}(t) = Ay(t) + DF(u(t; x))y(t), \quad t \geq s, \quad y(s) = h.$$

As proved in the previous lemma, if $h \in L^r(\mathcal{O}; \mathbb{R}^n)$, then $U_x(t, s)h \in L^p(\mathcal{O}; \mathbb{R}^n)$, for any $s < t$, $\mathbb{P}$-a.s., and there exists a continuous increasing function $\mu_{r,p}(t)$ such that

$$\sup_{x \in H} |U_x(t, s)h|_p \leq \mu_{r,p}(t - s)(t - s)^{-\frac{d(p-r)}{2pr}} |h|_r, \qquad \mathbb{P} - \text{a.s.}$$

This implies that

$$\sup_{x \in H} |z^h(t; x)|_p \leq \mu_{r,p}(t)t^{-\frac{d(p-r)}{2pr}} |h|_r + \int_0^t \mu_{r,p}(t - s)(t - s)^{-\frac{d(p-r)}{2pr}} |\theta(s)|_r \, ds. \quad (4.2.12)$$

Theorem 4.2.4. *Assume the Hypotheses 4.1 and 4.2 and fix $k < (2 + 4/d) \wedge (r + 1)$. Then for any $p \geq 2$ the solution $u(\cdot; x)$ is k-times differentiable in the mean of order p, with respect to the initial datum $x \in H$.*

Proof. Step 1: We show that the functional $\mathcal{F}$ fulfills the Hypothesis C.2, with $\Lambda = H$, $E = \mathcal{H}_p(T, H)$ and $G = \mathcal{C}_{2p}(T, H)$. Thanks to the Proposition C.0.4, this implies that the solution $u(\cdot; x)$ is twice differentiable in the mean of order p, at any $x \in H$ and along any directions $h_1, h_2 \in H$, and $D^2u(\cdot; x)(h_1, h_2)$ satisfies the following equation

$$\begin{cases} \dfrac{dz}{dt}(t) &= Az(t) + DF(u(t; x))z(t) + D^2F(u(t; x)) \left(D_x u(t; x)h_1, D_x u(t; x)h_2 \right), \\[2ex] z(0) &= 0. \end{cases}$$
$$(4.2.13)$$

According to (4.2.6) and (4.2.7), if $j \geq 1$ and $i + j \geq 2$, we have

$$\frac{\partial^{i+j} \mathcal{F}}{(\partial u \, \partial x)^{(\alpha, \beta)}}(x, u) = \frac{\partial^{i+j} \mathcal{F}}{\partial u^{i_1} \partial x^{j_1} \cdots \partial u^{i_h} \partial x^{j_h}}(x, u) = 0,$$

for all (α, β) such that $|\alpha| = i_1 + \cdots + i_h = i$ and $|\beta| = j_1 + \cdots + j_h = j$ and for all $u \in \mathcal{H}_p(T, H)$ and $x \in H$. In particular the mappings

$$\frac{\partial^{i+j} \mathcal{F}}{(\partial u \, \partial x)^{(\alpha, \beta)}} : H \times \mathcal{H}_p(T, H) \to \mathcal{L}\left(H^j \times (\mathcal{H}_p(T, H))^i; \mathcal{H}_p(T, H) \right),$$

are all continuous.

Now, for $u \in \mathcal{H}_p(T, H)$ and $u_1, u_2 \in \mathcal{H}_{2p}(T, H)$ fixed, we consider the mapping

$$\mathbb{R} \to \mathcal{H}_p(T, H), \quad \lambda \mapsto \frac{\partial \mathcal{F}}{\partial u}(x, u + \lambda u_1)u_2, \quad x \in H.$$

We have already seen that for $t \geq 0$

$$\left(\frac{\partial \mathcal{F}}{\partial u}(x, u + \lambda u_1)u_2\right)(t) = \int_0^t e^{(t-s)A} DF(u(s) + \lambda u_1(s))u_2(s)\, ds,$$

and then, if we define

$$\frac{\partial^2 \mathcal{F}}{\partial u^2}(x, u)(u_1, u_2) = \frac{d}{d\lambda}\left(\frac{\partial \mathcal{F}}{\partial u}(x, u + \lambda u_1)u_2\right)\Bigg|_{\lambda=0},$$

we have

$$\left(\frac{\partial^2 \mathcal{F}}{\partial u^2}(x, u)(u_1, u_2)\right)(t) = \int_0^t e^{(t-s)A} D^2F(u(s))(u_1(s), u_2(s))\, ds.$$

Indeed, if we set

$$\Delta_\lambda(s) = \frac{1}{\lambda}\left[DF(u(s) + \lambda u_1(s)) - DF(u(s))\right]u_2(s) - D^2F(u(s))(u_1(s), u_2(s)),$$

we have

$$\sup_{t \in [0,T]}\left|\frac{1}{\lambda}\left(\frac{\partial \mathcal{F}}{\partial u}(x, u + \lambda u_1)u_2 - \frac{\partial \mathcal{F}}{\partial u}(x, u)u_2\right)(t) - \frac{\partial^2 \mathcal{F}}{\partial u^2}(x, u)(u_1, u_2)(t)\right|_H$$

$$= \sup_{t \in [0,T]}\left|\int_0^t e^{(t-s)A}\Delta_\lambda(s)\, ds\right|_H \leq c \sup_{t \in [0,T]}\int_0^t (t-s)^{-\frac{d}{4}}|\Delta_\lambda(s)|_1\, ds$$

$$\leq c \sup_{t \in [0,T]}\left(\int_0^T (T-s)^{-\frac{d\delta}{4}}\, ds\right)^{1/\delta}\left(\int_0^T |\Delta_\lambda(s)|_1^{\frac{\delta}{\delta-1}}\, ds\right)^{\frac{\delta-1}{\delta}},$$

$$\tag{4.2.14}$$

where δ is some constant greater than 1 such that $d\delta/4 < 1$ (and this is possible, as $d \leq 3$).

As we have seen in the subsection 4.1.3, if $y \in H$, then the mapping $DF(\cdot)y : L^1(\mathcal{O}; \mathbb{R}^n) \to L^1(\mathcal{O}; \mathbb{R}^n)$ has directional derivatives at any point $x \in L^1(\mathcal{O}; \mathbb{R}^n)$ and along any direction $z \in H$. This means that for any $s \in [0, t]$

$$\lim_{\lambda \to 0} |\Delta_\lambda(s)|_1 = 0, \qquad \mathbb{P} - \text{a.s.}$$

Moreover, for any $\xi \in \mathcal{O}$ and $s \in [0, t]$ it holds

$$\Delta_\lambda(s)(\xi) = \int_0^1 \left[D_\sigma^2 f(\xi, u(s)(\xi) + \theta\lambda u_1(s)(\xi)) - D_\sigma^2 f(\xi, u(s)(\xi))\right] d\theta\, (u_1(s)u_2(s))(\xi),$$

and then it follows

$$|\Delta_\lambda(s)(\xi)| \leq c\, |u_1(s)(\xi)||u_2(s)(\xi)|.$$

As we are assuming $u_1, u_2 \in C_{2p}(T, H)$, this easily implies that

$$\mathbb{E}\left(\int_0^T |\Delta_\lambda(s)|_1^{\frac{\delta}{\delta-1}} \, ds\right)^{\frac{(\delta-1)p}{\delta}} \tag{4.2.15}$$

$$\leq T^{\frac{(\delta-1)p}{\delta}} \mathbb{E} \langle u_1 \rangle_2^p \langle u_2 \rangle_2^p \leq T^{\frac{(\delta-1)p}{\delta}} \|u_1\|_{C_{2p}(T,H)}^p \|u_2\|_{C_{2p}(T,H)}^p$$

and due to (4.2.14) from the dominated convergence theorem, we can conclude that

$$\frac{\partial^2 \mathcal{F}}{\partial u^2}(x,u)(u_1,u_2) = \lim_{\lambda\to 0} \frac{1}{\lambda}\left(\frac{\partial \mathcal{F}}{\partial u}(x, u+\lambda u_1)u_2 - \frac{\partial \mathcal{F}}{\partial u}(x,u)u_2\right)$$

$$= \int_0^\cdot e^{(\cdot-s)A} D^2 F(u(s))(u_1(s), u_2(s)) \, ds, \quad \text{in } \mathcal{H}_p(T, H).$$

Clearly, thanks to (4.2.14) and (4.2.15) it is immediate to check that for any $x \in H$ and $u \in \mathcal{H}_p(T, H)$ the linear operator $\partial^2 \mathcal{F}/\partial u^2(x, u)$ is bounded from $C_{2p}(T, H) \times C_{2p}(T, H)$ into $\mathcal{H}_p(T, H)$ and if $u_1, u_2 \in C_{2p}(T, H)$ the mapping

$$\frac{\partial^2 \mathcal{F}}{\partial u^2}(\cdot, \cdot)(u_1, u_2) : H \times \mathcal{H}_p(T, H) \to \mathcal{H}_p(T, H)$$

is continuous. Actually, if $x \in H$ and $u \in \mathcal{H}_p(T, H)$ and if $\{x_n\} \subset H$ and $\{u_n\} \subset \mathcal{H}_p(T, H)$ are two sequences converging respectively to x and u, we have

$$\frac{\partial^2 \mathcal{F}}{\partial u^2}(x, u)(u_1, u_2)(t) - \frac{\partial^2 \mathcal{F}}{\partial u^2}(x_n, u_n)(u_1, u_2)(t)$$

$$= \int_0^t e^{(t-s)A}\left(D^2 F(u(s)) - D^2 F(u_n(s))\right)(u_1(s), u_2(s)) \, ds.$$

Thus,

$$\left\|\frac{\partial^2 \mathcal{F}}{\partial u^2}(x, u)(u_1, u_2) - \frac{\partial^2 \mathcal{F}}{\partial u^2}(x_n, u_n)(u_1, u_2)\right\|_{\mathcal{H}_p(T,H)}^p$$

$$\leq c\mathbb{E} \sup_{t\in[0,T]} \left(\int_0^t (t-s)^{-\frac{d}{4}}\left|\left(D^2 F(u(s)) - D^2 F(u_n(s))\right)(u_1(s), u_2(s))\right|_1 \, ds\right)^p.$$

By proceeding as in the first step of the proof of the Proposition 4.2.1, and by using once more the dominated convergence theorem, we can conclude that

$$\lim_{n\to+\infty} \left\|\frac{\partial^2 \mathcal{F}}{\partial u^2}(x, u)(u_1, u_2) - \frac{\partial^2 \mathcal{F}}{\partial u^2}(x_n, u_n)(u_1, u_2)\right\|_{\mathcal{H}_p(T,H)} = 0.$$

Moreover, for any $x \in H$ the mapping

$$\frac{\partial^2 \mathcal{F}}{\partial u^2}(x, \cdot) : \mathcal{H}_p(T, H) \to \mathcal{L}^2(\mathcal{C}_{2p}(T, H); \mathcal{H}_p(T, H))$$

is bounded on bounded subsets. In fact, from (4.2.15) it is clear that it is bounded in the whole space $\mathcal{H}_p(T, H)$.

Finally, in order to conclude, we have to show that

$$\lim_{\lambda \to 0} \frac{u(\cdot; x + \lambda h) - u(\cdot; x)}{\lambda} = D_x u(\cdot; x) h, \qquad \text{in } \mathcal{C}_{2p}(H),$$

for any $x, h \in H$. We have

$$\sup_{t \in [0,T]} \left| \frac{u(t; x + \lambda h) - u(t; x)}{\lambda} - D_x u(\cdot; x) h \right|_H^p$$

$$= \sup_{t \in [0,T]} \left| \int_0^t e^{(t-s)A} \left(\frac{F(u(s; x + \lambda h)) - F(u(s; x))}{\lambda} - DF(u(s; x)) D_x u(s; x) h \right) ds \right|_H^p$$

$$\leq T^{p-1} \int_0^T \left| \frac{F(u(s; x + \lambda h)) - F(u(s; x))}{\lambda} - DF(u(s; x)) D_x u(s; x) h \right|_H^p ds.$$

Clearly, for almost all $s \in [0, T]$

$$\lim_{h \to 0} \left| \frac{F(u(s; x + \lambda h)) - F(u(s; x))}{\lambda} - DF(u(s; x)) D_x u(s; x) h \right|_H = 0, \qquad \mathbb{P} - \text{a.s.}$$

Moreover, due to (4.2.5) and to the boundedness of DF

$$\left| \frac{F(u(s; x + \lambda h)) - F(u(s; x))}{\lambda} - DF(u(s; x)) D_x u(s; x) h \right|_H \leq \nu_2(T) |h|_H, \qquad \mathbb{P} - \text{a.s.}$$

so that we can conclude from the dominated convergence theorem.

Step 2: We give a sketch of the proof of the existence of derivatives of order $2 < k < (4/d + 2) \wedge (r + 1)$.

For any $x \in H$, $u \in \mathcal{H}_p(T, H)$ and $u_i \in \mathcal{K}_{kp}^k(T, H)$, with $i = 1, \ldots, k$, the mapping

$$\mathbb{R} \to \mathcal{H}_p(T, H), \qquad \lambda \mapsto \frac{\partial^{k-1} \mathcal{F}}{\partial u^{k-1}}(x, u + \lambda u_1)(u_2, \ldots, u_k)$$

is differentiable and if we define

$$\frac{\partial^k \mathcal{F}}{\partial u^k}(x, u)(u_1, \ldots, u_k) = \frac{d}{d\lambda} \left(\frac{\partial^{k-1} \mathcal{F}}{\partial u^{k-1}}(x, u + \lambda u_1)(u_2, \ldots, u_k) \right) \bigg|_{\lambda=0},$$

we have

$$\left(\frac{\partial^k \mathcal{F}}{\partial u^k}(x, u)(u_1, \dots, u_k)\right)(t) = \int_0^t e^{(t-s)A}\left(D^k F(u(s))(u_1(s), \dots, u_k(s))\right) ds.$$

For the proof one has to proceed as in the first step for the second order derivative.

All continuity and boundedness properties of the operator $\partial^k \mathcal{F}/\partial u^k$ follow from the same arguments used in the first step for $\partial^2 \mathcal{F}/\partial u^2$. For example, let us show that

$$\frac{\partial^k \mathcal{F}}{\partial u^k}(x, u) \in \mathcal{L}^k(\mathcal{K}^k_{kp}(T, H), \mathcal{H}_p(T, H)),$$

for any $x \in H$ and $u \in \mathcal{H}_p(T, H)$. We have

$$\left|\frac{\partial^k \mathcal{F}}{\partial u^k}(x, u)(u_1, \dots, u_k)(t)\right|_H$$

$$\leq c \int_0^t (t-s)^{-\frac{d}{4}}\left|D^k F(u(s))(u_1(s), \dots, u_k(s))\right|_1 ds$$

$$\leq c \int_0^t (t-s)^{-\frac{d}{4}} \prod_{i=1}^k |u_i(s)|_k \, ds \leq c \prod_{i=1}^k \langle u_i \rangle_k \int_0^t (t-s)^{-\frac{d}{4}} s^{-\frac{d(k-2)}{4}} ds.$$

Therefore, since we are assuming $k < 4/d + 2$, the integral above is convergent and we get

$$\left\|\frac{\partial^k \mathcal{F}}{\partial u^k}(x, u)(u_1, \dots, u_k)\right\|^p_{\mathcal{H}_p(T, H)} = \mathbb{E} \sup_{t \in [0, T]}\left|\frac{\partial^k \mathcal{F}}{\partial u^k}(x, u)(u_1, \dots, u_k)(t)\right|^p_H$$

$$\leq c \prod_{i=1}^k \left(\mathbb{E} \langle u_i \rangle^{kp}_k\right)^{\frac{1}{k}} \sup_{t \in [0, T]}\left|\int_0^t (t-s)^{-\frac{d}{4}} s^{-\frac{d(k-2)}{4}} ds\right|^p = c_{k,p}(T) \prod_{i=1}^k \|u_i\|^p_{\mathcal{K}^k_{kp}(T, H)}.$$

$$\square$$

Remark 4.2.5. Assume that $k \geq 4/d + 2$. Due to the Lemma 4.2.2, it is immediate to check that for any $0 \leq s \leq t \leq T$

$$\left|e^{(t-s)A} D^k F(u(s; x))(D_x u(s; x)h_1, \dots, D_x u(s; x)h_k)\right|_H$$

$$\leq c(t-s)^{-\frac{d}{4}}|D_x u(s; x)h_1 \cdots D_x u(s; x)h_k|_1$$

$$\leq c(t-s)^{-\frac{d}{4}} s^{-\frac{d(k-2)}{4}} \prod_{i=1}^k \langle D_x u(t; x)h_i \rangle_k \leq c(t-s)^{-\frac{d}{4}} s^{-\frac{d(k-2)}{4}} \mu^k_{k,k}(T) \prod_{i=1}^k |h_i|_H,$$

and this singularity is not integrable near 0. Then, in order to get the k-th derivative in the mean of order p for the solution $u(t; x)$, we can not proceed as in previous proposition. This difficulty seems to be not only technical. Actually, if $u(t; x)$ were k-times differentiable, then at least formally, $D_x^k u(t; x)(h_1, \ldots, h_k)$ should be a mild solution of

$$\frac{d}{dt}z(t) = Az(t) + DF(u(t; x))z(t) + \theta(t), \quad t \geq s, \quad z(s) = 0,$$

for a certain process $\theta(t)$ in which appears, together with other terms,

$$D^k F(u(t; x)) (D_x u(t; x)h_1, \ldots, D_x u(t; x)h_k).$$

Thus the term

$$\int_0^t e^{(t-s)A} D^k F(u(s; x)) (D_x u(s; x)h_1, \ldots, D_x u(s; x)h_k) \, ds$$

is well defined in H if there exist some p_i with $p_1^{-1} + \cdots + p_k^{-1} = 1$ such that $D_x u(t; x)h_i \in L^{p_i}(\mathcal{O}; \mathbb{R}^n)$, for any $t > 0$ and

$$\mathbb{E} \int_0^t (t - s)^{-\frac{d}{4}} \prod_{i=1}^{k} |D_x u(s; x)h_i|_{p_i} \, ds < \infty.$$

But this is not the case, as $k \geq 2 + 4/d$.

Proposition 4.2.6. *Assume the Hypotheses 4.1 and 4.2, and fix $t \geq 0$ and $1 < j < (4/d+2)\wedge(r+1)$. Then $D_x^j u(t; x)(h_1, \ldots, h_j) \in L^p(\mathcal{O}; \mathbb{R}^n)$, for any $x, h_1, \ldots, h_j \in H$ and for any $p \geq 1$ such that*

$$\frac{1}{p} > \frac{dj - 6}{6d}.$$

Moreover, the following estimates hold

$$\sup_{x \in H} \left| D_x^j u(t; x)(h_1, \ldots, h_j) \right|_p \leq \nu_{j,p}(t) \prod_{i=1}^{j} |h_i|_H, \qquad \mathbb{P} - a.s. \qquad (4.2.16)$$

where $\nu_{j,p}$ are suitable increasing continuous functions.

Proof. In the Theorem 4.2.4 we have seen that $D_x^2 u(\cdot; x)(h_1, h_2)$ is the mild solution of the equation (4.2.13). Thus, according to the Remark 4.2.3 we have

$$D_x^2 u(t; x)(h_1, h_2) = \int_0^t U_x(t, s)\theta(s) \, ds$$

where

$$\theta(s) = D^2 F(u(s; x)) (D_x u(s; x)h_1, D_x u(s; x)h_2).$$

Then, thanks to (4.2.12) with $r = 2$

$$\left|D_x^2 u(t;x)(h_1,h_2)\right|_p \leq \int_0^t \mu_{2,p}(t-s)(t-s)^{-\frac{d(p-2)}{4p}} |\theta(s)|_H \, ds, \qquad \mathbb{P} - \text{a.s.}$$

For any $s \in [0,t]$, we have

$$|\theta(s)|_H^2 = \int_{\mathcal{O}} \left|D_\sigma^2 f(\xi, u(s;x)(\xi))(D_x u(s;x)h_1)(\xi)(D_x u(s;x)h_2)(\xi)\right|^2 \, d\xi,$$

and then

$$|\theta(s)|_H \leq c \, |D_x u(s;x)h_1|_4 \, |D_x u(s;x)h_2|_4 \, .$$

Now, from the Lemma 4.2.2 we have that if $h_i \in H$, then $D_x u(s;x)h_i \in \mathcal{K}_q^4(H)$, for any $q \geq 1$ and

$$|D_x u(s;x)h_i|_4 \leq s^{-\frac{d}{8}} \langle D_x u(\cdot;x)h_i\rangle_4 \leq s^{-\frac{d}{8}} \mu_{2,4}(s)|h_i|_H,$$

so that we get

$$\left|D_x^2 u(t;x)(h_1,h_2)\right|_p \leq \int_0^t \mu_{2,p}(t-s)(t-s)^{-\frac{d(p-2)}{4p}} s^{-\frac{d}{4}} \mu_{2,4}^2(s) \, ds|h_1|_H|h_2|_H.$$

This implies that

$$\sup_{x \in H} \left|D_x^2 u(t;x)(h_1,h_2)\right|_p \leq \nu_{2,p}(t)|h_1|_H|h_2|_H \qquad \mathbb{P} - \text{a.s.}$$

with

$$\nu_{2,p}(t) = \int_0^t \mu_{2,p}(t-s)(t-s)^{-\frac{d(p-2)}{4p}} \mu_{2,4}^2(s)s^{-\frac{d}{4}} \, ds,$$

and this is meaningful for any $p \leq \infty$.

Concerning the third derivative, we have seen that $D_x^3 u(t;x)(h_1,h_2,h_3)$ is the unique mild solution of the problem

$$\frac{d}{dt}z(t) = Az(t) + DF(u(t;x))z(t) + \theta(t), \qquad z(0) = 0,$$

where

$$\theta(t) = \frac{1}{4} \sum_{\sigma \in S_3} D^2 F(u(t;x)) \left(D_x^2 u(t;x)(h_{\sigma(1)}, h_{\sigma(2)}), D_x u(t;x)h_{\sigma(3)}\right)$$

$$+ D^3 F(u(t;x)) \left(D_x u(t;x)h_1, D_x u(t;x)h_2, D_x u(t;x)h_3\right),$$

where S_3 is the set of permutations in $\{1,2,3\}$. Thus, if we proceed as for the second derivative, according to the Remark 4.2.3 we have that

$$\left|D_x^3 u(t;x)(h_1,h_2,h_3)\right|_p \leq \int_0^t \mu_{q,p}(t-s)(t-s)^{-\frac{d(p-q)}{2qp}} |\theta(s)|_q \, ds,$$

for any r such that

$$2 \geq q > \frac{dp}{2p+d}. \qquad (4.2.17)$$

Now, since

$$|\theta(s)|_q \leq c \sum_{\sigma \in S_3} \left| D_x^2 u(s;x)(h_{\sigma(1)}, h_{\sigma(2)}) \right|_\infty \left| D_x u(s;x) h_{\sigma(3)} \right|_q$$

$$+ c \left| D_x u(s;x) h_1 \right|_{3q} \left| D_x u(s;x) h_2 \right|_{3q} \left| D_x u(s;x) h_3 \right|_{3q},$$

from (4.2.5), (4.2.10) and (4.2.16) for $j = 2$, it follows

$$|\theta(s)|_q \leq c \left(\nu_{2,\infty}(s)\nu_2(s) + \mu_{2,3q}^3(s) s^{-\frac{d(3q-2)}{4q}} \right) \prod_{i=1}^{3} |h_i|_H.$$

Then, if we fix any q which fulfills (4.2.17) such that $d(3q - 2)/4q < 1$, the function $\nu_{3,p}$ defined by

$$\nu_{3,p}(t) = c \int_0^t \mu_{q,p}(t - s)(t - s)^{-\frac{d(p-q)}{2qp}} \left(\nu_{2,\infty}(s)\nu_2(s) + \mu_{2,3q}^3(s) s^{-\frac{d(3q-2)}{4q}} \right) ds,$$

is meaningful and (4.2.16) follows for the third derivative. Notice that the condition $d(3q - 2)/4q < 1$ together with (4.2.17) easily implies that $p \leq \infty$ if $d = 1$, $p < \infty$ if $d = 2$ and $p < 6$ if $d = 3$.

Finally, in dimension $d = 1$, if $f(\xi, \cdot)$ is assumed to be of class C^5 (that is $r = 5$)) there exist also the fourth and fifth derivatives. They are respectively the unique mild solutions of the fourth and the fifth variation equations which can be written as

$$\frac{d}{dt} z(t) = Az(t) + DF(u(t;x))z(t) + \theta_j(t), \qquad z(0) = 0, \qquad j = 4, 5,$$

where $\theta_j(t)$ is a process depending on all lower order derivatives. As before, since $d = 1$, we have

$$\left| D_x^j u(t;x)(h_1, \ldots, h_j) \right|_p \leq c \int_0^t \mu_{q,p}(t - s)(t - s)^{-\frac{p-q}{2qp}} |\theta_j(s)|_r \, ds,$$

for any q which fulfills (4.2.17). Thus we have only to estimate $|\theta_j(s)|_q$, for $j = 4, 5$. We only show how to estimate the worst term

$$\left| D^5 F(u(s;x)) \left(D_x u(s;x) h_1, \ldots, D_x u(s;x) h_5 \right) \right|_q$$

arising in $|\theta_5(s)|_q$. Due to (4.2.10), we have

$$\left| D^5 F(u(s;x)) \left(D_x u(s;x) h_1, \ldots, D_x u(s;x) h_5 \right) \right|_q$$

$$\leq c \prod_{i=1}^{5} |D_x u(s;x) h_i|_{5q} \leq \mu_{2,5q}^5(s) s^{-\frac{5q-2}{4q}} \prod_{i=1}^{5} |h_i|_H.$$

Thus, if we fix $q = 1$ we have $(p-q)/2qp = (p-1)/2p < 1$ and $(5q-2)/4q = 3/4 < 1$, so that the integral

$$\int_0^t \mu_{1,p}(t-s)(t-s)^{-\frac{p-1}{2p}} \mu_{2,5}^5(s)s^{-\frac{3}{4}} \, ds$$

converges. The estimates of all other terms are even easier.   □

We conclude this section by showing that the highest order derivative of $u(\cdot;x)$ is continuous with respect to $x \in H$. This fact is needed in order to have the continuity of the derivative of the semigroup.

Proposition 4.2.7. *For any $x_0 \in H$ and $t \geq 0$*

$$\lim_{x \to x_0} \sup_{|h_i|_H \leq 1} \left| \left(D_x^j u(t;x) - D_x^j u(t;x_0) \right) (h_1, \ldots, h_j) \right|_H = 0, \qquad \mathbb{P} - a.s. \qquad (4.2.18)$$

where $j = 5$ if $d = 1$, $j = 3$ if $d = 2$ and $j = 2$ if $d = 3$.

Proof. Case $d = 1$. If we set $z(t) = D_x^5 u(t;x)(h_1, \ldots, h_5) - D_x^5 u(t;x_0)(h_1, \ldots, h_5)$, we have

$$\frac{dz}{dt}(t) = Az(t) + DF(u(t;x))z(t)$$

$$+ \left(DF(u(t;x)) - DF(u(t;x_0)) \right) D_x^5 u(t;x_0)(h_1, \ldots, h_5) + \theta_5(t), \quad z(0) = 0,$$

where $\theta_5(t)$ is a suitable process involving the derivatives of $u(t;x)$ and $u(t;x_0)$ up to the fourth order. According to the Remark 4.2.3 we have

$$z(t) = \int_0^t U_x(t,s) \left[\left(DF(u(s;x)) - DF(u(s;x_0)) \right) D_x^5 u(s;x_0)(h_1, \ldots, h_5) + \theta_5(s) \right] \, ds$$

and thanks to (4.2.12) and (4.2.16)

$$|z(t)|_H \leq \int_0^t \left(\nu_{5,\infty}(s) \left| DF(u(s;x)) - DF(u(s;x_0)) \right|_1 \prod_{i=1}^5 |h_i|_H + |\theta_5(s)|_1 \right)$$

$$\mu_{1,2}(t-s)(t-s)^{-\frac{1}{4}} \, ds.$$

Clearly, by using (4.2.5) we have

$$\left| DF(u(s;x)) - DF(u(s;x_0)) \right|_1 \leq c \left| u(s;x) - u(s;x_0) \right|_1 \leq c\nu_2(s) |x - x_0|_H$$

and then, in order to get (4.2.18), it is enough to prove that

$$\lim_{x \to x_0} \sup_{|h_i|_H \leq 1} \int_0^t \mu_{1,2}(t-s)(t-s)^{-\frac{1}{4}} |\theta_5(s)|_1 \, ds = 0.$$

We have

$$\theta_5(t) = \sum_{j=2}^{5} \Delta_F^j(s, x, x_0),$$

where

$$\Delta_F^j(s, x, x_0) = c_j \sum_{\pi \in S_5} \left[D^j F(u(s;x)) \left(D_x^{i_1} u(s;x) \left(h_{\pi(1)}, \ldots, h_{\pi(i_1)} \right) \right), \ldots, \right.$$

$$\left. D_x^{i_j} u(s;x) \left(h_{\pi(i_{j-1})}, \ldots, h_{\pi(i_j)} \right) \right) - D^j F(u(s;x_0)) \left(D_x^{i_1} u(s;x_0) \left(h_{\pi(1)}, \ldots, h_{\pi(i_1)} \right), \right.$$

$$\left. \ldots, D_x^{i_j} u(s;x_0) \left(h_{\pi(i_{j-1})}, \ldots, h_{\pi(i_j)} \right) \right) \right]$$

and where $i_1 + \cdots + i_j = 5$, with $i_1 \leq i_2 \leq \ldots \leq i_j$, S_5 is the set of permutations π of the set $\{1, \ldots, 5\}$ and c_j is a suitable constant. Hence we have to show that

$$\lim_{x \to x_0} \sup_{|h_i|_H \leq 1} \int_0^t \mu_{1,2}(t-s)(t-s)^{-\frac{1}{4}} |\Delta_F^j(s, x, x_0)|_1 \, ds = 0,$$

for each $j = 2, \ldots, 5$.

We prove this for $j = 5$; the other cases follow from similar arguments. We have

$$|\Delta_F^5(s, x, x_0)|_1 \leq \left| \left(D^5 F(u(s;x)) - D^5 F(u(s;x_0)) \right) (D_x u(s;x)h_1, \ldots, D_x u(s;x)h_5) \right|_1$$

$$+ \sum_{k=1}^{5} |\Gamma_k(s, x, x_0)|_1$$

where for each $k = 1, \ldots, 5$

$$\Gamma_k(s, x, x_0) = D^5 F(u(s;x_0)) \left(D_x u(s;x)h_1, \ldots, D_x u(s;x)h_{k-1}, \right.$$

$$\left. (D_x u(s;x_0) - D_x u(s;x))h_k, D_x u(s;x_0)h_{k+1}, \ldots, D_x u(s;x_0)h_5 \right).$$

Due to (4.2.10) and (4.2.16) we have

$$|\Gamma_k(s, x, x_0)|_1 \leq c \mu_{2,4}^4(s) s^{-\frac{1}{2}} \prod_{\substack{i=1 \\ i \neq k}}^{5} |h_i|_H \sup_{z \in H} \left| D_x^2 u(s;z)(x - x_0, h_k) \right|_\infty$$

$$\leq c \mu_{2,4}^4(s) s^{-\frac{1}{2}} \nu_{2,\infty}(s) |x - x_0|_H \prod_{i=1}^{5} |h_i|_H,$$

and then

$$\lim_{x \to x_0} \sup_{|h_i|_H \leq 1} \int_0^t \mu_{1,2}(t-s)(t-s)^{-\frac{1}{4}} \sum_{k=1}^{5} |\Gamma_k(s, x, x_0)|_1 \, ds = 0.$$

Now, thanks to (4.2.10)

$$\left| \left(D^5 F(u(s;x)) - D^5 F(u(s;x_0)) \right) \left(D_x u(s;x) h_1, \ldots, D_x u(s;x) h_5 \right) \right|_1$$

$$\leq \left| D^5 F(u(s;x)) - D^5 F(u(s;x_0)) \right|_6 \mu_{2,6}^5(s) s^{-\frac{5}{6}} \prod_{i=1}^{5} |h_i|_H,$$

so that, due to the continuity of $D^5 F$, for any $s > 0$ we obtain

$$\lim_{x \to x_0} \sup_{|h_i|_H \leq 1} \left| \left(D^5 F(u(s;x)) - D^5 F(u(s;x_0)) \right) \left(D_x u(s;x) h_1, \ldots, D_x u(s;x) h_5 \right) \right|_1 = 0,$$

$\mathbb{P}$-a.s. Moreover,

$$\left| \left(D^5 F(u(s;x)) - D^5 F(u(s;x_0)) \right) \left(D_x u(s;x) h_1, \ldots, D_x u(s;x) h_5 \right) \right|_1$$

$$\leq c \mu_{2,6}^5(s)(s) s^{-\frac{5}{6}} \prod_{i=1}^{5} |h_i|_H, \qquad \mathbb{P} - \text{a.s.}$$

and then, due to the dominated convergence theorem, we can conclude.

Case $d \geq 2$. The proof is analogous to the case $d = 1$. □

4.3   The transition semigroup

For any $\varphi \in B_b(H)$ and $t \geq 0$ we define

$$P_t \varphi(x) = \mathbb{E} \varphi(u(t;x)), \qquad x \in H,$$

where $u(t;x)$ is the unique mild solution of the equation (4.2.1). It is immediate to check that P_t is a contraction operator on $B_b(H)$ and due to the Chapman-Kolmogorov equation, P_t is a semigroup. Moreover it satisfies the Feller property, that is it maps $C_b(H)$ into itself. Indeed, in the Proposition 4.2.1, we have shown that for any $p \geq 1$ the solution $u(t;x)$ is differentiable in the mean of order p with respect to x and for any $t \geq 0$ and $h \in L^r(\mathcal{O}; \mathbb{R}^n)$ it holds

$$\sup_{x \in H} |D_x u(t;x) h|_r \leq \nu_r(t) |h|_r, \qquad \mathbb{P} - \text{a.s.}$$

Then, if we take $r = 2$

$$|u(t;x) - u(t;y)|_H \leq \nu_2(t) |x - y|_H, \qquad \mathbb{P} - \text{a.s.}$$

This easily implies that if φ is uniformly continuous then for any $T > 0$ and $\epsilon > 0$ there exists $\delta_{\epsilon,T} > 0$ such that

$$|x - y|_H < \delta_{\epsilon,T} \Rightarrow |P_t \varphi(x) - P_t \varphi(y)| \leq \mathbb{E} |\varphi(u(t;x)) - \varphi(u(t;y))|_H < \epsilon,$$

for any $t \in [0, T]$. In particular, the family of functions $\{ P_t \varphi \, ; \, t \in [0, T] \}$ is equi-uniformly continuous.

The next proposition shows that P_t is a weakly continuous semigroup on $C_b(H)$ (see Appendix B).

Proposition 4.3.1. *Under the Hypotheses 4.1 and 4.2, P_t is a weakly continuous semigroup on $C_b(H)$.*

Proof. Since P_t is a contraction on $C_b(H)$, the property 4 in the Definition B.1.2 is verified with $M = 1$ and $\omega = 0$. The property 1 has been proved above. As far as the property 2 is concerned, we have

$$P_t \varphi(x) - \varphi(x) = \mathbb{E} \left(\varphi \left(e^{tA} x + \int_0^t e^{(t-s)A} F(u(s; x)) \, ds + w^A(t) \right) - \varphi(x) \right)$$

and hence, if $\varphi \in C_b^{0,1}(H)$, we have

$$|P_t \varphi(x) - \varphi(x)|$$

$$\leq \|\varphi\|_{0,1}^H \left(|e^{tA} x - x|_H + \mathbb{E} |w^A(t)|_H + \mathbb{E} \left| \int_0^t e^{(t-s)A} F(u(s; x)) \, ds \right|_H \right).$$

Now, for any $t \geq 0$

$$\mathbb{E} |w^A(t)|_H^2 = \int_0^t \mathrm{Tr} \left[e^{sA} Q Q^* e^{sA^*} \right] ds,$$

so that due to the Hypothesis 4.1-1

$$\lim_{t \to 0} \mathbb{E} |w^A(t)|_H = 0.$$

For any $T > 0$ and $x \in H$ it holds

$$\sup_{t \in [0,T]} \mathbb{E} |u(t; x)|_H \leq c_T (1 + |x|_H),$$

and, as F has linear growth, this easily implies that

$$\mathbb{E} \left| \int_0^t e^{(t-s)A} F(u(s; x)) \, ds \right|_H \leq c \, t \, (1 + |x|_H).$$

In particular, if $K \subset H$ is compact, we have

$$\lim_{t \to 0^+} \sup_{x \in K} \mathbb{E} \left| \int_0^t e^{(t-s)A} F(u(s; x)) \, ds \right|_H = 0.$$

Moreover, for any $T > 0$ the family of functions

$$h_t : H \to H, \qquad x \mapsto e^{tA} x - x, \qquad t \in [0, T],$$

is equi-bounded and equi-continuous, so that, as e^{tA} is strongly continuous,

$$\lim_{t \to 0^+} \sup_{x \in K} |e^{tA}x - x|_H = 0.$$

Thus, we can conclude that if $\varphi \in C_b^{0,1}(H)$ and K is any compact set in H

$$\lim_{t \to 0^+} \sup_{x \in H} |P_t\varphi(x) - \varphi(x)| = 0, \tag{4.3.1}$$

that is $P_t\varphi$ is $\mathcal{K}$-convergent to φ, as t goes to 0 (see appendix B for the definition and main properties of $\mathcal{K}$-convergence). Since $C_b^{0,1}(H)$ is dense in $C_b(H)$ and P_t is a contraction operator, the limit (4.3.1) holds for general functions $\varphi \in C_b(H)$.

Finally, we show that the property 3 holds true, that is P_t is continuous with respect to the $\mathcal{K}$-convergence. Due to the properties 1 and 2, we have that the mapping

$$[0, +\infty) \times H \to \mathbb{R}, \quad (t, x) \mapsto P_t\varphi(x)$$

is continuous, for any $\varphi \in C_b(H)$. This implies that, for any $T > 0$ and for any compact $K \subset H$ the family of measures

$$\{ P_t(x, \cdot) ; t \in [0, T], x \in K \}$$

is tight. Actually, if $\{(t_n, x_n)\} \subset [0, T] \times K$, there exists $\{(t_{n_k}, x_{n_k})\} \subseteq \{(t_n, x_n)\}$ and $(t_0, x_0) \in [0, T] \times K$ such that (t_{n_k}, x_{n_k}) converges to (t_0, x_0), as k goes to infinity. Then, as the mapping $(t, x) \mapsto P_t\varphi(x)$ is continuous, for any $\varphi \in C_b(H)$, we obtain that $P_t(x_{n_k}, \cdot)$ converges weakly to $P_t(x_0, \cdot)$ and, according to the Prohorov theorem, tightness follows. By proceeding as in the proof of the Proposition 1.4.1, this allows us to conclude that if $\{\varphi_n\} \subset C_b(H)$ is a sequence $\mathcal{K}$-convergent to $\varphi \in C_b(H)$, then $\{P_t\varphi_n\}$ is $\mathcal{K}$-convergent to $P_t\varphi$. $\qquad\square$

4.4 Differentiability of the transition semigroup

In the Theorem 4.2.4 it has been proved that for any $p \geq 1$ the solution $u(\cdot; x)$ of the system (4.2.1) is k-times differentiable in the mean of with respect to x, where k is an integer strictly less than $(4/d+2) \wedge (r+1)$. Thus, as $P_t\varphi(x) = \mathbb{E}\varphi(t; x)$, for any $\varphi : H \to \mathbb{R}$ and for any $x \in H$ and $t \geq 0$, if φ is differentiable, it is possible to differentiate under the sign of integral. Hence, by using the estimates (4.2.5) and (4.2.16) and the Proposition 4.2.7 we get the following result

Proposition 4.4.1. *Under the Hypotheses 4.1 and 4.2, for any $t \geq 0$ and for any integer $k < (4/d+2) \wedge (r+1)$*

$$\varphi \in C_b^k(H) \Rightarrow P_t\varphi \in C_b^k(H).$$

Moreover

$$\|P_t\varphi\|_k^H \leq c \|\varphi\|_k^H. \tag{4.4.1}$$

In this section our aim is proving that the semigroup P_t has a smoothing effect. Namely, we want to show that, for any $t > 0$, P_t maps $B_b(H)$ into $C_b^k(H)$, where k equals 5 if $d = 1$, equals 3 if $d = 2$ and equals 2 if $d = 3$ (clearly under the assumption that F is regular enough). Furthermore, we want to prove that for any $j = 1, \ldots, k$ the following estimate holds

$$\|D^j (P_t \varphi)\|_0^H \leq c \, (t \wedge 1)^{-\frac{j(1+\epsilon)}{2}} \|\varphi\|_0^H, \quad t > 0,$$

where ϵ is the constant introduced in the Hypothesis 4.1-2.

To this purpose we need the following preliminary result.

Lemma 4.4.2. *Assume that*

$$j \leq \begin{cases} 5 \wedge r & \text{if } d = 1, \\[2mm] 3 \wedge r & \text{if } d = 2, \\[2mm] 2 & \text{if } d = 3. \end{cases}$$

Then for any $x, h_1, \ldots, h_j \in H$ and $t > 0$ we have that $D_x^j u(t; x)(h_1, \ldots, h_j) \in D(Q^{-1})$ and

$$\sup_{x \in H} \int_0^t |Q^{-1} D_x^j u(s; x)(h_1, \ldots, h_j)|_H^2 \, ds \leq c(t) \, t^{1-\epsilon} \prod_{i=1}^j |h_i|_H, \quad \mathbb{P} - a.s. \quad (4.4.2)$$

for some continuous increasing function $c(t)$. Moreover,

$$\lim_{x \to x_0} \sup_{|h_i|_H \leq 1} \int_0^t |Q^{-1} \left[D_x^j u(s; x) - D_x^j (u(s; x_0)) \right] (h_1, \ldots, h_j)|_H^2 \, ds = 0. \quad (4.4.3)$$

Proof. The first derivative $D_x u(t; x) h$ along the direction $h \in H$ coincides with the unique mild solution $z^h(t; x)$ of the problem

$$\frac{d}{dt} z(t) = Az(t) + DF(u(t; x)) z(t), \quad z(0) = h.$$

It is easy to prove that $z(t) \in D((-A)^\delta)$, for any $t > 0$ and for any $\delta < 1$. Besides, possibly approximating $z(t)$ by means of more regular solutions (we will use in more details these approximations in the chapter 6, see the Proposition 6.2.2), we have

$$\frac{1}{2} \frac{d}{dt} |z(t)|_H^2 + \langle (-A) z(t), z(t) \rangle_H = \langle DF(u(t; x)) z(t), z(t) \rangle_H.$$

Hence, by integrating with respect to t and by using the boundedness of DF, we get

$$|z(t)|_H^2 + 2 \int_0^t |(-A)^{1/2} z(s)|_H^2 \, ds \leq |h|_H^2 + c \int_0^t |z(s)|_H^2 \, ds.$$

Due to (4.2.5), with $q = 2$, this yields

$$\int_0^t |(-A)^{1/2} z(s)|_H^2 \, ds \leq c \, (1 + c \, t \, \nu_2(t)) \, |h|_H^2, \quad \mathbb{P} - \text{a.s.} \tag{4.4.4}$$

Now, according to the interpolatory inequality given in (4.1.8) (with $\delta_1 = 0$, $\delta = \epsilon/2$ and $\delta_2 = 1/2$), from the Hölder inequality, we have

$$\int_0^t |(-A)^{\epsilon/2} z(s)|_H^2 \, ds \leq \int_0^t |(-A)^{1/2} z(s)|_H^{2\epsilon} |z(s)|_H^{2(1-\epsilon)} \, ds$$

$$\leq \left(\int_0^t |(-A)^{1/2} z(s)|_H^2 \, ds \right)^\epsilon \left(\int_0^t |z(s)|_H^2 \, ds \right)^{1-\epsilon},$$

so that, thanks to (4.2.16) and (4.4.4), we can conclude that

$$\int_0^t |(-A)^{\epsilon/2} D_x u(s; x) h|_H^2 \, ds \leq c \left(1 + t \, \nu_2^2(t) \right)^\epsilon \nu_2^{2(1-\epsilon)}(t) t^{1-\epsilon} |h|_H^2, \quad \mathbb{P} - \text{a.s.}$$

Finally, recalling that due to the Hypothesis 4.1-2 $Q^{-1} = \Gamma_\epsilon (-A)^{\epsilon/2}$, for a suitable bounded linear operator Γ_ϵ, it follows that $D_x u(t; x) h \in D(Q^{-1})$ for any $t > 0$ and

$$\sup_{x \in H} \int_0^t |Q^{-1} D_x u(s; x) h|_H^2 \, ds$$

$$\leq \|\Gamma_\epsilon\|^2 \sup_{x \in H} \int_0^t |(-A)^{\epsilon/2} D_x u(s; x) h|_H^2 \, ds \leq c(t) t^{1-\epsilon} |h|_H^2.$$

Higher order derivatives $D_x^j u(t; x)(h_1, \ldots, h_j)$, $j \geq 2$, coincide with the solutions $z_j(t; x)$ of the corresponding variation equations

$$\frac{d}{dt} z(t) = A z(t) + DF(u(t; x)) z(t) + \theta_j(t), \quad z(0) = 0,$$

for suitable processes $\theta_j(t)$ depending on all lower order derivatives. Then, as before for the first derivative, we have that $z_j(t) \in D((-A)^{1/2})$ for $t > 0$ and

$$|z_j(t)|_H^2 + 2 \int_0^t |(-A)^{1/2} z_j(s)|_H^2 \, ds \leq c \int_0^t |z_j(s)|_H^2 \, ds + 2 \int_0^t \left| \langle \theta_j(s), z_j(s) \rangle_H \right| \, ds.$$

By using (4.2.16) and the Hölder inequality, this implies that

$$\sup_{x \in H} \int_0^t |(-A)^{1/2} z_j(s)|_H^2 \, ds \leq c_j(t) \prod_{i=1}^j |h_i|_H^2, \tag{4.4.5}$$

for suitable continuous functions $c_j(t) > 0$, increasing with respect to t. We prove (4.4.5) for $j = 2$. We have

$$\theta_2(t) = D^2 F(u(t;x)) \left(D_x u(t;x)h_1, D_x u(t;x)h_2\right)$$

and then, thanks to (4.2.5) and (4.2.16) we easily have

$$\int_0^t |(-A)^{1/2} z_2(s)|_H^2 \, ds \le c \int_0^t |z_2(s)|_H^2 \, ds + c \int_0^t |\theta_2(s)|_H |z_2(s)|_H \, ds$$

$$c \int_0^t \left(\nu_{2,2}(s) + \nu_{2,\infty}(s)\nu_2^2(s)\right) ds |h_1|_H^2 |h_2|_H^2.$$

Hence, by the same interpolatory arguments used above for the first derivative, the estimate (4.4.2) follows for $j = 2$.

In the case $j > 2$, we only show how to treat the most delicate term

$$\int_0^t \langle \eta_j(s), z_j(s) \rangle_H \, ds,$$

where

$$\eta_j(s) = D^j F(u(s;x)) \left(D_x u(s;x)h_1, \ldots, D_x u(s;x)h_j\right).$$

Since $d < 3$, $z_j \in L^\infty(0,T; L^\infty(\mathcal{O}; \mathbb{R}^n))$, $\mathbb{P}$-a.s. and then we have

$$\left| \int_0^t \langle \eta_j(s), z_j(s) \rangle_H \, ds \right| \le \int_0^t |z_j(s)|_\infty |\eta_j(s)|_1 \, ds$$

$$\le \nu_{2,\infty}(t) \int_0^t \prod_{i=1}^j |D_x u(s;x)h_i|_j \, ds \le \nu_{2,\infty}(t) \int_0^t \mu_j^j(s) s^{-\frac{d(j-2)}{4}} \, ds \prod_{i=1}^j |h_i|_H.$$

Once we have this estimate, we proceed as before.

The limit (4.4.3) follows from arguments analogous to those used in the proof of the Proposition 4.2.7. $\qquad \square$

Proposition 4.4.3. *Assume the Hypotheses 4.1 and 4.2. If $\varphi \in B_b(H)$ then $P_t\varphi$ is differentiable for any $t > 0$ and*

$$\|D(P_t\varphi)\|_0^H \le c (t \wedge 1)^{-\frac{1+\epsilon}{2}} \|\varphi\|_0^H. \tag{4.4.6}$$

Moreover, if $\varphi \in C_b(H)$ we have

$$\langle D(P_t\varphi)(x), h \rangle_H = \frac{1}{t} \mathbb{E}\varphi(u(t;x)) \int_0^t \langle Q^{-1} D_x u(s;x)h, dw(s) \rangle_H. \tag{4.4.7}$$

Proof. As proved in the Theorem 4.2.4, $u(t; x)$ is k-times mean-square differentiable with respect to $x \in H$, for any $k < (4/d + 2) \wedge (r + 1)$. Since $d \leq 3$ and $r \geq 2$, this implies that $u(t; x)$ is at least three times mean-square differentiable. Hence, as for any $t \geq 0$ and $x \in H$

$$P_t \varphi(x) = \mathbb{E} \varphi(u(t; x)),$$

by differentiating under the sign of integral, we have that for any $\varphi \in C_b^2(H)$ the function $(t, x) \mapsto P_t \varphi(x)$ belongs to $C^{1,2}([0, +\infty) \times H)$ and by using Itô's formula we have

$$\varphi(u(t; x)) = P_t \varphi(x) + \int_0^t \langle D_x (P_{t-s} \varphi)(u(s; x)), Q \, dw(s) \rangle_H. \tag{4.4.8}$$

The proof of this fact is not trivial, as we are in infinite dimensions and the process $u(t; x)$ is not a semimartingale. For a detailed proof we refer to [104]. Now, recalling that (4.4.2) holds, we can multiply both sides of (4.4.8) by the term

$$\int_0^t \langle Q^{-1} D_x u(s; x) h, dw(s) \rangle_H$$

and by taking expectation we get

$$\mathbb{E} \varphi(u(t; x)) \int_0^t \langle Q^{-1} D_x u(s; x) h, dw(s) \rangle_H$$

$$= \mathbb{E} \int_0^t \langle Q^* D(P_{t-s} \varphi)(u(s; x)), Q^{-1} D_x u(s; x) h \rangle_H \, ds$$

$$= \left\langle D \int_0^t \mathbb{E} (P_{t-s} \varphi)(u(s; x)) \, ds, h \right\rangle_H = t \langle D(P_t \varphi)(x), h \rangle_H.$$

This means that (4.4.7) holds for any $\varphi \in C_b^2(H)$. Now we extend the formula (4.4.7) to functions $\varphi \in C_b(H)$. As in [104], for any $\varphi \in C_b(H)$ and $k \in \mathbb{N}$ we define

$$\varphi_k(x) = \int_{\mathbb{R}^k} \rho_k(\xi - \Pi_k x) \varphi(T_k \xi) \, d\xi, \qquad x \in H,$$

where $\{\rho_k\}$ is a sequence of non-negative smooth functions such that

$$\operatorname{supp} \rho_k \subset \{\xi \in \mathbb{R}^k \, : \, |\xi| \leq 1/k\}, \qquad \int_{\mathbb{R}^k} \rho_k(\xi) \, d\xi = 1$$

and the functions $\Pi_k : H \to \mathbb{R}^k$ and $T_k : \mathbb{R}^k \to H$ are respectively defined by $P_k(x) = (\langle x, e_1 \rangle_H, \ldots, \langle x, e_k \rangle_H)$ and $T_k(\xi) = \sum_{i=1}^k \xi_i e_i$, for some complete orthonormal basis $\{e_k\}$ in H. It is possible to show that φ_k is smooth and

$$\lim_{k \to +\infty} \varphi_k(x) = \varphi(x), \qquad x \in H, \qquad \|\varphi_k\|_0^H \leq \|\varphi\|_0^H. \tag{4.4.9}$$

Therefore, since we have the formula (4.4.7) for each φ_k, thanks to (4.4.9) by the dominated convergence theorem we can take the limit as $k \to +\infty$ and we get the formula (4.4.7) for φ.

In order to obtain the estimate (4.4.6), we remark that

$$|\langle D(P_t\varphi)(x), h\rangle_H| = \frac{1}{t}\left|\mathbb{E}\varphi(u(t;x))\int_0^t \langle Q^{-1}D_xu(s;x)h, dw(s)\rangle_H\right|$$

$$\leq \frac{1}{t}\|\varphi\|_0^H\left(\mathbb{E}\int_0^t |Q^{-1}D_xu(s;x)h|_H^2\,ds\right)^{1/2}$$

and then, due to (4.4.2) we easily get (4.4.6).

Moreover, if we denote by $P_t(x, \cdot)$ the law of $u(t;x)$, for any $x, y \in H$ the estimate (4.4.6) yields

$$\text{Var}\,(P_t(x, \cdot) - P_t(y, \cdot)) \leq c\,(t \wedge 1)^{-\frac{1+\epsilon}{2}}|x - y|_H.$$

Hence for any $\varphi \in B_b(H)$ and $t > 0$, $P_t\varphi$ is Lipschitz-continuous in H and, since $P_t\varphi = P_{t/2}\left(P_{t/2}\varphi\right)$, this implies that $P_t\varphi$ is differentiable for any $\varphi \in B_b(H)$ and (4.4.7) holds. $\qquad\square$

In the Proposition 4.4.3, by assuming that the reaction term $f(\xi, \cdot)$ is at least twice differentiable we have shown that $P_t\varphi$ is differentiable, for any $t > 0$ and $\varphi \in B_b(H)$, and $\|D(P_t\varphi)\|_0^H \leq c\,(t \wedge 1)^{-\frac{1+\epsilon}{2}}\|\varphi\|_0^H$. Moreover, we have proved the Bismut-Elworthy formula (4.4.7). The assumption that $f(\xi, \cdot)$ is at least twice differentiable is required in order to have the second order mean-square derivative of $u(t, x)$ with respect to x. Actually, this implies that $P_t : C_b^2(H) \to C_b^2(H)$ and then the formula (4.4.8) holds.

In the next proposition we show that even if we assume $f(\xi, \cdot)$ to be only of class C^1, nevertheless $P_t\varphi \in C_b^1(H)$, for any $\varphi \in B_b(H)$ and $t > 0$, and $D(P_t\varphi)$ is given by (4.4.7).

Proposition 4.4.4. *Assume that* $f : \overline{\mathcal{O}} \times \mathbb{R}^n \to \mathbb{R}^n$ *is measurable in* $\xi \in \overline{\mathcal{O}}$, *for all* $\sigma \in \mathbb{R}^n$, *and for almost all* $\xi \in \overline{\mathcal{O}}$ *the function* $f(\xi, \cdot)$ *is of class* C^1 *and*

$$\sup_{\xi \in \overline{\mathcal{O}}}\,\sup_{\sigma \in \mathbb{R}^n} |D_\sigma f(\xi, \sigma)| < \infty.$$

Then, under the Hypothesis 4.1, $P_t\varphi$ is differentiable, for any $t > 0$ and $\varphi \in B_b(H)$, and if $\varphi \in C_b(H)$ (4.4.7) is fulfilled.

Proof. If $\{\rho_k\}$ is the usual sequence of mollifiers in $\mathbb{R}^n$, for any $(\xi, \sigma) \in \overline{\mathcal{O}} \times \mathbb{R}^n$ we define

$$f_k(\xi, \sigma) = \int_{\mathbb{R}^n} \rho_k(\theta)f(\xi, \sigma - \theta)\,d\theta = \int_{\mathbb{R}^n} \rho_k(\sigma - \theta)f(\xi, \theta)\,d\theta, \qquad k \in \mathbb{N}.$$

For any $k \in \mathbb{N}$ we have

$$\frac{\partial f_k}{\partial \sigma_i}(\xi, \sigma) = \int_{\mathbb{R}^n} \rho_k(\theta) \frac{\partial f}{\partial \sigma_i}(\xi, \sigma - \theta) \, d\theta,$$

and

$$\frac{\partial^2 f_k}{\partial \sigma_j \, \partial \sigma_i}(\xi, \sigma) = \int_{\mathbb{R}^n} \frac{\partial \rho_k}{\partial \sigma_j}(\sigma - \theta) \frac{\partial f}{\partial \sigma_i}(\xi, \theta) \, d\theta,$$

so that $f_k(\xi, \cdot)$ is twice differentiable, with bounded derivatives. This means that if $u_k(\cdot; x)$ is the solution of the problem

$$du(t) = [Au(t) + F_k(u(t))] \, dt + Q \, dw(t), \qquad u(0) = x,$$

where F_k is the Nemytskii operator associated with the reaction term f_k, $u_k(\cdot; x)$ is twice differentiable in the mean of order p, for any $p \geq 1$. Thus, according to the Proposition 4.4.3, if we denote by P_t^k the corresponding transition semigroup, $P_t^k \varphi \in C_b^1(H)$, for any $t > 0$ and $\varphi \in B_b(H)$, and if $\varphi \in C_b(H)$

$$\left\langle D(P_t^k \varphi)(x), h \right\rangle_H = \frac{1}{t} \mathbb{E} \varphi(u_k(t; x)) \int_0^t \left\langle Q^{-1} D_x u_k(s; x) h, dw(s) \right\rangle_H.$$

Since $f_k(\xi, \cdot)$ and $D_\sigma f_k(\xi, \cdot)$ converge respectively to $f(\xi, \cdot)$ and $D_\sigma f(\xi, \cdot)$, uniformly on compact sets, it is not difficult to check that

$$\lim_{k \to +\infty} \sup_{t \in [0,T]} |u_k(t; x) - u(t; x)|_H = 0$$

and

$$\lim_{k \to +\infty} \int_0^t \left| Q^{-1} \left(D_x u_k(s; x) - D_x u(s; x) \right) h \right|_H^2 \, ds = 0,$$

so that, if $\varphi \in C_b(H)$

$$\lim_{k \to +\infty} P_t^k \varphi(x) = P_t \varphi(x), \qquad (t, x) \in [0, \infty) \times H,$$

and

$$\lim_{k \to +\infty} \left\langle D(P_t^k \varphi)(x), h \right\rangle_H = \frac{1}{t} \mathbb{E} \varphi(u(t; x)) \int_0^t \left\langle Q^{-1} D_x u(s; x) h, dw(s) \right\rangle_H.$$

As the mapping $x \in H \mapsto D_x u(s; x) h$ is continuous, this easily implies that $P_t \varphi$ is continuously differentiable and its derivative is given by (4.4.7). $\qquad \square$

By using the semigroup law and the previous proposition we can prove the existence of higher order derivatives.

Theorem 4.4.5. *Under the Hypotheses 4.1 and 4.2, for any $\varphi \in B_b(H)$ and $t > 0$ we have $P_t\varphi \in C_b^k(H)$, where $k = 5 \wedge r$ if $d = 1$, $k = 3 \wedge r$ if $d = 2$ and $k = 2$ if $d = 3$. Moreover for $0 \leq i \leq j \leq k$ it holds*

$$\|D^j(P_t\varphi)\|_0^H \leq c (t \wedge 1)^{-\frac{(j-i)(1+\epsilon)}{2}} \|\varphi\|_i^H. \tag{4.4.10}$$

Proof. From the semigroup law we have

$$P_t\varphi(x) = P_{t/2}(P_{t/2}\varphi)(x) = \mathbb{E} P_{t/2}\varphi(u(t/2; x)).$$

Then, if we define $\psi = P_{t/2}\varphi$, from (4.4.7) we get

$$\langle D(P_t\varphi)(x), h \rangle_H = \frac{2}{t} \mathbb{E} \psi(u(t/2; x)) \int_0^{t/2} \langle Q^{-1} D_x u(s; x) h, dw(s) \rangle_H. \tag{4.4.11}$$

Now, thanks to the previous proposition, ψ is differentiable and then, due to the Proposition 4.2.1 and the Theorem 4.2.4, we can differentiate under the sign of integral in (4.4.11). This implies that $P_t\varphi$ is twice differentiable and for any $h, k \in H$ it holds

$$\langle D^2(P_t\varphi)(x)h, k \rangle_H = I_1(t; x)(h, k) + I_2(t; x)(h, k), \tag{4.4.12}$$

where

$$I_1(t; x)(h, k) = \frac{2}{t} \mathbb{E} \langle D\psi\left(u(t/2; x)\right), D_x u(t/2; x)k \rangle_H \int_0^{t/2} \langle Q^{-1} D_x u(s; x) h, dw(s) \rangle_H$$

$$I_2(t; x)(h, k) = \frac{2}{t} \mathbb{E} \psi\left(u(t/2; x)\right) \int_0^{t/2} \langle Q^{-1} D_x^2 u(s; x)(h, k), dw(s) \rangle_H.$$

In order to prove (4.4.10) for $j = 2$ and $i = 0$, by using (4.2.5) and (4.4.2) we have

$$|I_1(t; x)(h, k)| \leq \frac{2}{t} \|D\psi\|_0^H \nu_2(t/2) \left(\mathbb{E} \int_0^{t/2} |Q^{-1} D_x u(s; x) h|^2 ds \right)^{1/2} |k|_H$$

$$\leq \frac{2}{t} \|D\psi\|_0^H \nu_2(t/2) c(t) t^{\frac{1-\epsilon}{2}} |h|_H |k|_H.$$

so that

$$\sup_{x \in H} |I_1(t; x)(h, k)| \leq c(t) t^{-\frac{1+\epsilon}{2}} \|D\psi\|_0^H |h|_H |k|_H. \tag{4.4.13}$$

In the same way, by using once more (4.4.2), for the second term we have

$$|I_2(t; x)(h, k)| \leq \frac{2}{t} \|\psi\|_0^H \left(\mathbb{E} \int_0^{t/2} |Q^{-1} D_x^2 u(s; x)(h, k)|^2 ds \right)^{1/2} \tag{4.4.14}$$

$$\leq c(t) \|\psi\|_0^H t^{-\frac{1+\epsilon}{2}} |h|_H |k|_H.$$

Now, recalling that

$$\|\psi\|_0^H = \|P_{t/2}\varphi\|_0^H \le \|\varphi\|_0^H$$

and

$$\|D\psi\|_0^H = \|D(P_{t/2}\varphi)\|_0^H \le c\,(t \wedge 1)^{-\frac{1+\epsilon}{2}}\|\varphi\|_0^H,$$

we obtain

$$\sup_{x \in H}\left|\langle D^2(P_t\varphi)(x)h, k\rangle_H\right| \le c(t)\,t^{-(1+\epsilon)}\|\varphi\|_0^H|h|_H|k|_k,$$

For $t \ge 1$ we have

$$\langle D^2(P_t\varphi)(x)h, k\rangle_H = \langle D^2(P_1(P_{t-1}\varphi))(x)h, k\rangle_H \le c(1)\|P_{t-1}\varphi\|_0^H|h|_H|k|_k,$$

and then, as P_t is a contraction, we have that $P_t\varphi \in C_b^2(H)$, for any $t > 0$ and $\varphi \in B_b(H)$ and the estimate (4.4.10) holds for $i = 0$ and for $j = 2$.

Now, assume that $d \le 2$. Since $u(t; x)$ is three times mean-square differentiable and the function $\psi = P_{t/2}\varphi$ is twice differentiable, due to the Lemma 4.4.2 we can differentiate twice each side of (4.4.11) and we get that $P_t\varphi$ is three times differentiable for $t > 0$. Finally, the estimate (4.4.10) for $i = 0$ and $j = 3$ follows by the same arguments as before, by taking into account of (4.4.2) for the derivatives of $u(t; x)$, up to the third order, and of the estimates proved for the first two derivatives of ψ. Higher order derivatives are obtained in a similar way. Notice that the continuity of the k-th derivative follows from the Proposition 4.2.7 and from the limit 4.4.3.

We conclude by proving the estimate (4.4.10) for $i > 0$. The case $i = j$ follows from (4.4.1). Thus, we have to consider the case $1 \le i < j$. If $\varphi \in C_b^1(H)$, due to (4.4.1) we have

$$\|D\psi\|_0^H = \|DP_{t/2}\varphi\|_0^H \le c\|\varphi\|_1^H$$

and then, from (4.4.12), (4.4.13) and (4.4.14) the estimate (4.4.10) follows for $i = 1$ and $j = 2$.

We conclude by showing how to get the estimate (4.4.10) for $i = 1$ and $j = 3$; the other cases are analogous. If $\varphi \in C_b^1(H)$ we can differentiate each side of (4.4.12) along any direction $h_3 \in H$ and we get

$$D^3(P_t\varphi)(x)(h_1, h_2, h_3) = \frac{2}{t}\left(J_1(t; x) + J_2(t; x) + J_3(t; x)\right),$$

where

$$J_1(t; x) = \mathbb{E}\psi(u(t/2; x)) \int_0^{t/2} \langle Q^{-1}D_x^3 u(s; x)(h_1, h_2, h_3), dw(s)\rangle_H$$

and

$$J_2(t;x) = \mathbb{E}\langle D\psi(u(t/2;x)), D_x u(t/2;x)h_2\rangle_H \int_0^{t/2} \langle Q^{-1}D_x^2 u(s;x)(h_1,h_3), dw(s)\rangle_H$$

$$+\mathbb{E}\langle D\psi(u(t/2;x)), D_x^2 u(t/2;x)(h_2,h_3)\rangle_H \int_0^{t/2} \langle Q^{-1}D_x u(s;x)h_1, dw(s)\rangle_H$$

$$+\mathbb{E}\langle D\psi(u(t/2;x)), D_x u(t/2;x)h_3\rangle_H \int_0^{t/2} \langle Q^{-1}D_x^2 u(s;x)(h_1,h_2), dw(s)\rangle_H$$

and finally,

$$J_3(t;x) = \mathbb{E}D^2\psi(u(t/2;x))(D_x u(t/2;x)h_3, D_x u(t/2;x)h_2)$$

$$\times \int_0^{t/2} \langle Q^{-1}D_x u(s;x)h_1, dw(s)\rangle_H.$$

By using (4.2.16) and recalling that $\psi = P_{t/2}\varphi$, thanks to (4.4.1) we easily get

$$\sup_{x\in H} |J_1(t;x)| + |J_2(t;x)| \le c\|\varphi\|_1^H |h_1|_H |h_2|_H |h_3|_H.$$

Concerning the last term J_3, from (4.4.12), by using again (4.2.16), (4.4.10) and (4.4.1) we have that for any $\varphi \in C_b^1(H)$

$$\|D^2(P_t\varphi)\|_0^H \le c(t\wedge 1)^{-\frac{1+\epsilon}{2}}\|\varphi\|_1^H.$$

Hence, with the same computations as before we have

$$\sup_{x\in H} |J_3(t;x)| \le ct^{-\epsilon}\|\varphi\|_1^H.$$

This allows us to conclude that

$$\|P_t\varphi\|_3^H \le c(t\wedge 1)^{-(1+\epsilon)}\|\varphi\|_1^H. \tag{4.4.15}$$

□

Chapter 5

Kolmogorov equations in Hilbert spaces

With the same notations used in the previous chapter, we can introduce the following second order infinite dimensional differential operator

$$\mathcal{L}(x, D) = \frac{1}{2} \text{Tr} \, [D^2 QQ^*] + \langle Ax + F(x), D \rangle_H, \quad x \in D(A).$$

$\mathcal{L}(x, D)$ is the diffusion operator corresponding to the system (4.0.1).

In this chapter we want to study existence, uniqueness and optimal regularity in Hölder spaces for the solutions of the parabolic and the elliptic problems associated with the operator $\mathcal{L}(x, D)$.

In the chapter 4 we have proved that the system (4.0.1) admits a unique solution $u(\cdot; x)$, for any initial datum $x \in H$. Thus we have introduced the transition semigroup P_t and we have proved that P_t has a regularizing effect. Here, by means of P_t we give a probabilistic representation of the solutions of the infinite dimensional parabolic and elliptic problems associated with $\mathcal{L}(x, D)$

$$\begin{cases} \dfrac{\partial y}{\partial t}(t, x) = \mathcal{L}(x, D) y(t, x), & t > 0 \quad x \in D(A) \\[2mm] y(0, x) = \varphi(x), & x \in H, \end{cases} \tag{5.0.1}$$

and

$$\lambda \psi(x) - \mathcal{L}(x, D) \psi(x) = \varphi(x), \quad x \in D(A), \tag{5.0.2}$$

with $\lambda > 0$. Actually, if for any $(t, x) \in [0, \infty) \times H$ we define

$$y(t, x) = P_t \varphi(x), \quad x \in H, \tag{5.0.3}$$

we show that y is the unique *classical* solution of the problem (5.0.1). This means that $y : [0, +\infty) \times H \to \mathbb{R}$ is bounded and continuous, the function $y(t, \cdot)$ is in

$C_b^2(H)$, for any $t > 0$, the operator $D^2y(t,x)QQ^*$ is of *trace-class*, for any $x \in H$, and the function

$$(0, +\infty) \times H \to \mathbb{R}, \qquad (t,x) \mapsto \mathrm{Tr}\left[D^2y(t,x)QQ^*\right]$$

is continuous. Moreover for any $x \in D(A)$ the function $y(\cdot, x)$ is differentiable in $(0, +\infty)$ and y fulfills the equation (5.0.1).

In the same way, if we define

$$\psi(x) = \int_0^{+\infty} e^{-\lambda t} P_t\varphi(x)\,dt, \quad x \in H \tag{5.0.4}$$

(and this is meaningful, as P_t is a weakly continuous semigroup, see the appendix B for the definition and main properties), we prove that ψ is the unique strict solution of the problem (5.0.2), for any $\varphi \in C_b^1(H)$. This means that $\psi \in C_b^2(H)$, the operator $D^2\psi(x)QQ^*$ is of *trace-class*, for any $x \in H$, and the mapping $x \mapsto \mathrm{Tr}\left[D^2\psi(x)QQ^*\right]$ is bounded and continuous. Moreover ψ fulfills the equation (5.0.2). We also show that if the datum φ is only continuous, then the function ψ is a *strong* solution, that is there exist two sequences $\{\varphi_n\}$ and $\{\psi_n\}$, converging respectively to φ and ψ in $C_b(H)$, such that ψ_n is the unique strict solution of the problem (5.0.2) with datum φ_n.

In the chapter 4 we have also proved that for any $\varphi \in B_b(H)$ and $t > 0$

$$\left\| D^j\left(P_t\varphi\right) \right\|_0^H \le c\,(t \wedge 1)^{-\frac{(j-i)(1+\epsilon)}{2}} \|\varphi\|_i^H, \quad i \le j \le 3.$$

Then, by using interpolation in spaces of functions of infinite variables (see [12]) and the general method introduced by Lunardi in [93] which we have already used in the chapter 1, we get Schauder type estimates.

In the section 5.2 we study the parabolic problem. In order to show that the function y defined by (5.0.3) is the unique classical solution of (5.0.1) we have to prove that for any $t > 0$ and $x \in H$ the operator $D^2(P_t\varphi)(x)QQ^*$ is trace-class, for any $\varphi \in B_b(H)$. Notice that here, in addition to the difficulties arising from infinite sums, we do not assume Q to have a bounded inverse and then in order to use the Bismut-Elworthy formula we have also to give a meaning to the stochastic integrals where the unbounded operator Q^{-1} appears. To this purpose the crucial step will be proving that there exists a complete orthonormal system $\{e_k\}$ in H such that

$$\sup_{x \in H} \sum_{k=1}^{\infty} \int_0^t \left| Q^{-1}D_x u(s;x)QQ^*e_k \right|_H^2 ds \le c(t)\,t^\delta, \qquad \mathbb{P} - \text{a.s.} \tag{5.0.5}$$

for some constant $\delta > 0$ and some continuous increasing function $c(t)$, and for any bounded operator N

$$\sup_{x \in H} \sum_{k=1}^{\infty} \left| Q^{-1}D_x^2 u(t;x)(Ne_k, QQ^*e_k) \right|_H \le \|N\|\,c(t), \tag{5.0.6}$$

for a possibly different continuous increasing function $c(t)$ not depending on N.

Actually, since we prove the following Bismut-Elworthy formula for the second derivative of $P_t\varphi$

$$\langle D^2(P_t\varphi)(x)h, k\rangle_H$$

$$= \frac{1}{t}\mathbb{E}\,\langle D\varphi(u(t;x)), D_xu(t;x)h\rangle_H \int_0^t \langle Q^{-1}D_xu(s;x)k, dw(s)\rangle_H$$

$$+ \frac{1}{t}\mathbb{E}\varphi(u(t;x)) \int_0^t \langle Q^{-1}D_x^2u(s;x)(h,k), dw(s)\rangle_H,$$

where $D_xu(t;x)h$ and $D_x^2u(t;x)(h,k)$ are respectively the first and the second derivatives of $u(t;x)$ along the directions h and k, from the two inequalities (5.0.5) and (5.0.6), we can conclude that for any $\varphi \in B_b(H)$ and for any $t > 0$ and $x \in H$ the operator $D^2(P_t\varphi)(x)QQ^\star$ is of *trace-class* and

$$\sup_{x\in H}\left|\text{Tr}\left[D^2(P_t\varphi)(x)QQ^\star\right]\right| \leq c\,(t\wedge 1)^{-\alpha}\sup_{x\in H}|\varphi(x)|,$$

where α is a positive constant which we can determine explicitly. This follows if we assume that there exists a basis $\{e_k\}$ for H which fulfills some summability condition which will be described in detail later on. The existence and the uniqueness of the solution for (5.0.1) are first proved for regular initial data and then, by an approximating procedure, for any initial datum $\varphi \in C_b(H)$.

In the section 5.3 we study the elliptic problem. We prove that for any $\varphi \in C_b(H)$ there exists a unique strong solution. If $\varphi \in C_b^1(H)$, we show that the function ψ which is given by (5.0.4) is the unique strict solution. To this purpose we have first to verify that ψ belongs to $C_b^2(H)$ and $D^2\psi(x)QQ^\star$ is a trace-class operator for any $x \in H$. Notice that the study of the elliptic case is more delicate than the study of the parabolic case, as we have to check that the function

$$(0,\infty) \to \mathbb{R}, \qquad t \mapsto \text{Tr}\left[D^2(P_t\varphi)(x)QQ^\star\right]$$

is integrable near 0, for any $x \in H$. The general case of $\varphi \in C_b(H)$ follows, as $C_b^1(H)$ is dense in $C_b(H)$.

Finally, we conclude the chapter by giving Schauder estimates for the elliptic problem. More precisely, we prove that if in (5.0.1) the datum φ belongs to $C_b^\theta(H)$, then $\psi \in C_b^{2+\theta}(H)$ and

$$\|\psi\|_{C_b^{2+\theta}(H)} \leq c\,\|\varphi\|_{C_b^\theta(H)}.$$

5.1 Assumptions

The operator $\mathcal{L}$ is the diffusion generator corresponding to the system (4.0.1). With the notations introduced in the previous chapter, the system (4.0.1) can be written as

$$du(t) = [Au(t) + F(u(t))] \, dt + Q \, dw(t), \qquad u(0) = x,$$

where A is the realization in $H = L^2(\mathcal{O}; \mathbb{R}^n)$ of the second order differential operator $\mathcal{A} = (\mathcal{A}_1, \ldots, \mathcal{A}_n)$ and F is the Nemytskii operator associated with the reaction term f.

Concerning f we assume that

Hypothesis 5.1. *The function* $f : \overline{\mathcal{O}} \times \mathbb{R}^n \to \mathbb{R}^n$ *is measurable with respect to* $\xi \in \overline{\mathcal{O}}$, *for all* $\sigma \in \mathbb{R}^n$. *Moreover the function* $f(\xi, \cdot)$ *is of class* C^3, *for almost all* $\xi \in \overline{\mathcal{O}}$ *and it holds*

$$\sup_{\xi \in \overline{\mathcal{O}}} \sup_{\sigma \in \mathbb{R}^n} \left| \frac{\partial^j f}{\partial \sigma^j}(\xi, \sigma) \right| < \infty, \qquad 0 < j \leq 3.$$

Here we assume that the open set $\mathcal{O}$ and the operators A and Q satisfy some conditions which are more restrictive than those introduced in the section 4.1. Of course, all notations and preliminary results given in the chapter 4 for A and Q are still valid.

Hypothesis 5.2.   *1. There exists* $\delta > 0$ *such that for any* $t \geq 0$

$$\int_0^t s^{-\delta} \mathrm{Tr} \left[e^{sA} QQ^\star e^{sA^\star} \right] ds < +\infty. \tag{5.1.1}$$

2. There exists $\epsilon < 1$ *such that*

$$\mathrm{Range}\,((-A)^{-\epsilon/2}) \subset \mathrm{Range}\,(Q^{-1}). \tag{5.1.2}$$

As we have seen in the previous chapter, the hypotheses above are required in order to have the existence of mild solutions for the system (4.0.1) with continuous trajectories and in order to have smoothing properties of the corresponding transition semigroup P_t. In what follows, in order to prove that the operator $D^2(P_t\varphi)(x)QQ^\star$ is in $\mathcal{L}_1(H)$, for any $t > 0$ and $x \in H$, in addition to the Hypotheses 5.1 and 5.2 we shall assume the following conditions.

Hypothesis 5.3. *There exists a complete orthonormal basis* $\{e_k\}$ *in* H *such that*

$$\sum_{k=1}^{\infty} |e^{tA} QQ^\star e_k|_H \leq ct^{-\beta}, \qquad t \in (0, 1], \tag{5.1.3}$$

for some

$$\beta < 1 \wedge \left(2 - \frac{d + 2\epsilon}{4} \right). \tag{5.1.4}$$

Moreover, there exists some $\gamma < 1 - \epsilon$ *such that*

$$\|e^{tA} QQ^\star\|_2 \leq ct^{-\gamma/2}, \qquad t \in (0, 1]. \tag{5.1.5}$$

Remark 5.1.1.   1. It is immediate to check that the condition (5.1.5) implies the condition (5.1.1).

2. Assume that $Ae_k = -\alpha_k e_k$, where $\{e_k\}$ is a complete orthonormal system in H and for any $k \in \mathbb{N}$

$$\alpha_k \asymp k^{2/d}. \qquad (5.1.6)$$

If we define $Qe_k = \alpha_k^{-\epsilon/2} e_k$, for some $\epsilon < 1$ as in (5.1.2), we have

$$\sum_{k=1}^{\infty} |e^{tA} QQ^* e_k|_H = \sum_{k=1}^{\infty} e^{-t\alpha_k} \alpha_k^{-\epsilon}.$$

By easy calculations it is possible to show that for any $\beta > 0$ and $k \in \mathbb{N}$

$$e^{-t\alpha_k} \alpha_k^{-\epsilon} \leq \left(\frac{\beta}{\epsilon}\right)^{\beta} \alpha_k^{-(\epsilon+\beta)} t^{-\beta}, \qquad t > 0.$$

Hence we have

$$\sum_{k=1}^{\infty} |e^{tA} QQ^* e_k|_H \leq c_{\beta} t^{-\beta} \sum_{k=1}^{\infty} \alpha_k^{-(\epsilon+\beta)}$$

and, since we are assuming (5.1.6), we can conclude that (5.1.3) is fulfilled for any

$$\beta > \frac{d}{2} - \epsilon.$$

Notice that the condition above is compatible with (5.1.4). Actually it is sufficient that there exists some $\epsilon < 1$ such that

$$\frac{d}{2} - \epsilon < 1 \quad \text{and} \quad \frac{d}{2} - \epsilon < 2 - \frac{d + 2\epsilon}{4},$$

which is verified for

$$\epsilon > \left(\frac{3}{2}d - 4\right) \vee \left(\frac{d}{2} - 1\right).$$

Under the same hypotheses for A and Q, the condition (5.1.5) reads

$$\sum_{k=1}^{\infty} e^{-2t\alpha_k} \alpha_k^{-2\epsilon} \leq c_{\gamma} t^{-\gamma}$$

and with the same arguments used above it is possible to show that it is verified for

$$2\epsilon + \gamma > \frac{d}{2}.$$

Such a condition is compatible with $\gamma < 1 - \epsilon$, once one takes $\epsilon > d/2 - 1$.

5.2   The trace-class property of $D^2(P_t\varphi)QQ^\star$

In the chapter 4 we have proved that, when $d \leq 3$, $P_t\varphi$ is at least two times differentiable with bounded derivatives, for any $\varphi \in B_b(H)$ and $t > 0$. Our next step is to prove that if $\varphi \in C_b(H)$ the operator $D^2(P_t\varphi)(x)QQ^\star$ belongs to $\mathcal{L}_1(H)$, for any $t > 0$ and $x \in H$.

Before proving this fact, we need some preliminary results.

Lemma 5.2.1. *Under the Hypotheses 5.1, 5.2 and 5.3, if $\{e_k\}$ is the basis of H introduced in the Hypothesis 5.3, there exists a continuous increasing function $c(t)$ such that*

$$\sup_{x \in H} \sum_{k=1}^{\infty} \int_0^t \left|Q^{-1}D_x u(s;x)QQ^\star e_k\right|_H^2 ds \leq c(t)\, t^{1-\epsilon-\gamma}, \qquad \mathbb{P} - a.s. \qquad (5.2.1)$$

Moreover the series converges uniformly with respect to $x \in H$.

Proof. We set $v_k(t) = D_x u(t;x)QQ^\star e_k$. As seen in the previous chapter, the process $v_k(t)$ is the unique mild solution of the problem

$$\frac{dv}{dt}(t) = Av(t) + DF(u(t;x))v(t), \qquad v(0) = QQ^\star e_k,$$

that is

$$v_k(t) = e^{tA}QQ^\star e_k + \int_0^t e^{(t-s)A}DF(u(s;x))v_k(s)\, ds. \qquad (5.2.2)$$

Since the function $\partial f/\partial\sigma(\xi,\cdot)$ is bounded, uniformly with respect to $\xi \in \mathcal{O}$, we have

$$|DF(u(s;x))v_k(s)|_H \leq c\,|v_k(s)|_H, \qquad (5.2.3)$$

so that

$$|v_k(t)|_H \leq \left|e^{tA}QQ^\star e_k\right|_H + c \int_0^t |v_k(s)|_H\, ds.,$$

Thanks to (5.1.3), this implies that for any $n \in \mathbb{N}$

$$\sum_{k=1}^{n} |v_k(t)|_H \leq c\,t^{-\beta} + c \int_0^t \sum_{k=1}^{n} |v_k(s)|_H\, ds.$$

Since the integral equation

$$\alpha(t) = c\,t^{-\beta} + c \int_0^t \alpha(s)\, ds \qquad (5.2.4)$$

admits a unique continuous solution $\alpha : (0, +\infty) \to (0, +\infty)$ such that

$$\sup_{t \in (0,1]} t^\beta \alpha(t) < \infty.$$

Actually, if we set $f(t) = t^\beta \alpha(t)$, the equation (5.2.4) reads

$$f(t) = c + t^\beta \int_0^t s^{-\beta} f(s)\, ds$$

and this problem admits a unique solution in $C([0, +\infty))$, as the mapping

$$\Gamma(f)(t) = c + t^\beta \int_0^t s^{-\beta} f(s)\, ds$$

is a contraction in $C([0, T_0])$, for T_0 sufficiently small. Thus, by comparison we have

$$\sum_{k=1}^n |v_k(t)|_H \le c(t)\, t^{-\beta}, \qquad \mathbb{P} - \text{a.s.}$$

for some continuous increasing function $c(t)$ and by taking the limit as n goes to infinity, we get

$$\sup_{x \in H} \sum_{k=1}^\infty |D_x u(t; x) QQ^* e_k|_H \le c(t)\, t^{-\beta}, \qquad \mathbb{P} - \text{a.s.} \tag{5.2.5}$$

Notice that the convergence of the series is uniform with respect to $x \in H$ as the integral equation verified by $\alpha(t)$ is independent of $x \in H$.

Now, recalling (4.1.7) and the condition (5.1.2), since $v_k(t)$ fulfills the equation (5.2.2) we have that $v_k(t) \in D(Q^{-1})$, for any $t > 0$, and

$$\left|Q^{-1}v_k(t)\right|_H \le \left|Q^{-1}e^{tA}QQ^* e_k\right|_H + \int_0^t \left|Q^{-1}e^{(t-s)A}DF(u(s;x))v_k(s)\right|_H ds$$

$$\le ct^{-\frac{\epsilon}{2}}\left|e^{\frac{t}{2}A}QQ^* e_k\right|_H + c\int_0^t (t-s)^{-\frac{\epsilon}{2}}\left|DF(u(s;x))v_k(s)\right|_H ds.$$

Thanks to (5.2.3), this yields

$$\left|Q^{-1}v_k(t)\right|_H^2 \le ct^{-\epsilon}\left|e^{\frac{t}{2}A}QQ^* e_k\right|_H^2 + ct\int_0^t (t-s)^{-\epsilon}|v_k(s)|_H^2 ds$$

and then, by using the condition (5.1.5) and (5.2.5), for any $n \in \mathbb{N}$ we have

$$\sum_{k=1}^n \left|Q^{-1}v_k(t)\right|_H^2 \le ct^{-(\epsilon+\gamma)} + c(t)\int_0^t (t-s)^{-\epsilon}s^{-\beta}\, ds.$$

Hence, as $\epsilon + \gamma < 1$ and $\beta, \epsilon < 1$, from the arbitrariness of n we can conclude that

$$\int_0^t \sum_{k=1}^\infty \left|Q^{-1}v_k(s)\right|_H^2 ds \le c(t)\, t^{1-\epsilon-\gamma}, \qquad \mathbb{P} - \text{a.s.}$$

for some continuous increasing function $c(t)$.

$\square$

Now we prove an analogous result for the second order derivative.

Lemma 5.2.2. *Let $N \in \mathcal{L}(H)$ and let $\{e_k\}$ be the basis of H introduced in the Hypothesis 5.3. Then, under the Hypotheses 5.1, 5.2 and 5.3, for any $t \geq 0$ we have*

$$\sup_{x \in H} \sum_{k=1}^{\infty} |Q^{-1}D_x^2 u(t;x)(QQ^\star e_k, Ne_k)|_H \leq \|N\|\, c(t), \qquad \mathbb{P} - a.s. \qquad (5.2.6)$$

for some continuous increasing function $c(t)$ independent of N. Moreover the convergence of the series is uniform with respect to $x \in H$.

Proof. We set $v_k(t) = D_x^2 u(t;x)(Ne_k, QQ^\star e_k)$. In the previous chapter we have proved that $v_k(t)$ is the unique solution of the problem

$$v(t) = \int_0^t e^{(t-s)A} DF(u(t;x))v(s)\, ds$$

$$+ \int_0^t e^{(t-s)A} D^2 F(u(s;x))\, (D_x u(s;x)QQ^\star e_k, D_x u(s;x)Ne_k)\, ds.$$

As $\partial^2 f/\partial\sigma^2(\xi, \cdot)$ is bounded, uniformly with respect $\xi \in \mathcal{O}$, by using (4.1.5) and the Hölder inequality, for any $p \in (1,2)$ we have

$$\left| e^{(t-s)A} D^2 F(u(s;x))\, (D_x u(s;x)QQ^\star e_k, D_x u(s;x)Ne_k) \right|_H$$

$$\qquad\qquad\qquad\qquad (5.2.7)$$

$$\leq c\,(t-s)^{-\frac{d(2-p)}{4p}} |D_x u(s;x)QQ^\star e_k|_H\, |D_x u(s;x)Ne_k|_{\frac{2p}{2-p}}.$$

As proved in the Lemma 4.2.2, for any $p \geq 1$ we have

$$\sup_{x \in H}\ \sup_{t \in (0,T]}\ t^{\frac{d(p-2)}{4p}} |D_x u(t;x)h|_p \leq \mu_{2,p}(T)|h|_H, \qquad \mathbb{P} - a.s.$$

and then, thanks to the boundedness of $DF(x)$, this implies that

$$|v_k(t)|_H \leq c \int_0^t |v_k(s)|_H\, ds$$

$$+ c(t)\, \mu_{2,\frac{2p}{2-p}}(t)\|N\| \int_0^t (t-s)^{-\frac{d(2-p)}{4p}} s^{-\frac{d(p-1)}{2p}} |D_x u(s;x)QQ^\star e_k|_H\, ds.$$

Therefore, by using (5.2.5), for any $n \in \mathbb{N}$ we have

$$\sum_{k=1}^{n} |v_k(t)|_H \leq c \int_0^t \sum_{k=1}^{n} |v_k(s)|_H\, ds + c(t)\|N\| \int_0^t (t-s)^{-\frac{d(2-p)}{4p}} s^{-\left(\beta+\frac{d(p-1)}{2p}\right)}\, ds.$$

As p can be chosen arbitrarily close to 1 and $\beta < 1$, we can apply the Gronwall Lemma and we can conclude that there exists a continuous increasing function $c(t)$, which is independent of $x \in H$ and $N \in \mathcal{L}(H)$, such that for any $t \geq 0$

$$\sum_{k=1}^{n} |v_k(t)|_H \leq \|N\| c(t), \qquad \mathbb{P} - \text{a.s.}$$

Moreover, as the estimate is independent of $n \in \mathbb{N}$ and $x \in H$, we conclude that

$$\sum_{k=1}^{\infty} |D_x^2 u(t;x) (QQ^* e_k, N e_k)|_H \leq \|N\| c(t), \qquad \mathbb{P} - \text{a.s.} \qquad (5.2.8)$$

and the convergence of the series is uniform with respect to $x \in H$.

Now, due to (4.1.7) and (5.1.2), it is easy to check that $v_k(t) \in D(Q^{-1})$, for any $t \geq 0$, and

$$|Q^{-1} v_k(t)|_H \leq c \int_0^t (t-s)^{-\frac{\epsilon}{2}} \left| e^{\frac{t-s}{2}A} DF(u(s;x)) v_k(s) \right|_H ds$$

$$+ c \int_0^t (t-s)^{-\frac{\epsilon}{2}} \left| e^{\frac{t-s}{2}A} D^2 F(u(s;x)) (D_x u(s;x) QQ^* e_k, D_x u(s;x) N e_k) \right|_H ds.$$

If we fix $p \geq 1$ such that

$$\frac{1}{p} < \frac{4 + d - 2\epsilon}{2d}, \qquad (5.2.9)$$

we have $\epsilon/2 + d(2-p)/(4p) < 1$ and then, by using (5.2.7), it follows

$$|Q^{-1} v_k(t)|_H \leq c \int_0^t (t-s)^{-\frac{\epsilon}{2}} |v_k(s)|_H ds$$

$$+ c \mu_{2, \frac{2p}{2-p}}(t) \|N\| \int_0^t (t-s)^{-\left(\frac{\epsilon}{2} + \frac{d(2-p)}{4p}\right)} s^{-\frac{d(p-1)}{2p}} |D_x u(s;x) QQ^* e_k|_H ds.$$

Since we are assuming that β fulfills (5.1.4), it is easy to check that there exists some $p \in (1, 2)$ such that (5.2.9) is verified and

$$\beta + \frac{d(p-1)}{2p} < 1.$$

Therefore, thanks to (5.2.5) and (5.2.8), for any $n \in \mathbb{N}$ we get

$$\sum_{k=1}^{n} |Q^{-1} v_k(t)|_H \leq c(t) \|N\| \int_0^t (t-s)^{-\frac{\epsilon}{2}} ds$$

$$+ c(t) \mu_{2, \frac{2p}{2-p}}(t) \|N\| \int_0^t (t-s)^{-\left(\frac{\epsilon}{2} + \frac{d(2-p)}{4p}\right)} s^{-\left(\frac{d(p-1)}{2p} + \beta\right)} ds,$$

which implies (5.2.6), due to the arbitrariness of $n \in \mathbb{N}$. $\qquad \square$

Before proving the main result of this section, we recall a useful fact whose proof can be found in Dunford-Schwartz [55]

Lemma 5.2.3. *Let* $T \in \mathcal{L}(H)$. *Assume that there exists* $K > 0$ *such that, for all finite rank linear bounded operators* N, *one has*

$$|\mathrm{Tr}\,(NT)| \leq K\|N\|.$$

Then $T \in \mathcal{L}_1(H)$ *and*

$$|\mathrm{Tr}\,T| \leq K.$$

Theorem 5.2.4. *Under the Hypotheses 5.1, 5.2 and 5.3, for any* $\varphi \in B_b(H)$ *and for any* $t > 0$ *and* $x \in H$, *the operator* $D^2(P_t\varphi)(x)QQ^\star$ *is in* $\mathcal{L}_1(H)$. *Moreover, the mapping*

$$(0, +\infty) \times H \to \mathbb{R}, \qquad (t, x) \mapsto \mathrm{Tr}\left[D^2(P_t\varphi)(x)QQ^\star\right],$$

is continuous and

$$\sup_{x \in H} \left|\mathrm{Tr}\left[D^2(P_t\varphi)(x)QQ^\star\right]\right| \leq c\,(t \wedge 1)^{-\left(1+\epsilon+\frac{7}{2}\right)}\|\varphi\|_0^H, \qquad t > 0. \qquad (5.2.10)$$

Proof. Due to the previous lemma, if we show that for any finite rank operator N

$$\sup_{x \in H} \left|\mathrm{Tr}\left[ND^2(P_t\varphi)(x)QQ^\star\right]\right| \leq c\,\|N\|\,(t \wedge 1)^{-\left(1+\epsilon+\frac{7}{2}\right)}\|\varphi\|_0^H, \qquad t > 0,$$

we are done. Thanks to the Bismut-Elworthy formula for the first derivative of $P_t\varphi$, if $\{e_k\}$ is the complete orthonormal system introduced in the Hypothesis 5.3, for any $\varphi \in C_b(H)$ we have

$$\langle D(P_t\varphi)(x), QQ^\star e_k\rangle_H = \frac{1}{t}\mathbb{E}\,\varphi(u(t; x)) \int_0^t \langle Q^{-1}D_x u(s; x)QQ^\star e_k, dw(s)\rangle_H.$$

Thus, if $\varphi \in C_b^1(H)$, we can differentiate each side with respect to $x \in H$ along the direction $N^\star e_k$ and we get

$$\langle D^2(P_t\varphi)(x)QQ^\star e_k, N^\star e_k\rangle_H$$

$$= \frac{1}{t}\mathbb{E}\,\langle D\varphi(u(t; x)), D_x u(t; x)N^\star e_k\rangle_H \int_0^t \langle Q^{-1}D_x u(s; x)QQ^\star e_k, dw(s)\rangle_H$$

$$+ \frac{1}{t}\mathbb{E}\,\varphi(u(t; x)) \int_0^t \langle Q^{-1}D_x^2 u(s; x)\,(QQ^\star e_k, N^\star e_k)\,, dw(s)\rangle_H.$$

This implies that for any $n \in \mathbb{N}$

$$\sum_{k=1}^{n} \left\langle D^2(P_t\varphi)(x)QQ^\star e_k, N^\star e_k \right\rangle_H$$

$$= \frac{1}{t}\mathbb{E}\left\langle D\varphi(u(t;x)), D_x u(t;x)N^\star \left(\sum_{k=1}^{n} \int_0^t \left\langle Q^{-1}D_x u(s;x)QQ^\star e_k, dw(s) \right\rangle_H e_k \right) \right\rangle_H$$

$$+ \frac{1}{t}\mathbb{E}\varphi(u(t;x)) \int_0^t \sum_{k=1}^{n} \left\langle Q^{-1}D_x^2 u(s;x)\left(QQ^\star e_k, N^\star e_k\right), dw(s) \right\rangle_H.$$

By using (4.2.5) with $q = 2$, we have

$$\left| \sum_{k=1}^{n} \left\langle D^2(P_t\varphi)(x)QQ^\star e_k, N^\star e_k \right\rangle_H \right|$$

$$\leq \frac{1}{t}\nu_2(t)\|\varphi\|_1^H\|N\| \left(\mathbb{E}\left| \sum_{k=1}^{n} \int_0^t \left\langle Q^{-1}D_x u(s;x)QQ^\star e_k, dw(s) \right\rangle_H e_k \right|_H^2 \right)^{1/2}$$

$$+ \frac{1}{t}\|\varphi\|_0^H \left(\mathbb{E}\left| \int_0^t \sum_{k=1}^{n} \left\langle Q^{-1}D_x^2 u(s;x)\left(QQ^\star e_k, N^\star e_k\right), dw(s) \right\rangle_H \right|^2 \right)^{1/2} = J_1(t) + J_2(t).$$

It is immediate to check that

$$\mathbb{E}\left| \sum_{k=1}^{n} \int_0^t \left\langle Q^{-1}D_x u(s;x)QQ^\star e_k, dw(s) \right\rangle_H e_k \right|_H^2$$

$$= \mathbb{E}\sum_{k=1}^{n} \int_0^t \left| Q^{-1}D_x u(s;x)QQ^\star e_k \right|_H^2 ds$$

and then, due to (5.2.1), we have

$$J_1(t) = \nu_2(t)\|\varphi\|_1^H\|N\|\sqrt{c(t)}\, t^{-\frac{(1+\epsilon+\gamma)}{2}}, \qquad t > 0.$$

Moreover, thanks to (5.2.6)

$$J_2(t) = \frac{1}{t}\|\varphi\|_0^H \left(\mathbb{E}\int_0^t \left| \sum_{k=1}^{n} Q^{-1}D_x^2 u(s;x)\left(QQ^\star e_k, N^\star e_k\right) \right|_H^2 ds \right)^{1/2}$$

$$\leq c(t)\|\varphi\|_0^H\|N\|\, t^{-1/2}.$$

Therefore, as $n \in \mathbb{N}$ is arbitrary and N is a finite rank operator, by using the Lemma 5.2.3 it follows that for any $\varphi \in C_b^1(H)$ the operator $D^2(P_t\varphi)(x)QQ^\star$ is in $\mathcal{L}_1(H)$ and

$$\sup_{x \in H} \left| \mathrm{Tr} \left[D^2(P_t\varphi)(x)QQ^\star \right] \right| \leq c(t) \|\varphi\|_1^H \, t^{-\frac{1+\epsilon+\gamma}{2}}, \qquad t > 0.$$

Moreover, due to the Lemmata 5.2.1 and 5.2.2 the convergence of the series is uniform with respect to $x \in H$ and then the mapping

$$(0, +\infty) \times H \to \mathbb{R}, \qquad (t, x) \mapsto \mathrm{Tr} \left[D^2(P_t\varphi)(x)QQ^\star \right]$$

is continuous. Notice that if $t > 1$ we have $P_t\varphi = P_1(P_{t-1}\varphi)$ so that, by using (4.4.10) with $j = i = 1$, we conclude

$$\left| \mathrm{Tr} \left[D^2(P_t\varphi)(x)QQ^\star \right] \right| \leq c \|\varphi\|_1^H \, (t \wedge 1)^{-\frac{1+\epsilon+\gamma}{2}}, \qquad t > 0. \tag{5.2.11}$$

Finally, let $\varphi \in B_b(H)$. By the semigroup law and (4.4.10), with $j = 1$ and $i = 0$, we have

$$\left| \mathrm{Tr} \left[D^2(P_t\varphi)(x)QQ^\star \right] \right| \leq c \|P_{t/2}\varphi\|_1^H \left(\frac{t}{2} \wedge 1 \right)^{-\frac{1+\epsilon+\gamma}{2}} \leq c(t) \|\varphi\|_0^H \, (t \wedge 1)^{-(1+\epsilon+\frac{\gamma}{2})},$$

and this implies (5.2.10). $\square$

5.3   The parabolic problem

In this section we shall consider the parabolic problem associated with the diffusion operator $\mathcal{L}$

$$\begin{cases} \dfrac{\partial y}{\partial t}(t, x) = \mathcal{L}(x, D)y(t, x), & x \in D(A) \quad t > 0, \\[2mm] y(0, x) = \varphi(x), & x \in H. \end{cases} \tag{5.3.1}$$

Definition 5.3.1. *A bounded continuous function $y : [0, +\infty[\to \mathbb{R}$ is said a classical solution of the problem (5.3.1) if*

1. *for any $t > 0$, the function $y(t, \cdot) \in C_b^2(H)$ and for any $x \in H$ the operator $D^2y(t, x) \in \mathcal{L}_1(H)$. Moreover the function*

$$(0, +\infty) \times H \to \mathbb{R}, \qquad (t, x) \mapsto \mathrm{Tr}\,[D^2y(t, x)QQ^\star]$$

    is continuous;

2. *for any $x \in D(A)$, the function $y(\cdot, x)$ is differentiable on $(0, +\infty)$;*

3. *the function y satisfies the equation (5.3.1).*

Definition 5.3.2. *A bounded continuous function* $y : [0, +\infty) \times H \to \mathbb{R}$ *is said a strict solution of the problem (5.3.1) if satisfies the conditions 1-3 in the Definition 5.3.1, with* $t > 0$ *and* $(0, +\infty)$ *replaced respectively by* $t \geq 0$ *and* $[0, +\infty)$.

First of all we show that if φ is regular enough then the function

$$y : [0, +\infty) \times H \to \mathbb{R}, \quad (t, x) \mapsto y(t, x) = P_t \varphi(x)$$

is a strict solution of the problem (5.3.1).

If $A \in \mathcal{L}(H)$ and $\operatorname{Tr} QQ^* < +\infty$, the unique mild solution $u(t; x)$ of the equation

$$du(t) = [Au(t) + F(u(t))] \, dt + Q \, dw(t), \quad u(0) = x$$

is a strong solution, that is

$$u(t; x) = x + \int_0^t (Au(s; x) + F(u(s; x))) \, ds + Q \, w(t).$$

(for a proof see for example [47, Theorem 5.29]). Thus $u(t; x)$ is a semimartingale and we can apply the Itô formula for Hilbert spaces valued processes. Then, if $\varphi \in C_b^2(H)$ we have

$$P_t \varphi(x) = \varphi(x)$$

$$+ \mathbb{E} \int_0^t \left(\langle Au(s; x) + F(u(s; x)), D\varphi(u(s; x)) \rangle_H + \frac{1}{2} \operatorname{Tr} [D^2 \varphi(u(s; x)) QQ^*] \right) ds.$$

$$(5.3.2)$$

This implies that

$$\lim_{t \to 0^+} \frac{P_t \varphi(x) - \varphi(x)}{t} = \frac{1}{2} \operatorname{Tr} [D^2 \varphi(x) QQ^*] + \langle Ax + F(x), D\varphi(x) \rangle_H$$

and by standard arguments (see for example [47, Theorem 9.16]) it follows that the function $y(t, x)$ is a strict solution of the problem (5.3.1).

On the other hand, here we do not assume A to be bounded and QQ^* to be of trace-class, and then we can not apply directly the Itô formula. Therefore we have to proceed by approximation, assuming further hypotheses for φ. To this purpose we need to introduce some notations and to prove a preliminary lemma, whose proof is due to Fuhrman and it can be found in [69]. Here for the sake of completeness we give a sketch of it.

For any $n \in \mathbb{N}$, we define $Q_n = P_n Q$, where P_n is the projection of H into the finite dimensional space H_n generated by $\{e_1, \dots, e_n\}$. It is immediate to check that $Q_n x$ converges to Qx and $Q_n Q_n^* x$ converges to $QQ^* x$, as n goes to infinity, for any fixed $x \in H$.

Moreover, for any $n \in \mathbb{N}$ we define

$$w^{A,n}(t) = \int_0^t e^{(t-s)A} Q_n \, dw(s).$$

Then, if $w^A(t)$ is the stochastic convolution defined in the subsection 4.1.2 we have the following approximation result.

Lemma 5.3.3. *Under the Hypothesis 5.2, for any $T > 0$ and $p \geq 1$ we have*

$$\lim_{n \to +\infty} \mathbb{E} \sup_{t \in [0,T]} |w^{A,n}(t) - w^A(t)|_H^p = 0. \tag{5.3.3}$$

In particular, there exists a subsequence, which we still denote by $w^{A,n}(t)$, such that

$$\lim_{n \to +\infty} \sup_{t \in [0,T]} |w^{A,n}(t) - w^A(t)|_H^p = 0, \qquad \mathbb{P} - a.s.$$

Proof. We shall use the *factorization method* which has been introduced in [43]. It is based on the following elementary formula

$$\frac{\pi}{\sin \alpha \pi} = \int_\sigma^t (t - \sigma)^{\alpha-1}(s - \sigma)^{-\alpha} \, ds, \qquad \alpha \in (0, 1).$$

Actually, by using the formula above we obtain

$$w^A(t) = \frac{\sin \alpha \pi}{\pi} \int_0^t e^{(t-\sigma)A} \left(\int_\sigma^t (t - s)^{\alpha-1}(s - \sigma)^{-\alpha} \, ds \right) Q \, dw(\sigma),$$

and then, recalling the stochastic Fubini Tonelli theorem,

$$w^A(t) = \frac{\sin \alpha \pi}{\pi} \int_0^t e^{(t-s)A}(t - s)^{\alpha-1} \left(\int_0^s e^{(s-\sigma)A}(s - \sigma)^{-\alpha} Q \, dw(\sigma) \right) ds.$$

Thus, in our case, if δ is the constant introduced in the Hypothesis 5.2 we have

$$w^{A,n}(t) = \frac{\sin \delta \pi}{\pi} \int_0^t e^{(t-s)A}(t - s)^{\delta-1} v_n(s) \, ds$$

and

$$w^A(t) = \frac{\sin \delta \pi}{\pi} \int_0^t e^{(t-s)A}(t - s)^{\delta-1} v(s) \, ds,$$

where the process $v_n(s)$ is defined by

$$v_n(s) = \int_0^s e^{(s-\sigma)A}(s - \sigma)^{-\delta} Q_n \, dw(\sigma)$$

and $v(s)$ is the same as $v_n(s)$, with Q_n replaced by Q. From the Hölder inequality, if we fix $p > 1/2\delta$, we easily have

$$|w^{A,n}(t) - w^A(t)|_H^{2p} \leq c_T \int_0^T |v_n(s) - v(s)|_H^{2p} ds, \quad \mathbb{P} - \text{a.s.}$$

Therefore, since $v_n(s)$ and $v(s)$ have continuous path are gaussian random variables, we have

$$\mathbb{E} \sup_{t \in [0,T]} |w^{A,n}(t) - w^A(t)|_H^{2p} \leq c_T \int_0^T \left(\mathbb{E}|v_n(s) - v(s)|_H^2 \right)^p ds.$$

It holds

$$\mathbb{E}|v_n(s) - v(s)|_H^2 = \int_0^s (s-\sigma)^{-2\delta} \text{Tr} \left[e^{(s-\sigma)A}(Q_n - Q)(Q_n - Q)^\star e^{(s-\sigma)A^\star} \right] d\sigma,$$

then, in order to prove (5.3.3) we need to show that

$$\lim_{n \to +\infty} \int_0^T \left(\int_0^s (s-\sigma)^{-2\delta} \text{Tr} \left[e^{(s-\sigma)A}(Q_n - Q)(Q_n - Q)^\star e^{(s-\sigma)A^\star} \right] d\sigma \right)^p ds = 0.$$

$$(5.3.4)$$

If $\{e_k\}$ is any orthonormal basis in H, we have

$$\text{Tr} \left[e^{(s-\sigma)A}(Q_n - Q)(Q_n - Q)^\star e^{(s-\sigma)A^\star} \right] = \sum_{k=1}^\infty \left| (Q_n - Q)e^{(s-\sigma)A} e_k \right|_H^2.$$

For any $k \in \mathbb{N}$

$$\lim_{n \to +\infty} \left| (Q_n - Q)e^{(s-\sigma)A} e_k \right|_H^2 = 0.$$

Moreover $|Q_n x|_H \leq |Qx|_H$, for any $x \in H$, and then

$$\text{Tr} \left[e^{(s-\sigma)A}(Q_n - Q)(Q_n - Q)^\star e^{(s-\sigma)A^\star} \right]$$

$$\leq 2 \sum_{k=1}^\infty \left| Qe^{(s-\sigma)A} e_k \right|_H^2 = 2 \text{Tr} \left[e^{(s-\sigma)A} QQ^\star e^{(s-\sigma)A^\star} \right].$$

Therefore, recalling (5.1.1), by the dominated convergence theorem we get (5.3.4). The limit (5.3.3) for general $p \geq 1$ follows from the Hölder inequality. □

Proposition 5.3.4. *Assume that $\varphi \in C_b^2(H)$ and $D^2\varphi(x)QQ^\star \in \mathcal{L}_1(H)$ for any $x \in H$. Moreover assume that the mapping $x \mapsto \text{Tr}[D^2\varphi(x)QQ^\star]$ is continuous and*

$$\sup_{x \in H} \left| \text{Tr}[D^2\varphi(x)QQ^\star] \right| < +\infty.$$

$$(5.3.5)$$

Then, under the Hypotheses 5.1, 5.2 and 5.3, the function $P_t\varphi$ is a strict solution of the problem (5.3.1).

Proof. Step 1: Assume that $A \in \mathcal{L}(H)$. For each $n \in \mathbb{N}$, we consider the problem

$$du(t) = [Au(t) + F(u(t))] \, dt + Q_n \, dw(t), \quad u(0) = x.$$

Let $u_n(t; x)$ be its strong (and then mild) solution. If we introduce the corresponding transition semigroup, $P_t^n \varphi(x) = \mathbb{E}\varphi(u_n(t; x))$, due to Itô's formula we have

$$P_t^n \varphi(x) = \varphi(x) + \mathbb{E} \int_0^t \langle Au_n(s; x) + F(u_n(s; x)), D\varphi(u_n(s; x)) \rangle_H \, ds$$

(5.3.6)

$$+ \frac{1}{2} \mathbb{E} \int_0^t \operatorname{Tr} [D^2\varphi(u_n(s; x)) Q_n Q_n^\star] \, ds.$$

Now $u_n(\cdot; x)$ converges to $u(\cdot; x)$, as $n \to +\infty$, in $C([0, T]; H)$, $\mathbb{P}$-a.s. Indeed we have

$$u_n(t; x) - u(t; x) = \int_0^t e^{(t-s)A} \left(F(u_n(s; x)) - F(u(s; x)) \right) \, ds + w^{A,n}(t) - w^A(t),$$

and since F is Lipschitz-continuous, for any $p \geq 1$ it holds

$$|u_n(t; x) - u(t; x)|_H^{2p} \leq c \, t^{2p-1} \int_0^t |u_n(s; x) - u(s; x)|_H^{2p} \, ds + c \, |w^{A,n}(t) - w^A(t)|_H^{2p}.$$

From the Gronwall lemma, for any $T > 0$, this yields

$$\sup_{t \in [0,T]} |u_n(t; x) - u(t; x)|_H^{2p} \leq c_T \sup_{t \in [0,T]} |w^{A,n}(t) - w^A(t)|_H^{2p}, \quad \mathbb{P} - \text{a.s.}$$

and thanks to the Lemma 5.3.3 we get

$$\lim_{n \to +\infty} \sup_{t \in [0,T]} |u_n(t; x) - u(t; x)|_H^{2p} = 0, \quad \mathbb{P} - \text{a.s.} \quad (5.3.7)$$

Next, we prove that for each $x \in H$ and $t > 0$

$$\lim_{n \to +\infty} \mathbb{E} \int_0^t \left(\operatorname{Tr} [D^2\varphi(u_n(s; x)) Q_n Q_n^\star] - \operatorname{Tr} [D^2\varphi(u(s; x)) QQ^\star] \right) \, ds = 0.$$

We have

$$\operatorname{Tr} [D^2\varphi(u_n(t; x)) Q_n Q_n^\star] - \operatorname{Tr} [D^2\varphi(u(t; x)) QQ^\star]$$

$$= \operatorname{Tr} \left[(D^2\varphi(u_n(t; x)) - D^2\varphi(u(t; x))) Q_n Q_n^\star \right] + \operatorname{Tr} [D^2\varphi(u(t; x))(Q_n Q_n^\star - QQ^\star)].$$

Concerning the first term,

$$\left| \operatorname{Tr} \left[(D^2\varphi(u_n(t; x)) - D^2\varphi(u(t; x))) Q_n Q_n^\star \right] \right|$$

$$= \left| \operatorname{Tr} \left[(D^2\varphi(u_n(t; x)) - D^2\varphi(u(t; x))) P_n QQ^\star P_n \right] \right|$$

$$\leq \left| \operatorname{Tr} \left[(D^2\varphi(u_n(t; x)) - D^2\varphi(u(t; x))) QQ^\star \right] \right|.$$

Now, we recall that we are assuming that the mapping $x \mapsto \mathrm{Tr}\,[D^2\varphi(x)QQ^*]$ is continuous and (5.3.5) holds. Then, from the dominated convergence theorem and (5.3.7) we get

$$\lim_{n \to +\infty} \mathbb{E} \int_0^t \left(\mathrm{Tr}\,\left[(D^2\varphi(u_n(s;x)) - D^2\varphi(u(s;x))) \, Q_n Q_n^* \right] \right) ds = 0.$$

As far as the second term is concerned, recalling that $Q_n Q_n^* x$ converges to $QQ^* x$, for any $x \in H$, since

$$\left| \mathrm{Tr}\,\left[D^2\varphi(u(t;x)) \, (P_n QQ^* P_n - QQ^*) \right] \right| \leq 2\,\mathrm{Tr}\,\left[D^2\varphi(u(t;x))QQ^* \right],$$

we easily have

$$\lim_{n \to +\infty} \mathrm{Tr}\,\left[D^2\varphi(u(t;x))(Q_n Q_n^* - QQ^*) \right] = 0, \qquad \mathbb{P}-\text{a.s.}$$

and from the dominated convergence theorem it follows

$$\lim_{n \to +\infty} \mathbb{E} \int_0^t \mathrm{Tr}\,\left[D^2\varphi(u(s;x))(Q_n Q_n^* - QQ^*) \right] ds = 0.$$

Finally, since A and F are uniformly continuous and $\varphi \in C_b^2(H)$, we have

$$\lim_{n \to +\infty} \langle Au_n(s;x) + F(u_n(s;x)), D\varphi(u_n(s;x)) \rangle_H$$

$$= \langle Au(s;x) + F(u(s;x)), D\varphi(u(s;x)) \rangle_H, \qquad \mathbb{P}-\text{a.s.}$$

uniformly in $[0,T]$. Then, as $P_t^n\varphi(x)$ converges to $P_t\varphi(x)$, by taking the limit in each side of (5.3.6), we conclude that (5.3.2) holds for a general diffusion term Q.

Now, if φ satisfies the conditions of the proposition, it is easy to check that $P_t\varphi$ satisfies the same conditions, for any $t \geq 0$. Then, as proved in [47, Theorem 9.16], by using the semigroup law, we can conclude that the function $y(t,x) = P_t\varphi(x)$ is differentiable with respect to $t \geq 0$ and

$$\begin{cases} \dfrac{\partial y}{\partial t}(t,x) = \mathcal{L}(x,D)y(t,x), & x \in H \ \ t > 0, \\[2mm] y(0,x) = \varphi(x), & x \in H. \end{cases}$$

Step 2: Now, let A be the infinitesimal generator of a strongly continuous semigroup. We introduce the *Yosida approximations* $A_n = nAR(n,A)$ of A and for each n we consider the approximating problems

$$du(t) = [A_n u(t) + F(u(t))]\, dt + Q\, dw(t), \qquad u(0) = x. \tag{5.3.8}$$

Since $A_n \in \mathcal{L}(H)$, by the previous step we have that the function $y_n(t, x) = \mathbb{E}\varphi(u_n(t; x))$ is differentiable with respect to $t > 0$ and

$$
\begin{cases}
\dfrac{\partial y_n}{\partial t}(t, x) = \mathcal{L}_n(x, D)y_n(t, x), & x \in H, \ t > 0 \\[2mm]
y_n(0, x) = \varphi(x), & x \in H,
\end{cases}
$$

where $\mathcal{L}_n(x, D)$ is the operator corresponding to A_n. Let $u_n(t; x)$, $D_x u_n(t; x)$ and $D_x^2 u_n(t; x)$ denote respectively the solution of the equation (5.3.8) and the first two derivatives with respect to initial datum. As shown in [47, Theorem 9.17], for any $x, h, k \in H$

$$
\begin{cases}
u_n(t; x) \to u(t; x) \\[2mm]
D_x u_n(t; x)h \to D_x u(t; x)h \\[2mm]
D_x^2 u_n(t; x)(h, k) \to D_x^2 u(t; x)(h, k),
\end{cases}
\tag{5.3.9}
$$

$\mathbb{P}$-a.s., as $n \to +\infty$, uniformly for $t \in [0, T]$. Moreover, since e^{tA_n} is a contraction semigroup for any $n \in \mathbb{N}$, by proceeding as in the proof of the Proposition 4.2.1 and of the Theorem 4.2.4, we have

$$
\sup_{x \in H} |D_x u_n(t; x)h|_H \leq \nu_2(t)|h|_H, \qquad \mathbb{P} - \text{a.s.}
\tag{5.3.10}
$$

and

$$
\sup_{x \in H} \left| D_x^2 u_n(t; x)(h, k) \right|_H \leq \nu_{2,2}(t)|h|_H|k|_H, \qquad \mathbb{P} - \text{a.s.}
\tag{5.3.11}
$$

where $\nu_2(t)$ and $\nu_{2,2}(t)$ are the functions introduced respectively in (4.2.5) and (4.2.16).

It is immediate to check that

$$
\langle Dy_n(t, x), h \rangle_H = \mathbb{E}\langle D\varphi(u_n(t; x)), D_x u_n(t; x)h \rangle_H,
$$

then, according to (5.3.9), (5.3.10) and (5.3.11), since $\varphi \in C_b^2(H)$ from the dominated convergence theorem it follows

$$
\lim_{n \to +\infty} \langle Dy_n(t, x), h \rangle_H = \langle Dy(t, x), h \rangle_H.
$$

This implies that for any $x \in D(A)$

$$
\lim_{n \to +\infty} \langle Dy_n(t, x), A_n x + F(x) \rangle_H = \langle Dy(t, x), Ax + F(x) \rangle_H.
$$

If we prove that

$$
\lim_{n \to +\infty} \text{Tr}\,[D^2 y_n(t, x)QQ^\star] = \text{Tr}\,[D^2 y(t, x)QQ^\star],
$$

we conclude that $y(t, x)$ is differentiable with respect to $t \geq 0$ and is a strict solution of (5.3.1). We have

$$\langle D^2 y_n(t, x) QQ^* h, k \rangle_H = \mathbb{E} \langle D\varphi(u_n(t; x)), D_x^2 u_n(t; x)(QQ^* h, k) \rangle_H$$

$$+ \mathbb{E} \langle D^2 \varphi(u_n(t; x)) D_x u_n(t; x) QQ^* h, D_x u_n(t; x) k \rangle_H.$$

For the first term we have

$$\langle D\varphi(u_n(t; x)), D_x^2 u_n(t; x)(QQ^* e_k, e_k) \rangle_H - \langle D\varphi(u(t; x)), D_x^2 u(t; x)(QQ^* e_k, e_k) \rangle_H$$

$$= \langle D\varphi(u_n(t; x)), D_x^2 u_n(t; x)(QQ^* e_k, e_k) - D_x^2 u(t; x)(QQ^* e_k, e_k) \rangle_H$$

$$+ \langle D\varphi(u_n(t; x)) - D\varphi(u(t; x)), D_x^2 u(t; x)(QQ^* e_k, e_k) \rangle_H$$

and then, due to (5.2.8) (with $N = I$) we get

$$\sum_{k=1}^{\infty} |\langle D\varphi(u_n(t; x)), D_x^2 u_n(t; x)(QQ^* e_k, e_k) \rangle_H$$

$$- \langle D\varphi(u(t; x)), D_x^2 u(t; x)(QQ^* e_k, e_k) \rangle_H|$$

$$\leq \|D\varphi\|_0^H \sum_{k=1}^{\infty} |D_x^2 u_n(t; x)(QQ^* e_k, e_k) - D_x^2 u(t; x)(QQ^* e_k, e_k)|_H$$

$$+ |D\varphi(u_n(t; x)) - D\varphi(u(t; x))|_H \, c(t).$$

For any $k \in \mathbb{N}$, due to (5.3.9) we have

$$\lim_{n \to +\infty} |D_x^2 u_n(t; x)(QQ^* e_k, e_k) - D_x^2 u(t; x)(QQ^* e_k, e_k)|_H = 0, \qquad \mathbb{P} - \text{a.s.}$$

and due to (5.2.8) we have

$$\sum_{k=1}^{\infty} |D_x^2 u_n(t; x)(QQ^* e_k, e_k) - D_x^2 u(t; x)(QQ^* e_k, e_k)|_H \leq 2 \, c(t).$$

Hence it follows

$$\lim_{n \to +\infty} \sum_{k=1}^{\infty} |D_x^2 u_n(t; x)(QQ^* e_k, e_k) - D_x^2 u(t; x)(QQ^* e_k, e_k)|_H = 0.$$

Moreover, recalling that $D\varphi \in C_b(H; H)$ and (5.3.9) holds, we have

$$\lim_{n \to +\infty} |D\varphi(u_n(t; x)) - D\varphi(u(t; x))|_H = 0, \qquad \mathbb{P} - \text{a.s.}$$

and by the dominated convergence theorem we have

$$\lim_{n \to +\infty} \sum_{k=1}^{\infty} \mathbb{E} \langle D\varphi(u_n(t;x)), D_x^2 u_n(t;x)(QQ^* e_k, e_k) \rangle_H$$

$$= \sum_{k=1}^{\infty} \mathbb{E} \langle D\varphi(u(t;x)), D_x^2 u(t;x)(QQ^* e_k, e_k) \rangle_H.$$

Finally, for the second term, by using once more (5.3.9) and (5.3.10) and recalling that $D^2\varphi$ is continuous and the mapping $D^2\varphi QQ^*$ belongs to $C_b(H; \mathcal{L}^1(H))$, thanks to (5.2.5) we can conclude that

$$\lim_{n \to +\infty} \sum_{k=1}^{\infty} \mathbb{E} \langle D^2\varphi(u_n(t;x)) D_x u_n(t;x) QQ^* e_k, D_x u_n(t;x) e_k \rangle_H$$

$$= \sum_{k=1}^{\infty} \mathbb{E} \langle D^2\varphi(u(t;x)) D_x u(t;x) QQ^* e_k, D_x u(t;x) e_k \rangle_H.$$

$\square$

Now, we can show that the function $y(t, x) = P_t\varphi(x)$ is a classical solution when φ is only bounded and continuous.

Theorem 5.3.5. *Assume the Hypotheses 5.1, 5.2 and 5.3. Then for any $\varphi \in C_b(H)$ there exists a unique classical solution $y : [0, +\infty) \times H \to \mathbb{R}$ to the problem (5.3.1) which is given by*

$$y(t, x) = P_t\varphi(x) = \mathbb{E}\varphi(u(t;x)),$$

for any $t \geq 0$ and $x \in H$.

Proof. As in the proof of the Proposition 4.4.3, we can construct a sequence $\{\varphi_k\} \subset C_b^2(H)$ which converges pointwise to φ and is dominated in the sup-norm by φ. Recalling how φ_k is defined, it is easy to check that $D^2\varphi_k(x)QQ^* \in \mathcal{L}_1(H)$ for any $x \in H$ and the mapping $x \mapsto \text{Tr}\,[D^2\varphi_k(x)QQ^*]$ belongs to $C_b(H)$ (in fact this is true for $\text{Tr}\,[D^2\varphi_k]$).

If we define $y_k(t, x) = P_t\varphi_k(x)$, as proved in the previous proposition, y_k is a strict solution of the problem

$$\begin{cases} \dfrac{\partial y_k}{\partial t}(t, x) = \mathcal{L}(x, D)y_k(t, x), & x \in D(A) \ \ t \geq 0, \\[2mm] y_k(0, x) = \varphi_k(x), & x \in H. \end{cases}$$

First, we show that for any $t > 0$ and $x \in H$

$$\lim_{k \to +\infty} \text{Tr}\,[D^2 y_k(t, x)QQ^*] = \text{Tr}\,[D^2 y(t, x)QQ^*]. \qquad (5.3.12)$$

By using the semigroup law and the trace-class property of $D^2y_k(t,x)QQ^\star$ and $D^2y(t,x)QQ^\star$, we have

$$\mathrm{Tr}\,[D^2y_k(t,x)QQ^\star] - \mathrm{Tr}\,[D^2y(t,x)QQ^\star]$$

$$= \lim_{h\to\infty} \sum_{j=1}^{h} \mathbb{E}\big\langle D^2y_{k,t}(u(t/2;x))D_xu(t/2;x)QQ^\star e_j, D_xu(t/2;x)e_j\big\rangle_H$$

$$+ \lim_{h\to\infty} \mathbb{E}\bigg\langle Dy_{k,t}(u(t/2;x)), \sum_{j=1}^{h} D_x^2u(t/2;x)(QQ^\star e_j, e_j)\bigg\rangle_H,$$

where $y_{k,t} = P_{t/2}(\varphi_k - \varphi)$. For any $h \in \mathbb{N}$ we have

$$\left|\sum_{j=1}^{h} \mathbb{E}\big\langle D^2y_{k,t}(u(t/2;x))D_xu(t/2;x)QQ^\star e_j, D_xu(t/2;x)e_j\big\rangle_H\right|$$

$$\leq \nu_2(t/2)\,\mathbb{E}\,\big\|D^2y_{k,t}(u(t/2x))\big\|_{\mathcal{L}(H)} \sum_{j=1}^{h} |D_xu(t/2;x)QQ^\star e_j|_H$$

$$\leq \nu_2(t/2)c(t/2)\,t^{-\beta}\,\mathbb{E}\,\big\|D^2y_{k,t}(u(t/2;x))\big\|_{\mathcal{L}(H)},$$

the last inequality following from (5.2.5). In the same way, by using (5.2.8), with $N = I$, we have

$$\left|\mathbb{E}\bigg\langle Dy_{k,t}(u(t/2;x)), \sum_{j=1}^{h} D_x^2u(t/2;x)(QQ^\star e_j, e_j)\bigg\rangle_H\right|$$

$$\leq c\,(t/2)\,\mathbb{E}\,|Dy_{k,t}(u(t/2;x))|_H.$$

Therefore we have

$$\big|\mathrm{Tr}\,[D^2y_k(t,x)QQ^\star] - \mathrm{Tr}\,[D^2y(t,x)QQ^\star]\big|$$

$$\leq c(t)\left(t^{-\beta}\mathbb{E}\,\big\|D^2y_{k,t}(u(t/2;x))\big\|_{\mathcal{L}(H)} + \mathbb{E}\,|Dy_{k,t}(u(t/2;x))|_H\right),$$

(5.3.13)

for some continuous increasing function $c(t)$. Next, recalling (4.4.7), as φ_k converges to φ pointwise and $\|\varphi_k\|_0^H \leq \|\varphi\|_0^H$, by the dominated convergence theorem we get

$$\lim_{k\to+\infty} Dy_{k,t}(x) = \lim_{k\to+\infty} DP_{t/2}(\varphi_k - \varphi)(x) = 0,$$

for any $x \in H$. Recalling that

$$\|Dy_{k,t}\|_0^H \leq c(t \wedge 1)^{-\frac{1+\epsilon}{2}}\|\varphi\|_0^H,$$

due to (4.4.12) this easily implies that

$$\lim_{k \to +\infty} D^2 y_{k,t}(x) = 0, \qquad x \in H,$$

so that, according to (5.3.13) we can conclude that (5.3.12) holds true. Moreover, by using again (4.4.7) and (4.4.10), for any $x \in D(A)$ and $t > 0$ we have

$$\lim_{k \to +\infty} \langle Ax + F(x), Dy_k(t,x) \rangle_H = \langle Ax + F(x), Dy(t,x) \rangle_H.$$

Hence, since for any $t \geq 0$ and $x \in H$ we have

$$\lim_{k \to +\infty} y_k(t,x) = y(t,x),$$

and, thanks to (5.2.10), it holds

$$\begin{cases} |\langle Ax + F(x), Dy_k(t,x) \rangle_H| \leq (|Ax|_H + |F(x)|_H)(t \wedge 1)^{-1/2}\|\varphi\|_0^H \\[2mm] |\text{Tr}[D^2 y_k(t,x)QQ^\star]| \leq c(t \wedge 1)^{-(1+\epsilon+\frac{7}{2})}\|\varphi\|_0^H, \end{cases}$$

it follows that $y(t,x)$ is differentiable with respect to t, for any $t > 0$ and $x \in D(A)$, and satisfies the Kolmogorov equation (5.3.1).

The uniqueness of the classical solution follows as in the finite dimensional case, from a standard method based on the Itô formula (see the proof of the Theorem 1.6.2 and also [47, Theorem 9.17]). □

5.4 The elliptic problem

We are here concerned with the elliptic problem

$$\lambda \psi(x) - \mathcal{L}(x,D)\psi(x) = \varphi(x), \qquad x \in D(A), \tag{5.4.1}$$

for $\lambda > 0$ and $\varphi \in C_b(H)$.

We recall that in the previous chapter we have verified that P_t is a weakly continuous semigroup. Then we can introduce its weak generator $L : D(L) \subseteq C_b(H) \to C_b(H)$. As proved in the appendix B, since P_t is a contraction on $C_b(H)$, then $\rho(L) \supset \{\text{Re}\,\lambda > 0\}$ and for any $\varphi \in C_b(H)$ and $x \in H$

$$R(\lambda, L)\varphi(x) = \int_0^{+\infty} e^{-\lambda t} P_t\varphi(x)\,dt, \qquad \text{Re}\,\lambda > 0.$$

In this section we are showing that the solution of the elliptic problem (5.4.1) can be represented in terms of the resolvent of L.

Definition 5.4.1. *A function ψ is called a* strict solution *of (5.4.1) if*

1. *$\psi \in C_b^2(H)$;*

2. *the operator $D^2\psi(x)QQ^\star$ is in $\mathcal{L}_1(H)$, for all $x \in H$, and the mapping $x \mapsto \mathrm{Tr}\,[D^2\psi(x)QQ^\star]$ is in $C_b(H)$;*

3. *ψ satisfies the equation the (5.4.1).*

Definition 5.4.2. *A function ψ is called a* strong solution *of (5.4.1) if there exist two sequences $\{\psi_k\}$ and $\{\varphi_k\}$ in $C_b(H)$ such that*

1. *for any $k \in \mathbb{N}$, ψ_k is a strict solution of the problem*

$$\lambda\psi_k(x) - \mathcal{L}(x, D)\psi_k(x) = \varphi_k(x);$$

2. *the sequences $\{\psi_k\}$ and $\{\varphi_k\}$ converge respectively to ψ and φ in $C_b(H)$, as k goes to infinity.*

Theorem 5.4.3. *Under the Hypotheses 5.1, 5.2 and 5.3, for any $\lambda > 0$ and $\varphi \in C_b(H)$ there exists a unique strong solution ψ of the equation (5.4.1) which is given by*

$$\psi(x) = R(\lambda, L)\varphi(x) = \int_0^{+\infty} e^{-\lambda t} P_t\varphi(x)\,dt, \quad x \in H.$$

Moreover, if $\varphi \in C_b^1(H)$, then ψ is a strict solution.

Proof. Step 1: First we show that if $\varphi \in C_b^1(H)$, then $R(\lambda, L)\varphi$ is a strict solution of (5.4.1).

To this purpose, we first show that $R(\lambda, L)\varphi \in C_b^2(H)$. For any $x, h \in H$, we have

$$R(\lambda, L)\varphi(x + h) - R(\lambda, L)\varphi(x)$$

$$= \int_0^{+\infty} e^{-\lambda t}\left(P_t\varphi(x + h) - P_t\varphi(x)\right)dt = \int_0^{+\infty} e^{-\lambda t}\left\langle D(P_t\varphi)(x), h\right\rangle_H dt$$

$$+ \int_0^{+\infty} e^{-\lambda t}\int_0^1 \left\langle D(P_t\varphi)(x + \theta h) - D(P_t\varphi)(x), h\right\rangle_H d\theta\,dt,$$

the last two integrals being meaningful, as the mappings

$$t \mapsto \left\langle D(P_t\varphi)(x), h\right\rangle_H, \quad \text{and} \quad t \mapsto \int_0^1 \left\langle D(P_t\varphi)(x + \theta h), h\right\rangle_H d\theta$$

are measurable and bounded on bounded sets. Now, since $\|P_t\varphi\|_1^H \leq c\|\varphi\|_1^H$ and $D(P_t\varphi)$ is continuous, for any $t \geq 0$, it follows

$$\lim_{h\to 0} \frac{1}{|h|_H} \left| \int_0^{+\infty} e^{-\lambda t} \int_0^1 \langle D(P_t\varphi)(x+\theta h) - D(P_t\varphi)(x), h \rangle_H \, d\theta \, dt \right| = 0.$$

Moreover, if we define

$$\langle D\left(R(\lambda, L)\varphi\right)(x), h \rangle_H = \int_0^{+\infty} e^{-\lambda t} \langle D(P_t\varphi)(x), h \rangle_H \, dt,$$

we have $D\left(R(\lambda, L)\varphi\right)(x) \in \mathcal{L}(H; \mathbb{R})$ and $\|D\left(R(\lambda, L)\varphi\right)\|_0^H \leq c\|\varphi\|_1^H/\lambda$, so that $R(\lambda, L)\varphi$ is differentiable with bounded derivative. As far as the second derivative is concerned, for any fixed $k \in H$ we have

$$\langle D\left(R(\lambda, L)\varphi\right)(x+k) - D\left(R(\lambda, L)\varphi\right)(x), h \rangle_H = \int_0^{+\infty} e^{-\lambda t} D^2(P_t\varphi)(x)(k, h) \, dt$$

$$+ \int_0^{+\infty} e^{-\lambda t} \int_0^1 \left(D^2(P_t\varphi)(x+\theta k) - D^2(P_t\varphi)(x)\right)(k, h) \, d\theta \, dt.$$

Notice that the last two integrals are well defined, as the mapping $(0, +\infty] \to \mathbb{R}$, $t \mapsto D^2(P_t\varphi)(x)(h, k)$, is measurable and integrable, thanks to the Theorem 4.4.5. Thus, as $\|P_t\varphi\|_2^H \leq c(t \wedge 1)^{-(1+\epsilon)/2}\|\varphi\|_1^H$, for any $t > 0$ we have

$$\lim_{h,k\to 0} \frac{1}{|h|_H|k|_H} \left| \int_0^{+\infty} e^{-\lambda t} \int_0^1 \left(D^2(P_t\varphi)(x+\theta k) - D^2(P_t\varphi)(x)\right)(k, h) \, d\theta \, dt \right| = 0,$$

and then, by setting

$$D^2\left(R(\lambda, L)\varphi\right)(x)(k, h) = \int_0^{+\infty} e^{-\lambda t} D^2(P_t\varphi)(x)(k, h) \, dt,$$

it follows that $D^2\left(R(\lambda, L)\varphi\right) \in \mathcal{L}(H)$ and $\|D^2\left(R(\lambda, L)\varphi\right)\|_0^H \leq c\lambda^{-1/2}\|\varphi\|_1^H$, which means that $R(\lambda, L)\varphi$ is twice differentiable with bounded second derivative. Finally we prove the continuity of the second derivative. For any $x, y \in H$ we have

$$\left|\left(D^2\left(R(\lambda, L)\varphi\right)(x) - D^2\left(R(\lambda, L)\varphi\right)(y)\right)(k, h)\right|$$

$$\leq \int_0^{+\infty} e^{-\lambda t} \left|\left(D^2(P_t\varphi)(x) - D^2(P_t\varphi)(y)\right)(k, h)\right| \, dt.$$

Besides, due to (4.4.1) for any $t > 0$ it holds

$$\|P_t\varphi\|_1^H \leq c\|\varphi\|_1^H, \quad \|P_t\varphi\|_3^H \leq c(t \wedge 1)^{-(1+\epsilon)}\|\varphi\|_1^H$$

and then, by interpolation (for a proof see the Proposition 5.5.1), for any $\alpha \in (0,1)$ we get

$$P_t : C_b^1(H) \to C_b^{2+\alpha}(H), \quad \|P_t\varphi\|_{2+\alpha}^H \le c\,(t \wedge 1)^{-\frac{(1+\epsilon)(1+\alpha)}{2}}\|\varphi\|_1^H.$$

If we fix $\alpha < (1-\epsilon)/(1+\epsilon)$, this implies that

$$\left|\left(D^2\left(R(\lambda,L)\varphi\right)(x) - D^2\left(R(\lambda,L)\varphi\right)(y)\right)(k,h)\right|$$

$$\le \|\varphi\|_1^H \int_0^{+\infty} e^{-\lambda t}(t \wedge 1)^{-\frac{(1+\epsilon)(1+\alpha)}{2}}\,dt\,|h|_H|k|_H|x-y|^\alpha,$$

so that $R(\lambda,L)\varphi \in C_b^{2+\alpha}(H)$.

Now, we prove that the operator $D^2\left(R(\lambda,L)\varphi\right)(x)QQ^\star$ belongs to $\mathcal{L}_1(H)$, for any $x \in H$. We fix a finite rank operator $N \in \mathcal{L}(H)$. Due to (5.2.11) we have

$$\left|\text{Tr}\left[ND^2R(\lambda,L)\varphi(x)QQ^\star\right]\right|$$

$$\le c\,\|N\| \int_0^{+\infty} e^{-\lambda t}(t \wedge 1)^{-\frac{1+\epsilon+\gamma}{2}}\,dt\,\|\varphi\|_1^H \le c(\lambda)\|\varphi\|_1^H\|N\|.$$

Thus, thanks to the Lemma 5.2.3 this implies that $D^2(R(\lambda,L)\varphi)(x)QQ^\star \in \mathcal{L}_1(H)$, for any $x \in H$, and

$$\left|\text{Tr}\left[D^2R(\lambda,L)\varphi(x)QQ^\star\right]\right| \le c(\lambda)\|\varphi\|_1^H. \tag{5.4.2}$$

Besides, for $x \in D(A)$ and $t \ge 0$ we have

$$\left|\langle D(P_t\varphi)(x), Ax + F(x)\rangle_H\right| \le c\,\left(|Ax|_H + |F(x)|_H\right)\|\varphi\|_1^H. \tag{5.4.3}$$

Therefore

$$\left|\langle D(R(\lambda,L)\varphi)(x), Ax + F(x)\rangle_H\right| \le \frac{c}{\lambda}\|\varphi\|_1^H. \tag{5.4.4}$$

Now, we are able to show that $R(\lambda,A)\varphi$ is a strict solution of the equation (5.4.1). According to the Theorem 5.3.5, for any $x \in D(A)$ and $t > 0$ we have

$$\frac{d}{dt}P_t\varphi(x) = \frac{1}{2}\text{Tr}\left[D^2(P_t\varphi)(x)QQ^\star\right] + \langle D(P_t\varphi)(x), Ax + F(x)\rangle_H,$$

so that, by (5.2.10) and (5.4.3) it holds

$$\left|\frac{d}{dt}P_t\varphi(x)\right| \le c\left((t \wedge 1)^{-\frac{1+\epsilon+\gamma}{2}} + (|Ax|_H + F(x))\right)\|\varphi\|_1^H.$$

Then, from (5.4.2) and (5.4.4) we get

$$\frac{1}{2}\text{Tr}\left[D^2(R(\lambda,L)\varphi)(x)QQ^\star\right] + \langle D(R(\lambda,L)\varphi)(x), Ax + F(x)\rangle_H$$

$$= \int_0^{+\infty} e^{-\lambda t}\frac{d}{dt}P_t\varphi(x)\,dt = -\varphi(x) + \lambda(R(\lambda,L)\varphi)(x),$$

which means that $R(\lambda, L)\varphi$ is a strict solution of (5.4.1).

Step 2: We show that if $\varphi \in C_b(H)$, then $R(\lambda, L)\varphi$ is the unique strong solution of (5.4.1). Indeed, let $\{\varphi_k\} \subset C_b^1(H)$ be a sequence convergent to φ in $C_b(H)$. If we define $\psi_k = R(\lambda, L)\varphi_k$, by the previous step we have that ψ_k is a strict solution of the problem

$$\lambda\psi_k(x) - \frac{1}{2}\text{Tr}\,[D^2\psi_k(x)QQ^*] - \langle D\psi_k(x), Ax + F(x)\rangle_H = \varphi_k(x), \quad x \in D(A).$$

Then, as ψ_k converges to ψ in $C_b(H)$, as k goes to infinity, it follows that ψ is a strong solution.

Finally, let us prove the uniqueness. If φ is a strict solution of the equation $\lambda\psi(x) - \mathcal{L}(x, D)\psi(x) = 0$, then it is immediate to check that the function

$$\psi_\lambda(t, x) = e^{\lambda t}\psi(x)$$

is a classical solution of the problem (5.3.1). Thanks to the Theorem 5.3.5, ψ_λ is unique, so that ψ is unique, as well. Uniqueness for general second order terms φ follows immediately by linearity. If ψ is a strong solution to (5.4.1), let $\{\psi_k\}$ and $\{\varphi_k\}$ be as in the Definition 5.4.2. By uniqueness of strict solutions we have

$$\psi_k = R(\lambda, L)\varphi_k, \quad k \in \mathbb{N},$$

so that, letting k going to infinity, we have $\psi = R(\lambda, L)\varphi$. $\qquad\square$

5.5   Schauder estimates

In what follows we shall need the following generalization to the infinite dimensional case of classical interpolation results (for a proof of these results see [14]).

Proposition 5.5.1. *For any $\theta \in (0, 1)$*

$$\left(C_b^\alpha(H), C_b^\beta(H)\right)_{\theta,\infty} = C_b^{\alpha+\theta(\beta-\alpha)}(H), \qquad 0 \leq \alpha < \beta \leq 1$$

$$\left(C_b^\beta(H), C_b^{2+\beta}(H)\right)_{1-\frac{\theta}{2}(1-\theta),\infty} \subset C_b^{2+\theta\beta}(H), \quad 0 \leq \beta \leq 1.$$

(5.5.1)

By using the inclusions above, we can prove this important preliminary result.

Proposition 5.5.2. *Let $\theta \in (0, 1)$ and $\alpha \in (\theta, 3 - (d - 2)^+]$. Then for any $t > 0$ we have*

$$\|P_t\|_{\mathcal{L}(C_b^\theta(H), C_b^\alpha(H))} \leq c\,(t \wedge 1)^{-\frac{(\alpha-\theta)(1+\epsilon)}{2}}.$$

(5.5.2)

Moreover, for any $\varphi \in C_b^\theta(H)$ it holds

$$\sup_{x \in H} \left|\text{Tr}\,[D^2(P_t\varphi)(x)QQ^*]\right| \leq c\,(t \wedge 1)^{-\frac{(2-\theta)(1+\epsilon)+\gamma}{2}}\|\varphi\|_\theta^H.$$

(5.5.3)

Proof. The estimate (5.5.2) follows from the Proposition 5.5.1 and from (4.4.1), by using the same interpolation arguments used in the proof of the Proposition 1.7.2. Notice that in finite dimension in the (5.5.1) we have equality, while in infinite dimension only one inclusion was proved. But in order to prove (5.5.2) it is enough. It remains to prove (5.5.3). Due to (5.2.11), according to the semigroup law, for any $\varphi \in C_b^1(H)$ we have

$$\left| \text{Tr} \left[D^2(P_t\varphi)(x)QQ^* \right] \right| \leq c\,(t \wedge 1)^{-\frac{1+\epsilon+\gamma}{2}} \|P_{t/2}\varphi\|_1^H.$$

For any $\varphi \in C_b^\theta(H)$ we have

$$\|P_{t/2}\varphi\|_1^H \leq c\,(t \wedge 1)^{-\frac{(1-\theta)(1+\epsilon)}{2}} \|\varphi\|_\theta^H,$$

and then (5.5.3) follows by interpolation. $\qquad\square$

By using the previous proposition and proceeding as in the proof of the first step of the Theorem 5.4.3, we get

Corollary 5.5.3. *Let $\lambda > 0$ and define $\psi = R(\lambda, L)\varphi$, with $\varphi \in C_b(H)$. The following statements hold true*

1. $\psi \in C_b^{1+\theta}(H)$, *for any $\theta \in (0,1)$;*

2. *If $\varphi \in C_b^\theta(H)$, with $\theta \in (2\epsilon/(1+\epsilon), 1)$, then $\psi \in C_b^2(H)$ and it holds*

$$\|\psi\|_2^H \leq \Gamma\left(\theta(1+\epsilon)/2 - \epsilon\right) \lambda^{-\left(\frac{\theta(1+\epsilon)}{2} + \epsilon\right)} \|\varphi\|_\theta^H,$$

   where Γ is the Euler function.

3. *If $\varphi \in C_b^\theta(H)$, with $\theta \in ((2\epsilon + \gamma)/(1+\epsilon), 1)$, then ψ is a strict solution of the equation (5.4.1).*

We are now able to prove the Schauder estimates for the elliptic problem (5.4.1) associated with the operator $\mathcal{L}(x, D)$.

Theorem 5.5.4. *Assume that the Hypotheses 5.1, 5.2 and 5.3 hold and fix $\lambda > 0$ and $\theta \in (0,1)$. Then, if the dimension d is less or equal to 2, the function $R(\lambda, L)\varphi$ is in $C^{2+\theta}(H)$ for any $\varphi \in C_b^\theta(H)$ and*

$$\|R(\lambda, L)\varphi\|_{2+\theta}^H \leq c_\lambda \|\varphi\|_\theta^H.$$

Proof. We apply the general method due to Lunardi [93], which we already used in the proof of the Theorem 1.7.4. Since for any $\alpha \in (\theta, 1)$

$$\left(C_b^\alpha(H), C_b^{2+\alpha}(H) \right)_{1-\frac{\alpha-\theta}{2}, \infty} \subset C_b^{2+\theta}(H),$$

it suffices to show that

$$R(\lambda, L)\varphi \in \left(C_b^\alpha(H), C_b^{2+\alpha}(H)\right)_{1-\frac{\alpha-\theta}{2},\infty}.$$

This follows from the very first definition of interpolation spaces. Indeed for any $t \in [0, 1]$ we have

$$R(\lambda, L)\varphi(x) = a_t(x) + b_t(x), \quad x \in H$$

where

$$a_t(x) = \int_0^t e^{-\lambda s} P_s\varphi(x)\, ds$$

$$b_t(x) = \int_t^{+\infty} e^{-\lambda s} P_s\varphi(x)\, ds.$$

By using (5.5.2) and proceeding as in the proof of the first step of the Theorem 5.4.3, we get that $a_t \in C_b^\alpha(H)$ and $b_t \in C_b^{2+\alpha}(H)$, for any $t > 0$, and

$$\|a(t)\|_{C_b^{2+\alpha}(H)} \leq c\, t^{-(\alpha-\theta)/2} \|f\|_{C_b^\theta(H)},$$

$$\|b(t)\|_{C_b^\alpha(H)} \leq c\, t^{1-(\alpha-\theta)/2} \|f\|_{C_b^\theta(H)}.$$

Hence we can conclude by using the same interpolation argument due to Lunardi used in the first chapter for the finite dimensional case. □

Chapter 6

Smooth dependence on data for the SPDE: the non-Lipschitz case (I)

In the previous two chapters we have been dealing with stochastic reaction-diffusion systems of the following type

$$
\begin{cases}
\dfrac{\partial u_k}{\partial t}(t,\xi) = \mathcal{A}_k(\xi, D)u_k(t,\xi) + f_k(\xi, u_1(t,\xi), \ldots, u_n(t,\xi)) + Q_k \dfrac{\partial^2 w_k}{\partial t \partial \xi}(t,\xi) \\[2mm]
u_k(0,\xi) = x_k(\xi), \quad t > 0, \ \xi \in \overline{\mathcal{O}} \\[2mm]
\mathcal{B}_k(\xi, D)u_k(t,\xi) = 0, \quad t > 0, \ \xi \in \partial\mathcal{O}, \qquad k = 1, \ldots, n.
\end{cases}
$$

$$(6.0.1)$$

In those two chapters the reaction term $f(\xi, \cdot)$ is assumed to have bounded derivatives, uniformly with respect to $\xi \in \overline{\mathcal{O}}$. In the present chapter and in the following ones we consider reaction terms having polynomial growth and satisfying suitable dissipativity conditions.

The system (6.0.1) can also be regarded as a system in the Banach space $E = C(\overline{\mathcal{O}}; \mathbb{R}^n)$ and in what follows we will show that for any $x \in E$ there exists a unique mild solution $u(t; x)$ for it. Thus, we can introduce the transition semigroup associated with the equation (6.0.1) by setting for any $t \geq 0$ and $\varphi \in B_b(E)$

$$
P_t\varphi(x) = \mathbb{E}\varphi(u(t; x)), \quad x \in E.
$$

Our aim is to study the regularizing effect of P_t in $B_b(E)$. Namely, we prove that if $f(\xi, \cdot)$ is of class C^r, for any fixed $\xi \in \overline{\mathcal{O}}$, then

$$
\varphi \in B_b(E) \Rightarrow P_t\varphi \in C_b^r(E).
$$

In particular the semigroup P_t is *strongly Feller* on E, that is it maps Borel functions into uniformly continuous functions, for any $t > 0$. Moreover, for any $\varphi \in C_b(E)$ and $t > 0$ we establish the following Bismut-Elworthy type formula for the first derivative of $P_t\varphi$

$$\langle D(P_t\varphi)(x), h \rangle_E = \frac{1}{t} \mathbb{E}\varphi(u(t;x)) \int_0^t \langle Q^{-1} D_x u(s;x)h, dw(s) \rangle_H, \qquad (6.0.2)$$

where $D_x u(s;x)h$ is the derivative of $u(t;x)$ along the direction $h \in E$ and Q^{-1} is the inverse of Q, which is not bounded in general, as we have seen in the chapter 4. We also prove the following estimates for any $0 \le i \le j \le r$

$$\sup_{x \in E} \left| D^j(P_t\varphi)(x) \right| \le c(t \wedge 1)^{-\frac{(j-i)(1+\epsilon)}{2}} \sup_{x \in E} \left| D^i\varphi(x) \right|,$$

where ϵ is the constant introduced in the Hypothesis 4.1-2.

We remark that in order to get (6.0.2), we have to overcome some difficulties. First of all, we have to prove that the solution $u(t;x)$ is mean-square differentiable in E. This is not trivial, since the Nemytskii operator F associated with the function f is not Lipschitz-continuous in E and we cannot apply the theorem of contractions depending on parameters (see the appendix C for a generalization of it). Another difficulty lies in the fact that as in the previous two chapters, even here we do not assume Q to have a bounded inverse. Thus, in order to give a meaning to the Itô integral appearing in the right side of (6.0.2), first we have to prove that $D_x u(s;x)h \in D(Q^{-1})$ for any $x, h \in E$ and $s > 0$ and then

$$\mathbb{E}\int_0^t \left| Q^{-1} D_x u(s;x)h \right|_H^2 ds < +\infty. \qquad (6.0.3)$$

Since Itô's calculus is not suited to the Banach spaces framework, in order to get the differentiability of P_t in the Banach space E we have to approximate it by means of semigroups P_t^α defined in the Hilbert space $H = L^2(\mathcal{O}; \mathbb{R}^n)$. As a matter of fact, the main idea we follow throughout this chapter is to approximate f by a sequence of functions $\{f_\alpha\}$ which are Lipschitz-continuous and to introduce the approximating equations relative to the Nemytskii operators F_α corresponding to f_α. Then, if we denote by $u_\alpha(t;x)$ the corresponding solutions, we can work with the approximating semigroup $P_t^\alpha\varphi(x) = \mathbb{E}\varphi(u_\alpha(t;x))$ in $C_b(H)$, where we can use the usual Itô calculus and all the results proved in the chapter 4. We get the expected results for $P_t^\alpha\varphi$, we find that some *a priori* estimates hold and finally we pass to the limit.

In the section 6.2 we study some properties of the solution of the system (6.0.1) in the Banach space E. Moreover, by approximating the reaction term f by means of a sequence of Lipschitz-continuous functions $\{f_\alpha\}$, we introduce the corresponding approximating problems and we show that the solutions $u_\alpha(t;x)$ converge to $u(t;x)$ in $L^2(\Omega; C([0,T]; E))$, for any fixed $T > 0$.

In the section 6.3 we study the differentiability with respect to the initial datum for the solution $u(t; x)$. Due to the results proved in the chapter 4, for any $\alpha > 0$ the process $u_\alpha(t; x)$ is differentiable with respect to $x \in H$. We show that the derivatives of $u_\alpha(t; x)$ converge in $L^2(\Omega; C([0, T]; E))$ to the solutions of the variation equations corresponding to the system (6.0.1). Hence we get the differentiability of $u(t; x)$ with respect to $x \in E$. Moreover, we show that for the derivatives of $u(t; x)$ some estimates uniform with respect to $x \in E$ hold true.

In the section 6.4 we prove that the derivatives of $u(t; x)$ belong to the domain of Q^{-1}, for $t > 0$, and the estimates (6.0.3) are satisfied.

We conclude the chapter with the section 6.5, where we apply the results proved in the previous sections and we show that P_t has a smoothing effect. Moreover we estimate the supremum norm of the derivatives of $P_t\varphi$, for any $\varphi \in B_b(E)$.

6.1   Assumptions and preliminary results

As in the previous chapters, we denote by H the Hilbert space $L^2(\mathcal{O}; \mathbb{R}^n)$, with the norm $|\cdot|_H$ and the inner product $\langle \cdot, \cdot \rangle_H$. Moreover, we denote by E the Banach space $C(\overline{\mathcal{O}}; \mathbb{R}^n)$, endowed with the norm $|\cdot|_E$ and the duality pairing $\langle \cdot, \cdot \rangle_E$ in $E \times E^\star$.

With the notations introduced in the section 4.1, we denote by $\mathcal{A}$ the realization in H of the operator

$$\mathcal{A} - (\rho + 1)I = (\mathcal{A}_1 - (\rho + 1)I, \ldots, \mathcal{A}_n - (\rho + 1)I),$$

with the boundary conditions given by $\mathcal{B} = (\mathcal{B}_1, \ldots, \mathcal{B}_n)$, where

$$B_k(\xi, D) = I \quad \text{or} \quad B_k(\xi, D) = \sum_{i,j=1}^d a_{ij}^k(\xi)\nu_j(\xi)\frac{\partial}{\partial \xi_i}, \quad \xi \in \partial\mathcal{O}. \qquad (6.1.1)$$

Thus in this chapter and in the following ones we are only dealing only with Dirichlet or conormal boundary conditions.

Whenever we need, we refer to the properties of $\mathcal{A}$ described in the section 4.1. In what follows we will denote again by A the realization in E of the operator $\mathcal{A} - (\rho + 1)I$, with the boundary conditions (6.1.1). As shown for example in [91], it is given by

$$\begin{cases} D(A) = \left\{ x \in \bigcap_{p \geq 1} W_{loc}^{2,p}(\mathcal{O}; \mathbb{R}^n) : x, \mathcal{A}x \in E \ \mathcal{B}x_{|\partial\mathcal{O}} = 0 \right\} \\ \\ Ax = \mathcal{A}x - (\rho + 1)x, \quad x \in D(A), \end{cases}$$

It generates an analytic semigroup e^{tA}. In the case of Dirichlet boundary conditions we have

$$\overline{D(A)}^E = \left\{ x \in E : x_{|\partial\mathcal{O}} = 0 \right\} \neq E$$

and, due to the Hille-Yosida theorem, this implies that e^{tA} is not strongly continuous on E. In the case of conormal boundary conditions $\overline{D(A)}^E = E$, so that e^{tA} is strongly continuous. In each of the two cases $\overline{D(A)}^E$ is a dense Borel set of H. We remark that since we are assuming that for any $k = 1, \ldots, n$

$$\inf_{\xi \in \overline{\mathcal{O}}} \sum_{i,j=1}^{d} a_{ij}^k(\xi) h_i h_j \geq \nu |h|^2, \quad h \in \mathbb{R}^d,$$

for some $\nu > 0$, then, if δ_x is the element in $\partial |x|_E$ described in (A.1.1), we have

$$\langle Ax, \delta_x \rangle_E \leq 0, \quad x \in E. \tag{6.1.2}$$

Now, let us define

$$\mathcal{G}x = (\mathcal{G}_1 x_1, \ldots, \mathcal{G}_n x_n), \tag{6.1.3}$$

where for any $k = 1, \ldots, n$

$$\mathcal{G}_k(\xi, D) = \sum_{i=1}^{d} \left(b_i^k(\xi) - \sum_{j=1}^{d} \frac{\partial a_{ij}^k}{\partial \xi_j}(\xi) \right) \frac{\partial}{\partial \xi_i}, \quad \xi \in \mathcal{O},$$

and by difference let us define

$$C = A - (\rho + 1)\mathcal{I} - \mathcal{G}. \tag{6.1.4}$$

The second order elliptic operators C generates a negative analytic semigroup e^{tC} in $L^p(\mathcal{O}; \mathbb{R}^n)$, for any $p \in (1, \infty]$, and also in E. The semigroup e^{tC} enjoys the properties (4.1.1), (4.1.2), (4.1.3) and (4.1.4). Moreover, due to the boundary conditions (6.1.1), it is self-adjoint in H. Notice that if C_k is the realization in $L^p(\mathcal{O})$ or in $C(\overline{\mathcal{O}})$ of the operator $A_k - (\rho + 1)\mathcal{I} - \mathcal{G}_k$, then

$$e^{tC}x = \left(e^{tC_1} x_1, \ldots, e^{tC_n} x_n \right).$$

For any $\delta \in \mathbb{R}$ we have that $D((-A)^\delta) = D((-C)^\delta)$ with equivalence of norms, that is

$$c_1 |(-A)^\delta x|_H \leq |(-C)^\delta x|_H \leq c_2 |(-A)^\delta x|_H, \tag{6.1.5}$$

for suitable positive constants c_1 and c_2 depending only on δ.

Concerning the realization of the operator $\mathcal{G}$, as the coefficients a_{ij}^k and b_i^k are assumed to be smooth, it is easy to check that $D(G^\star) = D((-C)^{1/2})$ and

$$c_1 |(-C)^{1/2}x|_H \leq |G^\star x|_H \leq c_2 |(-C)^{1/2}x|_H. \tag{6.1.6}$$

We shall assume that for any $k = 1, \ldots, n$ the following conditions hold.

Hypothesis 6.1.   *1. There exists a complete orthonormal basis $\{e_k\}$ in H which diagonalizes C and such that $e_k \in E$ and $\sup_{k \in \mathbb{N}} |e_k|_E < \infty$. Concerning the corresponding eigenvalues $\{-\alpha_k\}$, we have that*

$$\delta > \frac{d}{2} \implies \sum_{k=1}^{\infty} \alpha_k^{-\delta} < \infty. \tag{6.1.7}$$

2. The bounded linear operator $Q : H \to H$ is non-negative and diagonal with respect to the complete orthonormal basis $\{e_k\}$ which diagonalizes C. Moreover, if $\{\lambda_k\}$ is the corresponding set of eigenvalues, we have

$$\sum_{k=1}^{\infty} \frac{\lambda_k^2}{\alpha_k^{1-\rho}} < +\infty, \tag{6.1.8}$$

for some $\rho \in (0,1)$.

3. There exists $\epsilon < 1$ such that

$$\text{Range}\,((-C)^{-\frac{\epsilon}{2}}) \subset \text{Range}\,(Q). \tag{6.1.9}$$

Remark 6.1.1.   1. In the case of the Laplace operator Δ, both with Dirichlet and with Neumann boundary conditions, if the open set $\mathcal{O}$ is *strongly regular* (in the sense of Davies [50]) and if $\delta > d/2$ then

$$\sum_{k=1}^{\infty} \alpha_k^{-\delta} < \infty \iff \int_{\mathcal{O}} d(\xi)^{2\delta - d}\, d\xi < \infty,$$

where $d(\xi) = \min\{|\xi - \eta| \,;\, \eta \in \mathcal{O}^c\}$ (see [50, Theorem 1.9.6]). Thus, the condition (6.1.7) in the Hypothesis 6.1 is satisfied.

2. Let us fix $\nu_1, \nu_2 > 0$ and $\alpha \geq 0$ such that

$$\nu_1 \leq \lambda_k \alpha_k^{\alpha} \leq \nu_2, \qquad k \in \mathbb{N}.$$

If $d \leq 3$, then there exists $\alpha \geq 0$ such that Q satisfies the Hypothesis 6.1. Indeed, due to (6.1.7) we can check that for $\rho \in (0,1)$

$$\sum_{k=1}^{\infty} \frac{\lambda_k^2}{\alpha_k^{1-\rho}} < +\infty \iff (1 - \rho + 2\alpha) > d/2 \iff \alpha > (d/2 + \rho - 1)/2.$$

Moreover, (6.1.9) is satisfied by any $\alpha \leq \epsilon/2$. Therefore, if $d \leq 3$, it is possible to find some α such that Q verifies both (6.1.8) and (6.1.9). Notice that if $d = 1$, we can choose $\alpha = 0$, so that the non-degenerate case is covered.

Now, consider the Ornstein-Uhlenbeck system

$$
\begin{cases}
\dfrac{\partial z}{\partial t}(t,\xi) = \mathcal{A}z(t,\xi) + Q\dfrac{\partial^2 w}{\partial \xi \partial t}(t,\xi) \\[2mm]
z(0,\xi) = 0, \quad \xi \in \mathcal{O}, \quad z(t,\xi) = 0, \quad \xi \in \partial\mathcal{O} \qquad t \geq 0.
\end{cases}
\tag{6.1.10}
$$

As we recalled in the chapter 4, if for any $t \geq 0$

$$
\int_0^t \mathrm{Tr}\left[e^{sA} Q Q^* e^{sA^*} \right] ds < \infty,
$$

then the equation (6.1.10) has a unique mild solution $w^A(t)$ which is the mean-square continuous gaussian process with values in H given by

$$
w^A(t) = \int_0^t e^{(t-s)A} Q \, dw(s).
\tag{6.1.11}
$$

In the present chapter we will need $w^A(t)$ to be more regular. Namely, we will require that $w^A(t)$ is an E-valued process, having the p-th moment of its E-norm finite, for any p.

In [47] it is proved that if A and Q are diagonal with respect to the same basis and their eigenvalues satisfy the condition (6.1.8), then w^A has a version $w^A(t,\xi)$ which is α-Hölder continuous with respect to $t \geq 0$ and $\xi \in \overline{\mathcal{O}}$, P-a.s. for any $\alpha \in [0, 1/4)$. In particular $w^A(t)$ has an E-valued version with α-Hölder continuous paths. In the present situation we do not assume that A is self-adjoint. Nevertheless, the following regularity result holds.

Lemma 6.1.2. *Assume the Hypothesis 6.1. Then the process* $w^A : [0, +\infty) \times \overline{\mathcal{O}} \to \mathbb{R}^n$ *defined in (6.1.11) is continuous, P-a.s. Moreover, for any* $T > 0$ *and* $p \geq 2$

$$
\mathbb{E} \sup_{t \in [0,T]} |w^A(t)|_E^p < \infty.
\tag{6.1.12}
$$

Proof. The process $w^A(t)$ is the unique mild solution of the Ornstein-Uhlenbeck system (6.1.10), which may be rewritten in the following form

$$
dz(t) = [Cz(t) + Gz(t)] \, dt + Q \, dw(t), \quad z(0) = 0,
\tag{6.1.13}
$$

where C and G are the realization in H of the operators $\mathcal{C}$ and $\mathcal{G}$ defined respectively in (6.1.4) and (6.1.3). We are showing that for any $t > 0$ the equation (6.1.13) has a unique mild solution which belongs to $C([0,T]; E)$, P-a.s. To this purpose, for any $u \in C([0,T]; E)$ we define

$$
\Gamma(u)(t) = \int_0^t e^{(t-s)C} Gu(s) \, ds + \int_0^t e^{(t-s)C} Q \, dw(s) = \psi(u)(t) + w^C(t).
$$

If we prove that $\Gamma : C([0, T_0]; E) \to C([0, T_0]; E)$ is a contraction $\mathbb{P}$-a.s. for some T_0 sufficiently small, then the existence of a unique mild solution in $C([0, T_0]; E)$ follows by a fixed point argument.

For any $p \in [2, \infty)$, the domain of the operator G^* in $L^p(\mathcal{O}; \mathbb{R}^n)$ coincides with $D((-C)^{\frac{1}{2}})$ and for any $x \in L^p(\mathcal{O}; \mathbb{R}^n)$ and $t > 0$ it holds

$$|G^* e^{tC} x|_p \leq c_2 |(-C)^{1/2} e^{tC} x|_p \leq c e^{-t} (t \wedge 1)^{-\frac{1}{2}} |x|_p.$$

Now, if $u \in C([0, T]; W^{1,p}(\mathcal{O}; \mathbb{R}^n))$ and $z \in L^q(\mathcal{O}; \mathbb{R}^n)$, with $q = p/(p-1)$, we have

$$\left| \left\langle e^{(t-s)C} G u(s), z \right\rangle_{p,q} \right| = \left| \left\langle u(s), G^* e^{(t-s)C} z \right\rangle_{p,q} \right| \leq c e^{-(t-s)} ((t-s) \wedge 1)^{-\frac{1}{2}} |u(s)|_p |z|_q.$$

Due to the arbitrariness of z, by using (4.1.3) (actually the operator C fulfills the same properties as A) for any $\epsilon \in (0, 1)$ we get

$$|\psi(u)(t)|_{\epsilon, p} \leq c \int_0^t e^{-\frac{t-s}{2}} ((t-s) \wedge 1)^{-\frac{\epsilon+1}{2}} |u(s)|_p \, ds \leq c t^{\frac{1-\epsilon}{2}} |u|_{C([0,T]; L^p(\mathcal{O}; \mathbb{R}^n))},$$
(6.1.14)

so that $\psi(u)(t) \in W^{\epsilon, p}(\mathcal{O}; \mathbb{R}^n)$, for any $t \in [0, T]$. We claim that the mapping $\psi(u) : [0, T] \to W^{\epsilon, p}(\mathcal{O}; \mathbb{R}^n)$ is continuous. Indeed, if $u \in C([0, T]; W^{1,p}(\mathcal{O}; \mathbb{R}^n))$ and $0 < s < t$ we have

$$\psi(u)(t) - \psi(u)(s) = \int_s^t e^{(t-r)C} G u(r) \, dr + \int_0^s \left(e^{(t-s)C} - I \right) e^{(s-r)C} G u(r) \, dr$$

and then, by using (6.1.14), it easily follows that

$$\lim_{|t-s| \to 0} |\psi(u)(t) - \psi(u)(s)|_{\epsilon, p} = 0.$$

This means that $\psi(u) \in C([0, T]; W^{\epsilon, p}(\mathcal{O}; \mathbb{R}^n))$. Given $u \in C([0, T]; L^p(\mathcal{O}; \mathbb{R}^n))$, let $\{u_n\}$ be a sequence in $C([0, T]; W^{1,p}(\mathcal{O}; \mathbb{R}^n))$ converging to u in $C([0, T]; L^p(\mathcal{O}; \mathbb{R}^n))$. By (6.1.14) we have that $\{\psi(u_n)\}$ is a Cauchy sequence on $C([0, T]; W^{\epsilon, p}(\mathcal{O}; \mathbb{R}^n))$ and so it converges to a function $\psi(u)$ in $C([0, T]; W^{\epsilon, p}(\mathcal{O}; \mathbb{R}^n))$, which is independent of the choice of $\{u_n\}$. Therefore, if $u \in C([0, T]; E)$, the function $\psi(u)$ is well defined and belongs to $C([0, T]; W^{\epsilon, p}(\mathcal{O}; \mathbb{R}^n))$, for any $p \geq 1$. From the Sobolev embedding theorem we get that $\psi(u) \in C([0, T]; E)$ and

$$\sup_{t \in [0, T]} |\psi(u)(t)|_E \leq c T^\delta |u|_{C([0,T]; E)},$$
(6.1.15)

for some $\delta \in (0, 1/2)$.

The continuity in space and time of $w^C(t, \xi)$ is known, but we sketch a proof for completeness. Let us fix $T > 0$ and $\alpha \in (0, 1/2)$. In general, if

$$F(t) = \int_0^t e^{(t-s)A} (t-s)^{\alpha-1} f(s) \, ds, \quad t \in [0, T],$$

for $f \in L^p([0,T] \times \mathcal{O}; \mathbb{R}^n)$, it is possible to show that $F \in C([0,T]; W^{\epsilon, p}(\mathcal{O}; \mathbb{R}^n))$, for any $\epsilon < 2(p\alpha - 1)/p$. Moreover the following estimate holds

$$|F(t)|^p_{W^{\epsilon, p}(\mathcal{O}; \mathbb{R}^n)} \leq c_T |f|^p_{L^p([0,T] \times \mathcal{O}; \mathbb{R}^n)}.$$

Due to the Sobolev embedding theorem, if $p\alpha > 3/2$ this implies that F belongs to $C([0,T]; E)$ and

$$\sup_{\substack{t \in [0,T] \\ \xi \in \overline{\mathcal{O}}}} |F(t;\xi)|^p \leq c_T |f|^p_{L^p([0,T] \times \mathcal{O}; \mathbb{R}^n)}.$$

By using the factorization method, for any $\alpha \in (0,1)$ it holds

$$w^C(t) = \frac{\sin \pi \alpha}{\pi} \int_0^t (t-s)^{\alpha-1} e^{(t-s)C} v_\alpha(s) \, ds,$$

where

$$v_\alpha(s) = \int_0^s (s-r)^{-\alpha} e^{(s-r)C} Q \, dw(r).$$

Therefore, if we prove that $v_\alpha \in L^p([0,T] \times \mathcal{O}; \mathbb{R}^n)$, for p sufficiently large, it follows that $w^C \in C([0,T]; E)$ and

$$\sup_{t \in [0,T]} |w^C(t)|^p_E \leq c T^\delta |v_\alpha|^p_{L^p([0,T] \times \mathcal{O}; \mathbb{R}^n)}, \quad \mathbb{P}-\text{a.s.} \qquad (6.1.16)$$

for some $\delta > 0$ depending on α, p and d. By the Hypotheses 6.1 we have

$$v_\alpha(t, \xi) = \sum_{ki=1}^\infty \lambda_k \int_0^t (t-s)^{-\alpha} e^{-(t-s)\alpha_k} e_k(\xi) \, dw_k(s).$$

For all $(t,\xi) \in [0,T] \times \mathcal{O}$, the process $v_\alpha(t,\xi)$ is a centered gaussian random variable with covariance $\sigma_\alpha(t,\xi)$ given by

$$\sigma_\alpha(t,\xi) = \sum_{k=1}^\infty \lambda_k^2 \int_0^t (t-s)^{-2\alpha} e^{-2(t-s)\alpha_k} |e_k(\xi)|^2 \, ds.$$

Taking into account of (6.1.7) and (6.1.8), after some calculations we have that if $\alpha < \rho/2$

$$\sigma_\alpha(t,\xi) \leq \sum_{k=1}^\infty \lambda_k^2 \alpha_k^{2\alpha-1} < \infty.$$

Thus, from the gaussianity of v_α it follows

$$\mathbb{E}|v_\alpha(t,\xi)|^p \leq c \left(\mathbb{E}|v_\alpha(t,\xi)|^2 \right)^{p/2} \leq c \left(\sum_{k=1}^\infty \lambda_k^2 \alpha_k^{2\alpha-1} \right)^{p/2}.$$

This implies that

$$\mathbb{E} \int_0^T \int_{\mathcal{O}} |v_\alpha(t,\xi)|^p \, d\xi \, dt \leq cT|\mathcal{O}|, \tag{6.1.17}$$

where $|\mathcal{O}|$ is Lebesgue measure of the open set $\mathcal{O}$, so that $v_\alpha \in L^p([0,T] \times \mathcal{O}; \mathbb{R}^n)$, $\mathbb{P}$-a.s. and (6.1.16) follows. Thanks to (6.1.15) this implies that Γ maps $C([0,T],E)$ into itself.

Moreover, by (6.1.15) we have that Γ is a contraction in $C([0,T_0];E)$ for T_0 sufficiently small, so that there exists a unique mild solution for the Ornstein-Uhlenbeck system (6.1.13) in $[0,T_0]$, which may be extended in the whole interval $[0,T]$ by repeating these arguments in the intervals $[T_0, 2T_0]$, $[2T_0, 3T_0]$ etc.

Finally, by using (6.1.14) we have

$$|w^A(t)|_E \leq c \int_0^t ((t-s) \wedge 1)^{-\frac{\varepsilon+1}{2}} |w^A(s)|_E \, ds + |w^C(t)|_E,$$

and then, from the Gronwall lemma and (6.1.16) and (6.1.17), the estimate (6.1.12) follows. $\square$

6.1.1   The Nemytskii operator

In this chapter and in the following ones we shall assume that for any $k = 1, \ldots, n$ there exist two continuous functions $g_k : \overline{\mathcal{O}} \times \mathbb{R} \to \mathbb{R}$ and $h_k : \overline{\mathcal{O}} \times \mathbb{R}^n \to \mathbb{R}$ such that

$$f_k(\xi, \sigma_1, \ldots, \sigma_n) = g_k(\xi, \sigma_k) + h_k(\xi, \sigma_1, \ldots, \sigma_n),$$

for $\xi \in \overline{\mathcal{O}}$ and $\sigma = (\sigma_1, \ldots, \sigma_n) \in \mathbb{R}^n$. The functions g_k and h_k are assumed to enjoy the following conditions.

Hypothesis 6.2.   *1. There exists $r \geq 2$ such that for any $\xi \in \overline{\mathcal{O}}$ the function $h_k(\xi, \cdot)$ belongs to $C^r(\mathbb{R}^n; \mathbb{R})$ and has bounded derivatives up to the r-th order, uniformly with respect to $\xi \in \overline{\mathcal{O}}$. Moreover, the mapping $D_\sigma^j h_k : \overline{\mathcal{O}} \times \mathbb{R}^n \to \mathcal{L}^j(\mathbb{R}^n; \mathbb{R})$ is continuous, for any $j = 1, \ldots, r$.*

2. For any $\xi \in \overline{\mathcal{O}}$, the function $g_k(\xi, \cdot)$ belongs to $C^r(\mathbb{R})$ and there exists $m \geq 0$ such that

$$\sup_{\xi \in \overline{\mathcal{O}}} \sup_{t \in \mathbb{R}} \frac{|D_t^j g_k(\xi, t)|}{1 + |t|^{2m+1-j}} < \infty.$$

Moreover, the mapping $D_t^j g_k : \overline{\mathcal{O}} \times \mathbb{R} \to \mathbb{R}$ is continuous, for any $j = 1, \ldots, r$.

3. If $m \geq 1$, there exists $\beta \in \mathbb{R}$ such that for any $t \in \mathbb{R}$ and $\xi \in \overline{\mathcal{O}}$

$$D_t g_k(\xi, t) \leq \beta.$$

As in the chapter 4, we define the Nemytskii operator F associated with the function $(\xi, \sigma) \mapsto f(\xi, \sigma) + (\rho + 1)\sigma$ by setting

$$F(x)(\xi) = f(\xi, x(\xi)) + (\rho + 1)x(\xi), \quad \xi \in \mathcal{O},$$

for any $x : \overline{\mathcal{O}} \to \mathbb{R}^n$. It is immediate to check that F is not even well defined from H into itself if $m \geq 1$. Instead, F is well defined and continuous from $L^p(\mathcal{O}; \mathbb{R}^n)$ into $L^q(\mathcal{O}; \mathbb{R}^n)$, for p and q such that $p/q = 2m + 1$. In particular, if we set

$$p_\star = 2m + 2, \quad q_\star = \frac{2m + 2}{2m + 1},$$

F is continuous from $L^{p_\star}(\mathcal{O}; \mathbb{R}^n)$ into $L^{q_\star}(\mathcal{O}; \mathbb{R}^n)$. Since we are assuming $f(\xi, \cdot)$ to be of class C^r, for any $\xi \in \overline{\mathcal{O}}$, if $m \geq 1$ then F is r-times Fréchet differentiable from $L^{p_\star}(\mathcal{O}; \mathbb{R}^n)$ into $L^{q_\star}(\mathcal{O}; \mathbb{R}^n)$ and for any $x, h_1, \ldots, h_r \in L^{p_\star}(\mathcal{O}; \mathbb{R}^n)$ and $j = 1, \ldots, r$ we have

$$DF^j(x)(h_1, \ldots, h_j)(\xi) = D^j_\sigma f(\xi, x(\xi))(h_1(\xi), \cdots, h_j(\xi)) + \delta_{j1}(\rho + 1)h_1(\xi), \quad \xi \in \overline{\mathcal{O}}.$$
$$(6.1.18)$$

Due to the growth conditions verified by f, this implies that

$$\left| DF^j(x) \right|_{\mathcal{L}^j(L^{p_\star}(\mathcal{O}; \mathbb{R}^n); L^{q_\star}(\mathcal{O}; \mathbb{R}^n))} \leq c \left(1 + |x|_{p_\star}^{2m+1-j} \right), \quad x \in L^{p_\star}(\mathcal{O}; \mathbb{R}^n).$$

Moreover the functional F is well defined from E into itself and is r-times Fréchet differentiable. A formula analogous to (6.1.18) holds for its derivative, so that

$$\left| DF^j(x) \right|_{\mathcal{L}^j(E)} \leq c \left(1 + |x|_E^{2m+1-j} \right), \quad x \in E. \qquad (6.1.19)$$

This means that $D^j F$ is bounded on bounded sets of E and in particular F and its derivatives are locally Lipschitzcontinuous on E, up to the r-th order. Furthermore, from the Hypothesis 6.2-3 for any $x, h \in E$ we have

$$\langle DF(x)h, h \rangle_H = \int_{\mathcal{O}} \langle (D_\sigma f(\xi, x(\xi)) + (\rho + 1)) h(\xi), h(\xi) \rangle \, d\xi \leq c |h|_H^2, \qquad (6.1.20)$$

and by the mean-value theorem, this yields

$$\langle F(x) - F(y), x - y \rangle_H \leq c |x - y|_H^2. \qquad (6.1.21)$$

By using once more the Hypothesis 6.2-3 it is not difficult to show that for any $x, h \in E$

$$\langle F(x + h) - F(x), \delta_h \rangle_E \leq c |h|_E, \qquad (6.1.22)$$

where δ_h is the element of $\partial |h|_E$ defined in (A.1.1). This easily implies that

$$\sup_{x \in H} \langle DF(x)h, \delta_h \rangle_E \leq c |h|_E, \quad h \in E. \qquad (6.1.23)$$

The next stronger dissipativity condition will be crucial in order to prove some important estimates.

Hypothesis 6.3. *If $m \geq 1$, there exist $a, \gamma > 0$ and $c \in \mathbb{R}$ such that for any* $k = 1, \ldots, n$

$$\sup_{\xi \in \overline{\mathcal{O}}} \left(g_k(\xi, t+s) - g_k(\xi, t)\right) s \leq -a s^{2m+2} + c(1 + |t|^\gamma)|s|, \qquad s, t \in \mathbb{R}. \qquad (6.1.24)$$

From (6.1.24) we have that for any $\sigma, \rho \in \mathbb{R}^n$

$$\sup_{\xi \in \overline{\mathcal{O}}} \langle f(\xi, \sigma + \rho) - f(\xi, \rho), \sigma \rangle_{\mathbb{R}^n} \leq -a|\sigma|^{2m+2} + c\left(1 + |\rho|^{\frac{\gamma(2m+2)}{2m+1}}\right),$$

for some constants $a > 0$ and $c \in \mathbb{R}$ possibly different from those introduced in (6.1.24). This implies that

$$\langle F(x+h) - F(x), h \rangle_H \leq -a|h|_{p_*}^{p_*} + c\left(1 + |x|_{\gamma q_*}^{\gamma q_*}\right).$$

Finally, according to the definition of δ_h, the condition (6.1.24) implies that for any $x, h \in \mathbb{E}$

$$\langle F(x+h) - F(x), \delta_h \rangle_E \leq -a|h|_E^{2m+1} + c\left(|x|_E^\gamma + 1\right), \qquad (6.1.25)$$

where the constants $a > 0$ and $c \in \mathbb{R}$ are possibly different from those introduced in the Hypothesis 6.3.

Remark 6.1.3. For any $k = 1, \ldots, n$, let us define

$$g_k(\xi, t) = -c_k(\xi)t^{2m+1} + \sum_{j=0}^{2m} c_{kj}(\xi)t^j,$$

where c_k, c_{kj} are continuous bounded functions from $\overline{\mathcal{O}}$ into $\mathbb{R}$. If we assume that

$$\inf_{\xi \in \overline{\mathcal{O}}} c_k(\xi) > 0,$$

then it is possible to check that g_k fulfills the Hypotheses 6.2 and 6.3.

6.1.2 The approximating Nemytskii operators

Due to the Hypothesis 6.2, there exists $c \in \mathbb{R}$ such that for any $\xi \in \overline{\mathcal{O}}$ the mapping

$$\gamma(\xi, \cdot) : \mathbb{R}^n \to \mathbb{R}^n, \qquad \sigma \mapsto f(\xi, \sigma) - c\sigma$$

is dissipative. Then, proceeding as in the section 1.3 (see also the appendix A), for any $\alpha > 0$ we can define the function

$$\gamma_\alpha : \overline{\mathcal{O}} \times \mathbb{R}^n \to \mathbb{R}^n, \qquad (\xi, \sigma) \mapsto f(\xi, J_\alpha(\xi, \sigma)) - c J_\alpha(\xi, \sigma),$$

where for each $\xi \in \overline{\mathcal{O}}$

$$J_\alpha(\xi, \sigma) = (I - \alpha\gamma(\xi, \cdot))^{-1}(\sigma), \quad \sigma \in \mathbb{R}^n.$$

According to the results proved in the appendix A, section A.2, the function $J_\alpha(\xi, \cdot)$ is of class C^r, for any fixed $\xi \in \overline{\mathcal{O}}$. Now we check that the derivatives of J_α are continuous in both variables.

Lemma 6.1.4. *The function $D_\sigma^j J_\alpha : \overline{\mathcal{O}} \times \mathbb{R}^n \to \mathcal{L}^j(\mathbb{R}^n)$ is continuous, for any $\alpha > 0$ and $j = 0, 1, \ldots, r$.*

Proof. First we show that $J_\alpha(\xi, \sigma)$ is continuous in both variables. Let us fix two sequences $\{\xi_k\} \subset \overline{\mathcal{O}}$ and $\{\sigma_k\} \subset \mathbb{R}^n$, converging respectively to ξ_0 and σ_0. We have

$$|J_\alpha(\xi_k, \sigma_k) - J_\alpha(\xi_0, \sigma_0)| \leq |J_\alpha(\xi_k, \sigma_k) - J_\alpha(\xi_k, \sigma_0)| + |J_\alpha(\xi_k, \sigma_0) - J_\alpha(\xi_0, \sigma_0)|.$$

Due to (A.2.1) for any $k \in \mathbb{N}$ and $\alpha > 0$ we have

$$|J_\alpha(\xi_k, \sigma_k) - J_\alpha(\xi_k, \sigma_0)| \leq |\sigma_k - \sigma_0|.$$

Then, in order to prove the continuity of $J_\alpha(\xi, \sigma)$ we have to show that

$$\lim_{k \to +\infty} J_\alpha(\xi_k, \sigma_0) = J_\alpha(\xi_0, \sigma_0). \tag{6.1.26}$$

Suppose that (6.1.26) is not true. Then, if we define for any $k \in \mathbb{N}$

$$\rho_k = J_\alpha(\xi_k, \sigma_0), \quad \rho_0 = J_\alpha(\xi_0, \sigma_0),$$

then there exists $\epsilon > 0$ and a subsequence $\{\rho_{k_h}\}$ such that for any $h \in \mathbb{N}$

$$|(\rho_{k_h} - \alpha\gamma(\xi_0, \rho_{k_h})) - (\rho_0 - \alpha\gamma(\xi_0, \rho_0))| > \epsilon. \tag{6.1.27}$$

Actually, if this is not true, then $\rho_k - \alpha\gamma(\xi_0, \rho_k)$ converges to $\rho_0 - \alpha\gamma(\xi_0, \rho_0)$. Hence, as $J_\alpha(\xi_0, \cdot)$ is continuous, we have that

$$\lim_{k \to +\infty} \rho_k = \lim_{k \to +\infty} J_\alpha(\xi_0, \rho_k - \alpha\gamma(\xi_0, \rho_k)) = J_\alpha(\xi_0, \rho_0 - \alpha\gamma(\xi_0, \rho_0)) = \rho_0$$

and this is a contradiction, since we are assuming that ρ_k does not converge to ρ_0. Thus, by using the definition of ρ_k and ρ_0 and (6.1.27), we have

$$0 = |(\rho_{k_h} - \alpha\gamma(\xi_{k_h}, \rho_{k_h})) - (\rho_0 - \alpha\gamma(\xi_0, \rho_0))|$$

$$\geq |(\rho_{k_h} - \alpha\gamma(\xi_0, \rho_{k_h})) - (\rho_0 - \alpha\gamma(\xi_0, \rho_0))| - \alpha|\gamma(\xi_0, \rho_{k_h}) - \gamma(\xi_{k_h}, \rho_{k_h})|$$

$$\geq \epsilon - \alpha|\gamma(\xi_0, \rho_{k_h}) - \gamma(\xi_{k_h}, \rho_{k_h})|.$$

As γ is continuous, then γ is uniformly continuous on $\overline{\mathcal{O}} \times \{x \in \mathbb{R}^n \, ; \, |x| \leq c\}$, for any $c > 0$. Hence, since $|\rho_{k_h}| \leq c$, for any $h \in \mathbb{N}$, we have that

$$\lim_{k \to +\infty} |\gamma(\xi_0, \rho_{k_h}) - \gamma(\xi_{k_h}, \rho_{k_h})| = 0$$

and we get a contradiction.

To conclude, we have to prove that $D_\sigma J_\alpha(\xi, \sigma)$ is continuous on $\overline{\mathcal{O}} \times \mathbb{R}^n$. Indeed, the continuity of higher order derivatives follows by recurrence, by using the formulas (A.2.4) and (A.2.5) and analogous formulas for higher order derivatives of $J_\alpha(\xi, \sigma)$. We recall that

$$D_\sigma J_\alpha(\xi, \sigma) = (I - \alpha \, D_\sigma \gamma(\xi, J_\alpha(\xi, \sigma)))^{-1}$$

and then, since $D_\sigma f(\xi, \sigma)$ (and hence $D_\sigma \gamma(\xi, \sigma)$) is continuous in both variables and $J_\alpha(\xi, \sigma)$ is continuous, it follows that $D_\sigma J_\alpha(\xi, \sigma)$ is continuous, as well. □

As in the section 1.3, thanks to the result proved in the appendix A, for any $\alpha > 0$ the function

$$f_\alpha(\xi, \sigma) = \gamma_\alpha(\xi, \sigma) + c\sigma$$

is Lipschitz-continuous and for any $(\xi, \sigma) \in \overline{\mathcal{O}} \times \mathbb{R}^n$ and $\alpha > 0$

$$\langle D_\sigma f_\alpha(\xi, \sigma) h, h \rangle \leq c|h|^2, \quad h \in \mathbb{R}^n. \tag{6.1.28}$$

Moreover

$$\sup_{\xi \in \overline{\mathcal{O}}} |f_\alpha(\xi, \sigma) - f(\xi, \sigma)| \leq \alpha c(1 + |\sigma|^{4m+1}), \quad \sigma \in \mathbb{R}^n. \tag{6.1.29}$$

We are assuming that the function $f(\xi, \cdot)$ is of class C^r, for any $\xi \in \overline{\mathcal{O}}$, and the mappings $D_\sigma^j f$ are continuous in $\overline{\mathcal{O}} \times \mathbb{R}^n$. Then, due to the Lemma 6.1.4 we have that the functions $\gamma_\alpha(\xi, \cdot)$ are r-times differentiable and the functions $D_\sigma^j \gamma_\alpha$ are continuous on $\overline{\mathcal{O}} \times \mathbb{R}^n$. This implies that the same is true for the functions $D_\sigma^j f_\alpha(\xi, \sigma)$. Moreover, for any $j = 0, 1, \ldots, r$ and $R > 0$ it holds

$$\lim_{\alpha \to 0} \sup_{|\sigma| \leq R} |D_\sigma^j f_\alpha(\xi, \sigma) - D_\sigma^j f(\xi, \sigma)| = 0, \tag{6.1.30}$$

uniformly with respect to $\xi \in \overline{\mathcal{O}}$ and

$$\sup_{\xi \in \overline{\mathcal{O}}} \sup_{\sigma \in \mathbb{R}^n} \frac{|D_\sigma^j f_\alpha(\xi, \sigma)|}{1 + |\sigma|^{2m+1-j}} \leq c, \tag{6.1.31}$$

for a suitable constant c independent of α.

Now, for any $\alpha > 0$, we denote by F_α the Nemytskii operator associated with the function $f_\alpha + (\rho + 1)I$, that is

$$F_\alpha(x)(\xi) = f_\alpha(\xi, x(\xi)) + (\rho + 1)x(\xi), \quad \xi \in \overline{\mathcal{O}}.$$

As f_α is Lipschitz-continuous, then F_α is Lipschitz-continuous both as a functional in H and as a functional in E. Moreover, since $f_\alpha(\xi, \cdot) \in C^h(\mathbb{R}^n)$ and its derivatives $D^j_\sigma f_\alpha(\xi, \sigma)$ are continuous, it is possible to prove that $F_\alpha \in C^r(E; E)$ and the j-th Fréchet derivative of F_α, as a mapping from E to $\mathcal{L}^j(E)$, is given by

$$D^j F_\alpha(x)(y_1, \ldots, y_j)(\xi) = D^j_\sigma f_\alpha(\xi, x(\xi)) y_1(\xi) \cdots y_j(\xi) + \delta_{j1}(\rho + 1) y_1(\xi), \quad \xi \in \mathcal{O}.$$

Due to (6.1.31) this implies that there exists a constant c, which is independent of α, such that for $j = 0, \ldots, r$

$$|D^j F_\alpha(x)|_{\mathcal{L}^j(E)} \leq c \left(1 + |x|^{2m+1-j}_E\right), \quad x \in E, \tag{6.1.32}$$

so that all $D^j F_\alpha$ are bounded on bounded subsets of E, uniformly with respect to α. Finally, due to (6.1.28), proceeding as for F for any $x, h \in E$ we have

$$\langle F_\alpha(x + h) - F_\alpha(x), \delta_h \rangle_E \leq c |h|_E, \quad x, h \in E, \tag{6.1.33}$$

and for any $x, h \in H$

$$\langle F_\alpha(x + h) - F_\alpha(x), h \rangle_H \leq c |h|^2_H, \tag{6.1.34}$$

for some constant c independent of α.

Concerning the differentiability in H, F_α is not Fréchet differentiable, as f_α is not linear. Nevertheless, for any $j \leq r$, there exists the j-th directional derivative of F_α at any point $x \in H$ and along any directions $y_i \in L^{p_i}(\mathcal{O}; \mathbb{R}^n)$, with $p_1^{-1} + \ldots + p_j^{-1} = 1/2$ and it is given by

$$D^j_G F_\alpha(x)(y_1, \ldots, y_j)(\xi) = D^j_\sigma f_\alpha(\xi, x(\xi)) y_1(\xi) \cdots y_j(\xi), \quad \xi \in \overline{\mathcal{O}},$$

(for more details see the section 4.2). It is useful to remark that if $x, y_1, \ldots, y_j \in E$ then

$$D^j_G F_\alpha(x)(y_1, \ldots, y_j) = D^j F_\alpha(x)(y_1, \ldots, y_j).$$

Finally, thanks to (6.1.30), for any $R > 0$ and $j \leq r$ we have

$$\begin{cases} \displaystyle \lim_{\alpha \to 0} \sup_{|x|_E \leq R} |F_\alpha(x) - F(x)|_E = 0 \\[2mm] \displaystyle \lim_{\alpha \to 0} \sup_{\substack{|x|_E \leq R \\ |y_1|_E, \ldots, |y_j|_E \leq R}} |D^j F_\alpha(x)(y_1, \ldots, y_j) - D^j F(x)(y_1, \ldots, y_j)|_E = 0. \end{cases} \tag{6.1.35}$$

6.1.3 Functional spaces

As the Banach space E is continuously embedded into H, then $B_b(H)$ and $C^r_b(H)$ are continuously embedded respectively into $B_b(E)$ and $C^r_b(E)$. On the other hand the following result holds.

Proposition 6.1.5. *For any $\varphi \in C_b(E)$ there exists a sequence $\{\varphi_k\} \subset C_b(H)$ such that*

$$\begin{cases} \lim_{k \to +\infty} \varphi_k(x) = \varphi(x), & x \in E \\[2mm] \sup_{k \in \mathbb{N}} \|\varphi_k\|_0^H \leq \|\varphi\|_0^E. \end{cases} \qquad (6.1.36)$$

Proof. By using standard arguments of reflection, if the open set $\mathcal{O}$ is regular enough, it is possible to prove that there exists a bounded linear extension operator $P : L^2(\mathcal{O}; \mathbb{R}^n) \to L^2(\mathbb{R}^d; \mathbb{R}^n)$ such that for any $x \in E$ the function $Px \in E$. For any $x \in H$ and $k \in \mathbb{N}$ we define

$$x_k(\xi) = \frac{1}{|B_k(\xi)|} \int_{B_k(\xi)} Px(\zeta) \, d\zeta, \quad \xi \in \overline{\mathcal{O}},$$

where $B_k(\xi)$ is the ball of $\mathbb{R}^d$ of radius $1/k$ and centre ξ. Our aim is to show that if $\varphi \in C_b(E)$, then the function φ_k defined for any $k \in \mathbb{N}$ by

$$\varphi_k(x) = \varphi(x_k), \quad x \in H$$

belongs to $C_b(H)$ and satisfies (6.1.36). It is well known that $x_k \in E$, then φ_k is well defined. Moreover, for any $x \in H$ and $\xi \in \overline{\mathcal{O}}$,

$$|x_k(\xi)| \leq \frac{1}{|B_k(\xi)|} \int_{B_k(\xi)} |Px(\zeta)| \, d\zeta \leq c \, k^{d/2} |x|_H,$$

so that for any $x, y \in H$ we have

$$|x_k - y_k|_E = |(x - y)_k|_E \leq c \, k^{d/2} \, |x - y|_H. \qquad (6.1.37)$$

Then, as $\varphi \in C_b(E)$, we can conclude from (6.1.37) that φ_k is uniformly continuous on H. Finally, from the definition of φ_k we have

$$\|\varphi_k\|_0^H = \sup_{x \in H} |\varphi_k(x)| = \sup_{x \in H} |\varphi(x_k)| \leq \sup_{x \in E} |\varphi(x)| = \|\varphi\|_0^E.$$

Thus, $\varphi_k \in C_b(H)$ and $\|\varphi_k\|_0^H \leq \|\varphi\|_0^E$, for any $k \in \mathbb{N}$.

To conclude the proof we have to show that for any $x \in E$, $\varphi_n(x)$ converges to $\varphi(x)$, as k goes to infinity. Actually, since $\varphi \in C_b(E)$, this easily follows if we notice that x_k converges to x in E as k goes to infinity, for any $x \in E$. $\qquad \square$

6.2 Some a priori estimates for the solution

As in the section 4.2, for any $T > 0$ and $p \geq 1$, we denote by $\mathcal{H}_p(T, E)$ the Banach space of adapted processes in $C((0, T]; L^p(\Omega; E)) \cap L^\infty(0, T; L^p(\Omega; E))$, endowed with the norm

$$\|u\|_{\mathcal{H}_p(T, E)}^p = \sup_{t \in [0, T]} \mathbb{E} |u(t)|_E^p$$

(here we have taken the constant $\rho = 0$). $C_p(T, E)$ denotes the subspace of adapted processes $L^p(\Omega; C((0, T]; E) \cap L^\infty(0, T; E))$ endowed with the norm

$$\|u\|^p_{C_p(T,E)} = \mathbb{E} \sup_{t \in [0,T]} |u(t)|^p_E.$$

Clearly $C_p(T, E)$ is continuously embedded in $\mathcal{H}_p(T, E)$.

As in the chapter 4, we rewrite the equation (6.0.1) as the following stochastic differential equation in E

$$du(t) = (Au(t) + F(u(t)))\, dt + Q\, dw(t), \quad u(0) = x. \tag{6.2.1}$$

Definition 6.2.1. *An E-valued adapted process $u(t; x)$ is a mild solution of the problem (6.2.1) if*

$$u(t; x) = e^{tA}x + \int_0^t e^{(t-s)A} F(u(s; x))\, ds + w^A(t).$$

Proposition 6.2.2. *Under the Hypotheses 6.1 and 6.2, for any $x \in E$, the problem (6.2.1) has a unique mild solution $u(\cdot; x) \in C_p(T, E)$, for any $T > 0$ and $p \geq 1$. Moreover,*

$$|u(t; x)|_E \leq e^{ct}|x|_E + h(t), \quad \mathbb{P} - a.s. \tag{6.2.2}$$

where $h(t)$ is the process defined by

$$h(t) = c\, e^{ct} \int_0^t \left(1 + |w^A(s)|^{2m+1}_E\right) ds + \sup_{s \in [0,t]} |w^A(s)|_E, \tag{6.2.3}$$

and c is some positive constant.

Proof. The proof of the existence and uniqueness of a mild solution $u(\cdot; x)$ is well known and can be found in [47, Theorem 7.13]. Thus, we only prove the estimate (6.2.2).

We first remark that if we define

$$v(t; x) = u(t; x) - w^A(t), \quad t \geq 0,$$

then $v(t; x)$ is the unique mild solution of the problem

$$\frac{dv}{dt}(t) = Av(t) + F(v(t) + w^A(t)), \quad v(0) = x.$$

For any $\lambda \in \mathbb{N}$ we introduce the problem

$$\frac{dv}{dt}(t) = Av(t) + F(v(t) + w^A(t)), \quad v(0) = \lambda(\lambda - A)^{-1}x. \tag{6.2.4}$$

By using the same arguments used in [47, Theorem 7.13], the problem (6.2.4) has a unique solution $v_\lambda(t; x)$ which belongs to $C([0, +\infty[; E)$, $\mathbb{P}$-a.s. Now we fix $T > 0$ and we define the function

$$f_\lambda(t; x) = F(v_\lambda(t; x) + w^A(t)), \quad t \in [0, T], \quad x \in H.$$

As $f_\lambda(\cdot; x) \in C([0, T]; E)$ and $v_\lambda(0; x) = \lambda(\lambda - A)^{-1}x \in D(A)$, then v_λ is a strong solution of the problem (6.2.4). This means that there exists a sequence $\{v_{\lambda,n}\} \subset C^1([0, T]; E) \cap C([0, T]; D(A))$ such that

$$v_{\lambda,n} \to v_\lambda, \quad \frac{dv_{\lambda,n}}{dt} - Av_{\lambda,n} = f_{\lambda,n} \to f_\lambda, \quad \text{in } C([0, T]; E) \qquad (6.2.5)$$

as n goes to infinity (for a proof see [91, Proposition 4.1.8]).

Now, according to the Proposition A.1.3, for any $t \in [0, T]$ we have

$$\frac{d^-}{dt} |v_{\lambda,n}(t)|_E \leq \langle Av_{\lambda,n}(t), \delta_{\lambda,n}(t)\rangle_E$$

$$+ \left\langle F(v_{\lambda,n}(t) + w^A(t)) - F(w^A(t)), \delta_{\lambda,n}(t)\right\rangle_E + \left\langle F(w^A(t)), \delta_{\lambda,n}(t)\right\rangle_E \qquad (6.2.6)$$

$$+ \left\langle f_\lambda(t) - F(v_{\lambda,n}(t) + w^A(t)), \delta_{\lambda,n}(t)\right\rangle_E + \left\langle f_{\lambda,n}(t) - f_\lambda(t), \delta_{\lambda,n}(t)\right\rangle_E,$$

where $\delta_{\lambda,n}(t) = \delta_{v_{\lambda,n}(t)}$ is defined as in (A.1.1). Then, since e^{tA} is a contraction semigroup and (6.1.19) and (6.1.22) hold, we get

$$\frac{d^-}{dt} |v_{\lambda,n}(t)|_E \leq c |v_{\lambda,n}(t)|_E + |F(w^A(t))|_E + |f_\lambda(t) - F(v_{\lambda,n}(t) + w^A(t))|_E$$

$$+ |f_{\lambda,n}(t) - f_\lambda(t)|_E \leq c |v_{\lambda,n}(t)|_E + c (1 + |w^A(t)|_E^{2m+1})$$

$$+ \|f_\lambda - F(v_{\lambda,n} + w^A)\|_{C([0,t];E)} + \|f_{\lambda,n} - f_\lambda\|_{C([0,t];E)}.$$

This implies that

$$|v_{\lambda,n}(t)|_E \leq e^{ct} |v_{\lambda,n}(0)|_E + c \int_0^t e^{c(t-s)} (1 + |w^A(s)|_E^{2m+1})\, ds$$

$$+ e^{ct} t \left(\|f_\lambda - F(v_{\lambda,n} + w^A)\|_{C([0,t];E)} + \|f_{\lambda,n} - f_\lambda\|_{C([0,t];E)} \right).$$

By taking the limit as n goes to infinity, since F is continuous, (6.2.5) holds and $|\lambda(\lambda - A)^{-1}x|_E \leq |x|_E$, we have

$$|v_\lambda(t; x)|_E \leq e^{ct} |x|_E + c e^{ct} \int_0^t (1 + |w^A(s)|_E^{2m+1})\, ds. \qquad (6.2.7)$$

Due to (6.1.12), for any $t \geq 0$ we have

$$\|w^A\|_{C([0,t];E)} < \infty, \qquad \mathbb{P} - \text{a.s.}$$

and then, thanks to (6.2.7), the random variable $r(x)$ defined by

$$r(x) = \sup_{\lambda \in \mathbf{N}} \|v_\lambda(x)\|_{C([0,t];E)} + \sup_{\mu \in \mathbf{N}} \|v_\mu(x)\|_{C([0,t];E)} + \|w^A\|_{C([0,t];E)}.$$

is finite $\mathbb{P}$-a.s., for any $x \in E$. Hence, since F is locally Lipschitz-continuous, there exists a positive random variable $c_{r(x)} > 0$ which is finite $\mathbb{P}$-a.s. such that

$$|F(x) - F(y)|_E \leq c_{r(x)} |x - y|_E,$$

for any $x, y \in B_E(0, r(x)) = \{ h \in E \, ; \, |h|_E \leq r(x) \}$, so that

$$|v_\lambda(t;x) - v_\mu(t;x)|_E \leq \left| \left(\lambda(\lambda - A)^{-1} - \mu(\mu - A)^{-1} \right) e^{tA} x \right|_E$$

$$+ \int_0^t \left| e^{(t-s)A} \left(F(v_\lambda(s;x) + w^A(s)) - F(v_\mu(s;x) + w^A(s)) \right) \right|_E ds$$

$$\leq \left| \left(\lambda(\lambda - A)^{-1} - \mu(\mu - A)^{-1} \right) e^{tA} x \right|_E + c_{r(x)} \int_0^t |v_\lambda(s;x) - v_\mu(s;x)|_E \, ds.$$

This implies that

$$|v_\lambda(t;x) - v_\mu(t;x)|_E \leq e^{c_{r(x)}t} \left| \left(\lambda(\lambda - A)^{-1} - \mu(\mu - A)^{-1} \right) e^{tA} x \right|_E, \quad \mathbb{P} - \text{a.s.}$$

Therefore, if $t > 0$, it follows that the $\{v_\lambda(t;x)\}$ is a Cauchy sequence in E, so that it converges $\mathbb{P}$-a.s. in E to a process which is immediate to check that coincides with $v(t;x)$. Moreover, thanks to (6.2.7)

$$|v(t;x)|_E \leq e^{ct} |x|_E + c e^{ct} \int_0^t \left(1 + |w^A(s)|_E^{2m+1} \right) ds, \quad \mathbb{P} - \text{a.s.} \qquad (6.2.8)$$

and then, recalling that $u(t;x) = v(t;x) + w^A(t)$,

$$|u(t;x)|_E \leq e^{ct} |x|_E + c e^{ct} \int_0^t \left(1 + |w^A(s)|_E^{2m+1} \right) ds + \sup_{s \in [0,t]} |w^A(s)|_E.$$

$\square$

By using the stronger dissipativity condition for f described in the Hypothesis 6.3, we get a stronger estimate for $|u(t;x)|_E$, uniform with respect to $x \in E$. Such an estimate is one of the key points in the proof of all results which are following.

Theorem 6.2.3. *Let $u(\cdot; x)$ be the unique mild solution of (6.2.1). Then, under the Hypotheses 6.1, 6.2 and 6.3, for any $t > 0$ we have*

$$\sup_{x \in E} |u(t; x)|_E \leq k(t) t^{-1/2m}, \quad \mathbb{P} - a.s. \tag{6.2.9}$$

where the random variable $k(t)$ is defined by

$$k(t) = c \left(1 + \sup_{s \in [0,t]} |w^A(s)|_E^{\frac{\gamma}{2m+1} \vee 1} \right). \tag{6.2.10}$$

Proof. Let $\{v_{\lambda,n}\} \subset C^1([0,T]; E) \cap C([0,T]; D(A))$ be the approximating sequence of strict solutions introduced in the proof of the Proposition 6.2.2. Thanks to (6.1.25) and (6.2.6) we have

$$\frac{d^-}{dt} |v_{\lambda,n}(t)|_E \leq -a|v_{\lambda,n}(t)|_E^{2m+1} + c \left(|w^A(t)|_E^{\gamma} + 1 \right) + \left| F(w^A(t)) \right|_E$$

$$+ \left| f_\lambda(t) - F(v_{\lambda,n}(t) + w^A(t)) \right|_E + |f_{\lambda,n}(t) - f_\lambda(t)|_E,$$

so that

$$\frac{d^-}{dt} |v_{\lambda,n}(t)|_E \leq -a|v_{\lambda,n}(t)|_E^{2m+1} + c \left(\sup_{s \in [0,T]} |w^A(s)|_E^{\gamma \vee (2m+1)} + 1 \right)$$

$$+ \|f_\lambda - F(v_{\lambda,n} + w^A)\|_{C([0,t];E)} + \|f_{\lambda,n} - f_\lambda\|_{C([0,t];E)}.$$

Then, if we proceed as in the proof of the Proposition 1.2.7, by using the Lemma 1.2.6 and a comparison argument, for any $t > 0$ we get

$$|v_{\lambda,n}(t)|_E \leq c \left(1 + \sup_{s \in [0,T]} |w^A(s)|_E^{\frac{\gamma \vee (2m+1)}{2m+1}} + \gamma^{\lambda,n}(t) \right) t^{-\frac{1}{2m+1}}, \quad \mathbb{P} - \text{a.s.}$$

where

$$\gamma_{\lambda,n}(t) = \left(\|f_\lambda - F(v_{\lambda,n} + w^A)\|_{C([0,t];E)} + \|f_{\lambda,n} - f_\lambda\|_{C([0,t];E)} \right)^{1/(2m+1)}.$$

Therefore, by taking the limit as n goes to infinity, for any $t > 0$ we have that

$$|v_\lambda(t)|_E \leq c \left(1 + \sup_{s \in [0,t]} |w^A(s)|_E^{\frac{\gamma}{2m+1} \vee 1} \right) t^{-1/2m}, \quad \mathbb{P} - \text{a.s.}$$

and since $v_\lambda(t; x)$ converges $\mathbb{P}$-a.s. in E to $v(t; x)$, as λ goes to infinity, the same inequality holds for $|v(t; x)|_E$. This implies that for any $t > 0$

$$\sup_{x \in E} |u(t; x)|_E \leq \sup_{x \in E} |v(t; x)|_E + |w^A(t)|_E \leq k(t) t^{-1/2m}, \quad \mathbb{P} - \text{a.s.}$$

with $k(t)$ defined as in (6.2.10). $\qquad\square$

Remark 6.2.4. It is important to notice that, due to (6.1.12), for any $p \geq 1$ and $T > 0$

$$\mathbb{E} \sup_{t \in [0,T]} |k(t)|^p < +\infty. \tag{6.2.11}$$

6.2.1 The approximating problem

For any $\alpha > 0$ we consider the approximating problem

$$du(t) = [Au(t) + F_\alpha(u(t))] \, dt + Q dw(t), \quad u(0) = x, \tag{6.2.12}$$

where F_α is the approximating Nemytskii operator introduced in the subsection 6.1.2. We recall that F_α is Lipschitz-continuous in H and then the problem (6.2.12) admits a unique mild solution $u_\alpha(t; x) \in L^2(\Omega; C([0, +\infty[; H))$, for any $x \in H$. It is useful to note that if $x \in E$ then $u_\alpha(t; x)$ is an E-valued process, so that we can look at the problem (6.2.12) both as a problem in H and as a problem in E. Moreover, due to (6.1.33), we can adapt the proof of the Proposition 6.2.2 to this case and for each $\alpha > 0$ and $t \geq 0$ we have

$$|u_\alpha(t; x)|_E \leq e^{2ct}|x|_E + h(t), \quad \mathbb{P} - \text{a.s.} \tag{6.2.13}$$

where the random process h is the same as (6.2.3) (with c possibly different).

We prove the first approximation result.

Proposition 6.2.5. *Assume the Hypothesis 6.1 and 6.2. Then for any $R, T > 0$ we have*

$$\lim_{\alpha \to 0} \sup_{\substack{t \in [0,T] \\ |x|_E \leq R}} |u_\alpha(t; x) - u(t; x)|_E = 0, \quad \mathbb{P} - \text{a.s.} \tag{6.2.14}$$

According to (6.2.2) and (6.2.13), this means that $u_\alpha(\cdot; x)$ converges to $u(\cdot; x)$ in $C_p(T, E)$, as α goes to 0, uniformly with respect to $x \in \{y \in E : |y|_E \leq R\}$.

Proof. For any $\alpha > 0$ and $x \in E$, we set $z_\alpha(t; x) = u_\alpha(t; x) - u(t; x)$. The process $z_\alpha(t; x)$ is the unique mild solution of the problem

$$\frac{dz}{dt}(t) = Az(t) + F_\alpha(u_\alpha(t; x)) - F(u(t; x)), \quad z(0) = 0. \tag{6.2.15}$$

We can assume that $z_\alpha(t; x)$ is a strict solution of the problem (6.2.15), otherwise we proceed as in the proof of Proposition 6.2.2, approximating $z_\alpha(t; x)$ by means of a sequence of more regular processes. If $\delta_\alpha(t) = \delta_{z_\alpha(t;x)}$ is defined as in (A.1.1), due to the Proposition (A.1.3) we have

$$\frac{d^-}{dt}|z_\alpha(t)|_E \leq \langle Az_\alpha(t) + F_\alpha(u_\alpha(t; x)) - F(u(t; x)), \delta_\alpha(t) \rangle_E$$

$$\leq \langle F_\alpha(u_\alpha(t; x)) - F(u_\alpha(t; x)), \delta_\alpha(t) \rangle_E + \langle F(u_\alpha(t; x)) - F(u(t; x)), \delta_\alpha(t) \rangle_E$$

$$\leq |F_\alpha(u_\alpha(t; x)) - F(u_\alpha(t; x))|_E + c|z_\alpha(t)|_E,$$

and by the Gronwall lemma

$$|z_\alpha(t;x)|_E \leq \int_0^t e^{c(t-s)} |F_\alpha(u_\alpha(s;x)) - F(u_\alpha(s;x))|_E \, ds, \quad \mathbb{P} - \text{a.s.}$$

Due to (6.2.2) and (6.2.13), by using the first limit in (6.1.35) it follows

$$\lim_{\alpha \to 0} \sup_{\substack{t \in [0,T] \\ |x|_E \leq R}} |F_\alpha(u_\alpha(t;x)) - F(u_\alpha(t;x))|_E = 0, \quad \mathbb{P} - \text{a.s.}$$

and this implies (6.2.14). $\qquad\qquad\qquad\qquad\qquad\qquad\qquad\qquad\qquad\qquad$ $\square$

6.3 Differential dependence on initial data

For any $\alpha > 0$, $F_\alpha : E \to E$ is Lipschitz-continuous and r-times Fréchet differentiable. Then, by a standard contraction argument (see the appendix C), for any $p \geq 1$ the solution $u_\alpha(\cdot; x)$ of the approximating problem (6.2.12) is r times differentiable in E in the mean of order p with respect to $x \in E$. Moreover, for all $j = 1, \ldots, r$ the derivatives $D_x^j u_\alpha(\cdot; x)(h_1, \ldots, h_j)$ are solutions of the corresponding variation equations. Our aim is to prove that also the solution of the problem (6.2.1) is differentiable with respect to $x \in E$ and its derivatives are the limit in $C_p(T, E)$ of the derivatives of $u_\alpha(\cdot; x)$.

The first derivative $D_x u_\alpha(\cdot; x)h$ at the point $x \in E$ and along the direction $h \in E$ is the unique mild solution of the problem

$$\frac{d}{dt}v(t) = Av(t) + DF_\alpha(u_\alpha(t; x))v(t), \quad v(0) = h.$$

Then $D_x u_\alpha(\cdot; x)h \in L^p(\Omega; C((0, +\infty); E) \cap L^\infty_{\text{loc}}([0, +\infty); E))$, for any $p \geq 1$, and proceeding as in the proof of the Proposition 6.2.2, from (6.1.33) it follows that for any $\alpha > 0$

$$\sup_{x \in E} |D_x u_\alpha(t; x)h|_E \leq e^{ct}|h|_E, \quad \mathbb{P} - \text{a.s.} \tag{6.3.1}$$

Now, for any $x, h \in H$, we consider the first variation equation associated with the problem (6.2.1)

$$\frac{dv}{dt}(t) = Av(t) + DF(u(t; x))v(t), \quad v(0) = h. \tag{6.3.2}$$

According to (6.1.23), for any $x, y \in E$ we have

$$\sup_{s \in [0,t]} \langle DF(u(s; x))y, \delta_y \rangle_E \leq c(t) |y|_E, \quad \mathbb{P} - \text{a.s.}$$

where $c(t)$ is a suitable increasing function. Then it is easy to show that (6.3.2) has a unique mild solution in $L^p(\Omega; C((0, +\infty); E) \cap L^\infty_{\text{loc}}([0, +\infty); E))$, for any $p \geq 1$, which we denote by $v_1(t; x, h)$. Moreover, with the same arguments used before for $D_x u_\alpha(t; x)h$, we have

$$\sup_{x \in E} |v_1(t; x, h)|_E \le e^{ct} |h|_E, \quad \mathbb{P} - \text{a.s.} \tag{6.3.3}$$

Concerning the second derivative $D_x^2 u_\alpha(t; x)(h_1, h_2)$ of $u_\alpha(t; x)$, it is the unique mild solution of the problem

$$\begin{cases} \dfrac{dv}{dt}(t) = Av(t) + DF_\alpha(u_\alpha(t; x))v(t) \\[2mm] \qquad\qquad + D^2 F_\alpha(u_\alpha(t; x)) \left(D_x u_\alpha(t; x)h_1, D_x u_\alpha(t; x)h_2 \right) \\[2mm] v(0) = 0. \end{cases}$$

Hence, due to (6.1.32), (6.1.33), (6.2.13) and (6.3.1), we have that $D_x^2 u_\alpha(x)(h_1, h_2)$ belongs to $L^2(\Omega; C((0, +\infty); E))$ and for any $R, T > 0$ and $p \ge 1$

$$\sup_{\substack{t \in [0,T] \\ |x|_E \le R}} \left| D_x^2 u_\alpha(t; x)(h_1, h_2) \right|_E^p \le c |h_1|_E^p |h_2|_E^p, \quad \mathbb{P} - \text{a.s.} \tag{6.3.4}$$

where $c = c(R, T, p)$ is a random variable independent of α, having all moments finite.

Now, for $x, h_1, h_2 \in E$, we consider the second derivative equation

$$\begin{cases} \dfrac{dv}{dt}(t) = Av(t) + DF(u(t; x))v(t) + D^2 F(u(t; x)) \left(v_1(t; x, h_1), v_1(t; x, h_2) \right) \\[2mm] v(0) = 0. \end{cases}$$

From the estimates (6.1.19) and (6.1.22) and from the estimates (6.2.2) and (6.3.3) on $|u(t; x)|_E$ and $|v_1(t; x, h)|_E$ respectively, it is easy to verify that this problem has a unique mild solution $v_2(t; x, h_1, h_2)$ in $L^2(\Omega; C([0, +\infty[; E))$. Proceeding as for the second derivative $D_x^2 u_\alpha(t; x)(h_1, h_2)$, it is possible to show that for $v_2(t; x, h_1, h_2)$ an estimate analogous to (6.3.4) holds.

In a similar way, for any $j \le r$ and $x, h_1, \ldots, h_j \in E$, we will denote by $D_x^j u_\alpha(\cdot; x)(h_1, \ldots, h_j)$ the j-th derivative of $u_\alpha(\cdot; x)$. By recurrence it is possible to show that for any $R, T > 0$ and $p \ge 1$

$$\sup_{\substack{t \in [0,T] \\ |x|_E \le R}} \left| D_x^j u_\alpha(t; x)(h_1, \ldots, h_j) \right|_E^p \le c(T) \prod_{i=1}^{j} |h_i|_E^p, \quad \mathbb{P} - \text{a.s} \tag{6.3.5}$$

where $c(T) = c(j, R, T, p)$ is an increasing random process having all moments finite. Clearly, we can repeat the same arguments for the solution of the j-th derivative equation, with $j \leq r$, and we get

$$\sup_{\substack{t \in [0,T] \\ |x|_E \leq R}} |v_j(t; x, h_1, \ldots, h_j)|_E^p \leq c(T) \prod_{i=1}^{j} |h_i|_E^p, \quad \mathbb{P} - \text{a.s.} \tag{6.3.6}$$

where $c(T) = c(j, R, T, p)$ is an increasing random process having all moments finite.

Proposition 6.3.1. *Under the Hypotheses 6.1, 6.2 and 6.3, for any $j \leq r$, $R, T > 0$ and $p \geq 1$,*

$$\lim_{\alpha \to 0} D_x^j u_\alpha(t; x)(h_1, \ldots, h_j) = v_j(t; x, h_1, \ldots, h_j) \tag{6.3.7}$$

in $C_p(T, E)$, uniformly with respect to $x, h_i \in B_E(0, R)$, $i \leq j$.

Proof. Let us fix $x, h \in E$ and define $v_\alpha(t; x, h) = D_x u_\alpha(t; x)h - v_1(t; x, h)$. We have only to prove that

$$\lim_{\alpha \to 0} \sup_{\substack{t \in [0,T] \\ |x|_E, |h_i|_E \leq R}} |v_\alpha(t; x, h)|_E = 0, \quad \mathbb{P} - \text{a.s.} \tag{6.3.8}$$

Actually, if this is true, thanks to (6.3.5) and (6.3.6) we can use the dominated convergence theorem and (6.3.7) follows. We have

$$v_\alpha(t) = \int_0^t e^{(t-s)A} \left(DF_\alpha(u_\alpha(s; x)) D_x u_\alpha(s; x)h - DF(u(s; x)) v_1(s; x, h) \right) ds,$$

so that

$$|v_\alpha(t)|_E \leq \int_0^t |DF_\alpha(u_\alpha(s; x)) D_x u_\alpha(s; x)h - DF(u(s; x)) v_1(s; x, h)|_E \, ds.$$

This yields

$$|v_\alpha(t)|_E \leq \int_0^t |(DF_\alpha(u_\alpha(s; x)) - DF(u_\alpha(s; x))) D_x u_\alpha(s; x)h|_E \, ds$$

$$+ \int_0^t |(DF(u_\alpha(s; x)) - DF(u(s; x))) D_x u_\alpha(s; x)h|_E \, ds + \int_0^t |DF(u(s; x)) v_\alpha(s)|_E \, ds.$$

Due to (6.1.19), with $j = 1$, and (6.2.2), for any $y \in E$ we have

$$\sup_{s \in [0,t]} |DF(u(s; x))y|_E \leq c(t, x)|y|_E, \quad \mathbb{P} - \text{a.s.},$$

where $c(t; x)$ is a suitable random variable having finite moments of any order and increasing with respect to t. Then, for any $t \in [0, T]$ we get

$$|v_\alpha(t)|_E \leq \gamma_\alpha(t)(x, h) + c(t, x) \int_0^t |v_\alpha(s)|_E \, ds, \quad \mathbb{P} - \text{a.s.}$$

where

$$\gamma_\alpha(t)(x,h) = \int_0^t |(DF_\alpha(u_\alpha(s;x)) - DF(u_\alpha(s;x))) D_x u_\alpha(s;x)h|_E \, ds$$

$$+ \int_0^t |(DF(u_\alpha(s;x)) - DF(u(s;x))) D_x u_\alpha(s;x)h|_E \, ds.$$

Hence, due to the Gronwall lemma we have

$$|v_\alpha(t)|_E \leq \int_0^t \exp\left(\int_s^t c(r,x)\,dr\right) \sup_{|x|_E, |h|_E \leq R} \gamma_\alpha(s)(x,h)\,ds$$

$$\leq t e^{t\,c(t,x)} \sup_{\substack{s \in [0,t] \\ |x|_E, |h|_E \leq R}} \gamma_\alpha(s)(x,h), \qquad \mathbb{P} - \text{a.s.}$$

Now, according to (6.2.13) and to the second limit in (6.1.35), to (6.2.14) and to the continuity of DF, by using (6.3.1) it is not difficult to show that

$$\lim_{\alpha \to 0} \sup_{\substack{t \in [0,T] \\ |x|_E, |h|_E \leq R}} \gamma_\alpha(t)(x,h) = 0, \qquad \mathbb{P} - \text{a.s.}$$

This implies (6.3.8) and hence (6.3.7).

The proof for $j \geq 2$ is analogous. □

Remark 6.3.2. Since $f(\xi, \cdot)$ is assumed to be r-times differentiable, with the r-th derivative continuous, due to the Proposition 6.3.1 it is not difficult to verify that for any fixed $k_1, \ldots, k_r \in E$ the mapping

$$E \to C_p(T, E), \qquad x \mapsto v_r(\cdot; x, k_1, \ldots, k_r)$$

is continuous. The same conclusion holds if we replace v_r with $D_x^r u_\alpha$, for any $\alpha > 0$.

An important consequence of the previous convergence result is the differentiability of $u(\cdot; x)$ with respect to x.

Theorem 6.3.3. *If the Hypotheses 6.1, 6.2 and 6.3 hold, then for any $T > 0$ and $p \geq 1$, the solution $u(\cdot; x)$ of the equation (6.2.1) is r-times differentiable in the mean of order p with respect to x and, for any $h_1, \ldots, h_h \in E$ and $q \geq 1$,*

$$\sup_{\substack{t \in [0,T] \\ x \in E}} \left| D_x^j u(t;x)(h_1, \ldots, h_j) \right|_E^q \leq c(T) \prod_{i=1}^{j} |h_i|_E^q, \qquad \mathbb{P} - \text{a.s.} \qquad (6.3.9)$$

for an increasing random process $c(T) = c(q, j, T)$, $j = 1, \ldots, r$, having finite moments of any order.

Proof. Since $u_\alpha(\cdot; x)$ is differentiable in the mean of order p, then for any $\alpha > 0$ we have

$$u_\alpha(\cdot; x + h) - u_\alpha(\cdot; x) = D_x u_\alpha(\cdot; x)h + \int_0^1 \int_0^1 D_x^2 u_\alpha(\cdot; x + \rho\theta h)(h, h)\, d\rho\, d\theta.$$

According to the Propositions 6.2.5 and 6.3.1, we can take the limit in $\mathcal{H}_p(T, E)$ as $\alpha \to 0$ and we get

$$u(\cdot; x + h) - u(\cdot; x) = v_1(\cdot; x, h) + \int_0^1 \int_0^1 v_2(\cdot; x + \rho\theta h, h, h)\, d\rho\, d\theta.$$

Due to (6.3.3), the mapping

$$v_1(\cdot; x) : E \to \mathcal{H}_p(T, E), \quad h \mapsto v_1(x, h),$$

is clearly linear and continuous. Then, thanks to (6.3.6) we have

$$\lim_{|h|_E \to 0} \frac{1}{|h|_E} \left| \int_0^1 \int_0^1 v_2(\cdot; x + \rho\theta h, h, h)\, d\rho\, d\theta \right|_E = 0, \quad \text{in } \mathcal{H}_p(T, E),$$

so that $u(\cdot; x)$ is differentiable in the mean of order p and its derivative coincides with $v_1(\cdot; x, h)$. The estimate (6.3.9) for $j = 1$ has just been proved.

Second and higher order differentiability can be proved by using the same arguments used before. We remark that the existence of the r–th derivative follows from the Remark 6.3.2. Indeed, for any $t \geq 0$ and $x, k, k_1, \ldots, k_{r-1} \in E$ we have

$$D^{r-1} u_\alpha(t; x + k)(k_1, \ldots, k_{r-1}) - D^{r-1} u_\alpha(t; x)(k_1, \ldots, k_{r-1})$$

$$= D^r u_\alpha(t; x)(k_1, \ldots, k_{r-1}, k) + R_\alpha(t; x),$$

where

$$R_\alpha(t; x) = \int_0^1 (D^r u_\alpha(t; x + \theta k) - D^r u_\alpha(t; x))(k_1, \ldots, k_{r-1}, k)\, d\theta.$$

By taking the limit in $\mathcal{H}_p(T, E)$, as α goes to 0, we get

$$v_{r-1}(t; x + k, k_1, \ldots, k_{r-1}) - v_{r-1}(t; x, k_1, \ldots, k_{r-1})$$

$$= v_r(t; x, k_1, \ldots, k_{r-1}, k) + R(t; x)$$

and then, as the mapping $x \mapsto v_r(\cdot; x, k_1, \ldots, k_{r-1}, k)$ is continuous and (6.3.6) holds, due to the dominated convergence theorem we have that

$$\lim_{|k|_E \to 0} \frac{|R(\cdot; x)|}{|k|_E} = 0, \quad \text{in } \mathcal{H}_p(T, E).$$

This means that there exists the r-th derivative of $v_{r-1}(\cdot; x)$ and it coincides with $v_r(\cdot; x)$.

Now, let us prove (6.3.9) for $j = 2$. We can assume that $D_x^2 u(\cdot; x)(h_1, h_2)$ is a strict solution of the second derivative equation, otherwise we approximate it as in the proof of Proposition 6.2.2. It holds

$$\frac{d^-}{dt} |D_x^2 u(t; x)(h_1, h_2)|_E$$

$$\leq c|D_x^2 u(t; x)(h_1, h_2)|_E + \left|D^2 F(u(t; x)) \left(D_x u(t; x)h_1, D_x u(t; x)h_2\right)\right|_E$$

$$\leq c|D_x^2 u(t; x)(h_1, h_2)|_E + c\left(1 + |u(t; x)|_E^{2m-1}\right) e^{2ct}|h_1|_E|h_2|_E,$$

last inequality following from (6.1.19) and (6.3.9) for $j = 1$. Therefore, by using (6.2.9), due to the Gronwall lemma, for any $t > 0$ we have

$$\sup_{x \in E} |D_x^2 u(t; x)(h_1, h_2)|_E \leq \sup_{x \in E} ce^{ct} \int_0^t \left(1 + |u(s; x)|_E^{2m-1}\right) ds|h_1|_E|h_2|_E$$

$$\leq ce^{ct} \int_0^t \left(1 + k(s)^{2m-1}s^{-1+1/2m}\right) ds|h_1|_E|h_2|_E$$

$$\leq ce^{ct} \left(t + k(t)^{2m-1}t^{1/2m}\right) |h_1|_E|h_2|_E.$$

The result now follows from (6.2.11).

In order to prove (6.3.9) for general $j \leq r$, we proceed by recurrence, by using the estimates (6.3.9) for $i < j$. Actually, we have

$$\frac{d^-}{dt} |D_x^j u(t; x)(h_1, \ldots, h_j)|_E$$

$$\leq c|D_x^j u(t; x)(h_1, \ldots, h_j)|_E + c(t)\left(1 + |u(t; x)|_E^{2m+1-j}\right) \prod_{i=1}^{j} |h_i|_E$$

and then, by using (6.2.9) and (6.2.11), the inequality (6.3.9) is obtained for any $j \leq h$. $\qquad \square$

6.4 Further properties of the derivatives of the solution

Our aim is now proving that the derivatives of $u(\cdot; x)$ belong to the domain of Q^{-1} and satisfy suitable estimates, uniform with respect to $x \in E$.

Proposition 6.4.1. *Assume the Hypotheses 6.1, 6.2 and 6.3. Then for $t > 0$ and $j \leq r$ the processes $D_x^j u(t; x)(h_1, \ldots, h_j)$ take value in $D(Q^{-1})$ and*

$$\sup_{x \in E} \int_0^t |Q^{-1} D_x^j u(s; x)(h_1, \ldots, h_j)|_H^2 \, ds \leq c_j(t) \, t^{1-\epsilon} \prod_{i=1}^j |h_i|_E^2, \qquad \mathbb{P} - a.s. \quad (6.4.1)$$

for suitable increasing processes $c_j(t)$ having finite moments of any order.

Proof. As seen in the previous section, the first derivative $D_x u(t; x) h$ along the direction $h \in E$ is the unique mild solution of the problem

$$\frac{dv}{dt}(t) = Av(t) + DF(u(t; x))v(t), \quad v(0) = h.$$

Such an equation can be regarded as an equation in H and due to (4.1.7) it is easy to prove that for any $\delta < 1$ its mild solution $v(t) \in D((-A)^\delta)$, for any $t > 0$. Besides, possibly by approximating $v(t)$ by means of more regular solutions (see the proof of the Proposition 6.2.2), we have

$$\frac{1}{2} \frac{d}{dt} |v(t)|_H^2 + \langle (-A)v(t), v(t) \rangle_H = \langle DF(u(t; x))v(t), v(t) \rangle_H \,.$$

Hence, by integrating with respect to t and by using (6.1.20), we get

$$|v(t)|_H^2 + 2 \int_0^t |(-A)^{1/2} v(s)|_H^2 \, ds \leq |h|_H^2 + 2c \int_0^t |v(s)|_H^2 \, ds.$$

Due to (6.3.9), this yields

$$\int_0^t |(-A)^{1/2} v(s)|_H^2 \, ds \leq c(t) |h|_E^2, \qquad \mathbb{P} - a.s. \quad (6.4.2)$$

Now, by using the interpolatory inequality (4.1.8) (with $\delta_1 = 0$, $\delta = \epsilon/2$ and $\delta_2 = 1/2$) and by the Hölder inequality, we have

$$\int_0^t |(-A)^{\epsilon/2} v(s)|_H^2 \, ds \leq \int_0^t |(-A)^{1/2} v(s)|_H^{2\epsilon} |v(s)|_H^{2(1-\epsilon)} \, ds$$

$$\leq \left(\int_0^t |(-A)^{1/2} v(s)|_H^2 \, ds \right)^\epsilon \left(\int_0^t |v(s)|_H^2 \, ds \right)^{1-\epsilon},$$

so that, thanks to (6.3.9) and (6.4.2),

$$\int_0^t |(-A)^{\epsilon/2} D_x u(s; x)h|_H^2 \, ds \leq c(t) e^{2(1-\epsilon)ct} t^{1-\epsilon} |h|_E^2, \qquad \mathbb{P} - a.s.$$

Finally, recalling that due to the Hypothesis 6.1-3 we have $Q^{-1} = \Gamma_\epsilon(-A)^{\epsilon/2}$, for a bounded linear operator Γ_ϵ, it follows that $D_x u(t; x)h \in D(Q^{-1})$ for any $t > 0$ and

$$\sup_{x \in E} \int_0^t |Q^{-1} D_x u(s; x)h|_H^2 \, ds$$

$$\leq \|\Gamma_\epsilon\|^2 \sup_{x \in E} \int_0^t |(-A)^{\epsilon/2} D_x u(s; x)h|_H^2 \, ds \leq c(t) t^{1-\epsilon} |h|_E^2.$$

As proved in the previous section, the j-th derivative $D_x^j u(t; x)(h_1, \ldots, h_j)$, with $j \geq 2$, is the unique mild solution of the problem

$$\frac{dv}{dt}(t) = Av(t) + DF(u(t; x))v(t) + \theta_j(t), \quad v(0) = 0,$$

where $\theta_j(t)$ is a process depending on all lower order derivatives. Then, as before for the first derivative, the solution $v_j(t)$ belongs to $D((-A)^{1/2})$, for $t > 0$, and

$$|v_j(t)|_H^2 + 2 \int_0^t |(-A)^{1/2} v_j(s)|_H^2 \, ds \leq 2c \int_0^t |v_j(s)|_H^2 \, ds + 2 \int_0^t |\langle \theta_j(s), v_j(s) \rangle|_H \, ds.$$

By using (6.3.9) and the Hölder inequality, we can show that this implies that

$$\sup_{x \in E} \int_0^t |(-A)^{1/2} v_j(s)|_H^2 \, ds \leq c_j(t) \prod_{i=1}^j |h_i|_E^2, \quad \mathbb{P} - \text{a.s.} \qquad (6.4.3)$$

for some increasing process $c_j(t)$ such that $\mathbb{E}|c_j(t)|^p < \infty$, for any $p \geq 1$. We prove (6.4.3) for $j = 2$, as the proof for $j > 2$ is identical. We have

$$\theta_2(t) = D^2 F(u(t; x)) (D_x u(t; x)h_1, D_x u(t; x)h_2)$$

and then, by (6.2.9) and (6.3.9)

$$\int_0^t |(-A)^{1/2} v_2(s)|_H^2 \, ds$$

$$\leq 2c \int_0^t |v_2(s)|_H^2 \, ds + c \int_0^t e^{2cs} \left(1 + |u(s; x)|_E^{2m-1}\right) |v_2(s)|_H \, ds |h_1|_E |h_2|_E$$

$$\leq 2ct \sup_{s \in [0,t]} |v_2(s)|_H^2 + c e^{2ct} \left(t + k(t)^{2m-1} t^{1/2m}\right) \sup_{s \in [0,t]} |v_2(s)|_H |h_1|_E |h_2|_E.$$

Therefore, by using (6.2.11) and (6.3.9) for $j = 2$, the estimate (6.4.3) easily follows. Thanks to the same interpolatory arguments used above for the first derivative, (6.4.1) follows also for $j = 2$. $\qquad \square$

Remark 6.4.2. The same arguments used in the proof of the previous proposition can be used in order to show that for any $j \leq r$ and $t > 0$ the derivatives $D_x^j u_\alpha(t; x)(h_1, \ldots, h_j)$ take value in $D(Q^{-1})$, for any $\alpha > 0$ and $x, h_1, \ldots, h_r \in E$. Furthermore for any $R > 0$ there exists an increasing process $c_R(t) > 0$ having finite moments of any order, such that

$$\sup_{|x|_E \leq R} \int_0^t |Q^{-1} D_x^j u_\alpha(s; x)(h_1, \ldots, h_j)|_H^2 \, ds \leq c_R(t) t^{1-\epsilon} \prod_{i=1}^j |h_i|_E^2, \qquad \mathbb{P}-\text{a.s.} \quad (6.4.4)$$

Proposition 6.4.3. *Under the Hypotheses 6.1, 6.2 and 6.3, for any $x, h \in E$*

$$\lim_{\alpha \to 0} \mathbb{E} \int_0^t |Q^{-1} (D_x u_\alpha(s; x) h - D_x u(s; x) h)|_H^2 \, ds = 0, \quad t \geq 0. \qquad (6.4.5)$$

Proof. If we set $z_\alpha(t) = D_x u_\alpha(s; x) h - D_x u(s; x) h$, we have

$$Q^{-1} z_\alpha(t) = \int_0^t \Gamma(t - s) \left[DF_\alpha(u_\alpha(s; x)) D_x u_\alpha(s; x) h - DF(u(s, x)) D_x u(s; x) h \right] ds,$$

where $\Gamma(t) = Q^{-1} e^{tA} = Q^{-1}(-A)^{-\epsilon/2}(-A)^{\epsilon/2} e^{tA} = \Gamma_\epsilon(-A)^{\epsilon/2} e^{tA}$. Then, according to the Hypothesis 6.1-3 we have that $\|\Gamma(t)\| \leq c\, t^{-\epsilon/2}$ and it easily follows

$$|Q^{-1} z_\alpha(t)|_H^2$$

$$\leq c\, t^{1-\epsilon} \int_0^t |DF_\alpha(u_\alpha(s; x)) D_x u_\alpha(s; x) h - DF(u(s; x)) D_x u(s; x) h|_H^2 \, ds.$$

By using the Proposition 6.3.1 and the estimates (6.2.2), (6.3.1) and (6.3.9), due to (6.1.35) it is possible to show that

$$\lim_{\alpha \to 0} \int_0^t |DF_\alpha(u_\alpha(s; x)) D_x u_\alpha(s; x) h - DF(u(s; x)) D_x u(s; x) h|_H^2 \, ds = 0,$$

$\mathbb{P}$-a.s. and then by the dominated convergence theorem (6.4.5) follows. $\square$

Proposition 6.4.4. *The process $Q^{-1} D_x^j u(t; x)(h_1, \ldots, h_j)$ is mean-square differentiable in H along any direction $h_{j+1} \in E$, for any $j < r$ and $x, h_1, \ldots, h_r \in E$, and*

$$D_x \left(Q^{-1} D_x^j u(t; x)(h_1, \ldots, h_j) \right) h_{j+1} = Q^{-1} D_x^{j+1} u(t; x)(h_1, \ldots, h_{j+1}).$$

Proof. We give a proof for $j = 1$. The case $j > 1$ is analogous.
For any $\delta > 0$ and $x, h, k \in E$ we have

$$Q^{-1} D_x u(t; x + \delta k) h - Q^{-1} D_x u(t; x) h$$

$$= \int_0^t \Gamma(t - s) \left[DF(u(s; x + \delta k)) D_x u(s; x + \delta k) h - DF(u(s, x)) D_x u(s; x) h \right] ds.$$

Now, $DF(u(s;x))D_x u(s;x)h$ is differentiable in E along any direction $k \in E$ and its derivative is given by

$$D^2 F(u(t;x))(D_x u(t;x)h, D_x u(t;x)k) + DF(u(t;x))D_x^2 u(t;x)(h,k).$$

Hence, since $\Gamma(t)$ is continuous in H for any $t > 0$ and $\|\Gamma(t)\| \le ct^{-\epsilon/2}$, we can conclude that

$$\lim_{\delta \to 0} \frac{1}{\delta} \left(Q^{-1} D_x u(t; x + \delta k)h - Q^{-1} D_x u(t; x)h \right)$$

$$= \int_0^t \Gamma(t-s) D^2 F(u(s;x))(D_x u(s;x)h, D_x u(s;x)k)\, ds$$

$$+ \int_0^t \Gamma(t-s) DF(u(s;x)) D_x^2 u(s;x)(h,k)\, ds = Q^{-1} D_x^2 u(t;x)(h,k)$$

and the limit is in $\mathcal{H}_p(T,H)$, for any $p \ge 1$.   □

Remark 6.4.5. The same statement of Proposition 6.4.3 can be proved for the derivatives of $u_\alpha(t;x)$. Moreover, since $u_\alpha(t;x)$ is at least twice differentiable in H (see chapter 4), for any $x, h \in H$ the process $Q^{-1} D_x u_\alpha(t;x)h$ is differentiable in H and for any $k \in H$

$$D_x \left(Q^{-1} D_x u_\alpha(t;x)h \right) k = Q^{-1} D_x^2 u_\alpha(t;x)(h,k).$$

6.5   Smoothing properties of the transition semigroup

For any $\varphi \in B_b(E)$ and $t \ge 0$ we define

$$P_t \varphi(x) = \mathbb{E}(\varphi(u(t;x))), \qquad x \in E.$$

P_t is the Markov transition semigroup associated with the problem (6.2.1). Due to the Theorem 6.3.3, it is easy to show that P_t maps $C_b^k(E)$ into itself, for any $k \le r$. In particular P_t is a Feller semigroup on E, that is it maps $C_b(E)$ into $C_b(E)$.

For any $x \in H$, the approximating problem (6.2.12) has a unique mild solution $u_\alpha(t;x)$ and if $x \in E$ the solution is an E-valued process. Then the associated transition semigroup P_t^α, $t \ge 0$, can be also studied in $B_b(E)$. In the chapter 4 regularity properties of P_t^α have been studied for its restriction to $B_b(H)$. It has been proved that $C_b(H)$ is an invariant subspace for P_t^α and P_t^α has a smoothing effect. That is

$$\varphi \in B_b(H) \Rightarrow P_t^\alpha \varphi \in C_b^j(H), \quad t > 0,$$

for some j depending on the ultracontractivity property of the operator A. Actually, even if $f \in C^\infty(\mathbb{R})$, we only have that $j = 5$ if $d = 1$, $j = 3$ if $d = 2$ and $j = 2$ if

$d = 3$. Our aim now is studying the smoothing properties of P_t in $B_b(E)$. To this purpose, we remark that for any $\varphi \in C_b(E)$ and $R, T > 0$

$$\lim_{\alpha \to 0} \sup_{\substack{t \in [0,T] \\ |x|_E \leq R}} |P_t^\alpha \varphi(x) - P_t \varphi(x)| = 0. \tag{6.5.1}$$

Indeed

$$P_t^\alpha \varphi(x) - P_t \varphi(x) = \mathbb{E}(\varphi(u_\alpha(t; x)) - \varphi(u(t; x))),$$

and then, as (6.2.14) holds and $\varphi \in C_b(E)$, we get (6.5.1). Therefore, in order to prove the smoothing properties of P_t we use analogous properties of P_t^α and then we proceed by approximation. We recall that in the chapter 4 we have proved that under the Hypotheses 6.1 and 6.2, when $d \leq 3$ the function $P_t^\alpha \varphi$ is at least twice differentiable, for any $\varphi \in B_b(H)$ and $t > 0$. Moreover, if $\varphi \in C_b(H)$ it holds

$$\langle D(P_t^\alpha \varphi)(x), h \rangle_H = \frac{1}{t} \mathbb{E} \varphi(u_\alpha(t; x)) \int_0^t \langle Q^{-1} D_x u_\alpha(s; x) h, dw(s) \rangle_H. \tag{6.5.2}$$

Now, if we set $\psi_\alpha = P_{t/2}^\alpha \varphi$, thanks to the Remark 6.4.5 and the semigroup law, we can differentiate each side of (6.5.2) with respect to x in H and we obtain

$$\langle D^2(P_t^\alpha \varphi)(x)k, h \rangle_H$$

$$= \frac{2}{t} \mathbb{E} \langle D\psi_\alpha(u_\alpha(t/2; x)), D_x u_\alpha(t/2; x)k \rangle_H \int_0^{t/2} \langle Q^{-1} D_x u_\alpha(s; x)h, dw(s) \rangle_H$$

$$+ \frac{2}{t} \mathbb{E} \psi_\alpha(u_\alpha(t/2; x)) \int_0^{t/2} \langle Q^{-1} D_x^2 u_\alpha(s; x)(h, k), dw(s) \rangle_H. \tag{6.5.3}$$

Theorem 6.5.1. *Assume the Hypotheses 6.1, 6.2 and 6.3. Then* $P_t \varphi \in C_b^r(E)$, *for any* $\varphi \in B_b(E)$ *and* $t > 0$, *and*

$$\|P_t \varphi\|_j^E \leq c (t \wedge 1)^{-\frac{j(1+\epsilon)}{2}} \|\varphi\|_0^E, \quad j = 1, \ldots, r. \tag{6.5.4}$$

Moreover, if $\varphi \in C_b(E)$ *for any* $x, h \in E$ *we have*

$$\langle D(P_t \varphi)(x), h \rangle_E = \frac{1}{t} \mathbb{E} \varphi(u(t; x)) \int_0^t \langle Q^{-1} D_x u(s; x)h, dw(s) \rangle_H. \tag{6.5.5}$$

Proof. Let $\varphi \in C_b(H)$. For any $\alpha > 0$ and $x, h \in E$ we have

$$P_t^\alpha \varphi(x + h) - P_t^\alpha \varphi(x)$$

$$= \langle D(P_t^\alpha \varphi)(x), h \rangle_H + \int_0^1 \langle D(P_t^\alpha \varphi)(x + \theta h) - D(P_t^\alpha \varphi)(x), h \rangle_H \, d\theta, \tag{6.5.6}$$

where $D(P_t^\alpha \varphi)$ is defined by (6.5.2). Thanks to (6.2.13), (6.2.14) and (6.4.5) it is not difficult to prove that

$$\lim_{\alpha \to 0} \langle (D(P_t^\alpha \varphi)(x), h \rangle_H = \frac{1}{t} \mathbb{E}\varphi(u(t; x)) \int_0^t \langle Q^{-1} D_x u(s; x)h, dw(s) \rangle_H,$$

and the limit is uniform with respect to x, h in $B_E(0, R)$, for any $R > 0$. Moreover, from (6.4.4) we have

$$|\langle D(P_t^\alpha \varphi)(x), h \rangle_H| \leq \frac{1}{t} \|\varphi\|_0^H \left(\mathbb{E} \int_0^t |Q^{-1} D_x u_\alpha(s; x)h|_H^2 \, ds \right)^{1/2}$$

$$\leq c_{|x|_E}(t) \|\varphi\|_0^H t^{-\frac{1+\epsilon}{2}} |h|_E. \tag{6.5.7}$$

Then, by the dominated convergence theorem, thanks to (6.5.1) we can take the limit in (6.5.6) as α goes to 0, and we get

$$P_t\varphi(x + h) - P_t\varphi(x) = \frac{1}{t}\mathbb{E}\varphi(u(t; x)) \int_0^t \langle Q^{-1} D_x u(s; x)h, dw(s) \rangle_H + R_t(x, h),$$

for a suitable remainder $R_t(x, h)$. Under our hypotheses, $P_t^\alpha \varphi$ is at least twice differentiable, and

$$\int_0^1 \langle D(P_t^\alpha \varphi)(x + \theta h) - D(P_t^\alpha \varphi)(x), h \rangle_H \, d\theta$$

$$= \int_0^1 \theta \int_0^1 \langle D^2(P_t^\alpha \varphi)(x + \rho\theta h)h, h \rangle_H \, d\rho \, d\theta.$$

Now, since $\langle D^2(P_t^\alpha \varphi)(x)h, k \rangle_H$ is given by (6.5.3), thanks to (6.3.1) we have that

$$|\langle D^2(P_t^\alpha \varphi)(x)h, k \rangle_H| \leq \frac{2}{t} \|D\psi_\alpha\|_0^H e^{ct} |k|_E \left(\mathbb{E} \int_0^t |Q^{-1} D_x u_\alpha(s; x)h|_H^2 \, ds \right)^{1/2}$$

$$+ \frac{2}{t} \|\psi_\alpha\|_0^H \left(\mathbb{E} \int_0^t |Q^{-1} D_x^2 u_\alpha(s; x)(h, k)|_H^2 \, ds \right)^{1/2}$$

and then, recalling (6.4.4) and (6.5.7), we get

$$|\langle D^2(P_t^\alpha \varphi)(x)h, k \rangle_H| \leq c_{|x|_E}(t)(t \wedge 1)^{-(1+\epsilon)} \|\varphi\|_0^H |h|_E |k|_E.$$

Since

$$R_t(x, h) = \lim_{\alpha \to 0} \int_0^1 \theta \int_0^1 \langle D^2(P_t^\alpha \varphi)(x + \rho\theta h)h, h \rangle_H \, d\rho \, d\theta,$$

this implies that

$$\frac{|R_t(x,h)|}{|h|_E} \leq c_{|x|_E}(t)(t \wedge 1)^{-(1+\epsilon)} \|\varphi\|_0^H |h|_E \to 0,$$

as $|h|_E \to 0$, so that $P_t\varphi$ is differentiable in E, for any $t > 0$ and $\varphi \in C_b(H)$, and

$$\langle D(P_t\varphi)(x), h \rangle_E = \frac{1}{t} \mathbb{E}\varphi(u(t;x)) \int_0^t \langle Q^{-1}D_x u(s;x)h, dw(s) \rangle_H.$$

Moreover, proceeding as for (6.5.7), due to (6.4.1) we get the estimate (6.5.4) uniform with respect to $x \in E$.

Now, let $\varphi \in C_b(E)$ and let $\{\varphi_k\} \subset C_b(H)$ be the approximating sequence introduced in the Proposition 6.1.5. For any $k \in \mathbb{N}$ and $t > 0$, we have that $P_t\varphi_k$ is differentiable in E. Besides, since $u(t;x) \in E$, for any $t \geq 0$ and $x \in E$, we have

$$\begin{cases} P_t\varphi_k(x) \to P_t\varphi(x) \\ \\ \langle D(P_t\varphi_k)(x), h \rangle_E \to \frac{1}{t}\mathbb{E}\varphi(u(t;x)) \int_0^t \langle Q^{-1}D_x u(s;x)h, dw(s) \rangle_H, \end{cases}$$

as k goes to infinity, for any $x, h \in E$. Then, proceeding as before, we get that $P_t\varphi$ is differentiable in E and (6.5.4) and (6.5.5) hold. Finally, in order to get that $P_t\varphi$ is differentiable for any $\varphi \in B_b(E)$ and $t > 0$ we proceed as in the proof of the Propositions 1.5.1 and 4.4.3, by considering the variation of $P_t(x, \cdot)$, for each $t > 0$ and $x \in E$, and by using the semigroup law.

Higher order differentiability follows by recurrence from the formula (6.5.5), by using the Theorem 6.3.3 and the Proposition 6.4.4 and the semigroup law.

In order to prove the estimate (6.5.4) for $j \leq r$, we use the estimates (6.3.9) and (6.5.4) and the estimate (6.5.4), which we have already proved for $i < j$.   □

Remark 6.5.2. In the proof of the first order differentiability of $P_t\varphi$ and of the estimate (6.5.4), with $j = 1$, it is sufficient to assume that the reaction term $f(\xi, \cdot)$ is of class C^1. Actually, due to the Proposition 4.4.4, $P_t^\alpha : B_b(H) \to C_b^1(H)$, even if $f_\alpha(\xi, \cdot)$ is only C^1. Then, as the differentiability of $P_t\varphi$ is obtained from the differentiability of $P_t^\alpha\varphi$, it is not difficult to conclude.

Chapter 7

Smooth dependence on data for the SPDE: the non-Lipschitz case (II)

In the previous chapter we have studied the regularizing properties of the transition semigroup associated with the stochastic reaction-diffusion system

$$du(t) = [Au(t) + F(u(t))] \, dt + Q \, dw(t), \qquad u(0) = x, \qquad (7.0.1)$$

in the Banach space E of continuous functions. In this chapter we study the same problem, but in the Hilbert space H of square integrable functions.

Under the same assumptions introduced in the section 6.1, for any $x \in H$ the problem (7.0.1) has a unique solution $u(\cdot; x)$, in a generalized sense that we are specifying later on. Then we can introduce the *Markov transition semigroup* corresponding to the system (7.0.1), regarded as a system in H, by setting

$$P_t\varphi(x) = \mathbb{E}\varphi(u(t; x)), \qquad x \in H,$$

for each $t \geq 0$ and $\varphi \in B_b(H)$.

Our purpose is showing that the semigroup P_t has a smoothing effect in H. Namely, we are proving that for any $t > 0$

$$\varphi \in B_b(H) \Rightarrow P_t\varphi \in C_b^1(H)$$

and the following estimate holds

$$\sup_{x \in H} |\langle D(P_t\varphi)(x), h \rangle_H| \leq c \, (t \wedge 1)^{-\frac{1+\epsilon}{2}} \|\varphi\|_0^H |h|_H, \quad h \in H, \qquad (7.0.2)$$

where ϵ is the constant introduced in the Hypothesis 6.1-3. Notice that when A is the realization in H of the Laplace operator with Dirichlet or conormal boundary conditions, ϵ may be taken equal zero if $d = 1$ and strictly less than one if $d \leq 3$, so that the singularity at $t = 0$ arising in the right hand side of (7.0.2) turns out to be integrable. Moreover we are proving that if $\varphi \in C_b(H)$, then a Bismut-Elworthy type formula holds for the derivative of $P_t\varphi$. In fact, we show that for any $x, h \in H$ and $t > 0$

$$\langle D(P_t\varphi)(x), h \rangle_H = \frac{1}{t} \mathbb{E} \varphi(u(t;x)) \int_0^t \langle Q^{-1}v(s;x,h), dw(s) \rangle_H, \qquad (7.0.3)$$

where the process $v(t; x, h)$ is not the mean-square derivative of the solution $u(t; x)$ along the direction h, in general, as one should expect. Actually, it can be proved that $v(t; x, h)$ is a *generalized* solution of the first variation equation associated with the problem (7.0.1). As in the previous chapter, in order to give a meaning to the formula (7.0.3), we have to show that $v(s; x, h) \in \text{Range}(Q)$ for each $s > 0$ and

$$\mathbb{E} \int_0^t |Q^{-1}v(s;x,h)|_H^2 \, ds < \infty.$$

The regularizing property of the transition semigroup P_t is rather crucial in the study of the ergodicity of the system as well as in the study of some parabolic and elliptic Hamilton-Jacobi-Bellman problems and the associated stochastic optimal control problems. The former case will be extensively studied in the chapter 8 and the latter case will be studied in the chapters 9 and 10.

In the section 7.1 we show that the system (7.0.1) admits a unique *generalized* solution $u(\cdot; x)$ with values in H and we prove some estimates for such solution. Hence, we introduce the corresponding transition semigroup and we prove that it is a weakly continuous semigroup on $C_b(H)$.

In the section 7.2 we prove some approximation results for the solution $u(\cdot; x)$ and for the *generalized* solution of the first variation equation corresponding to (7.0.1).

By using such approximation results and the differentiability of the semigroup P_t in E described in the previous chapter, in the last section we show the smoothing effect of P_t in H, and we give both the estimate (7.0.2) and the formula (7.0.3).

7.1   The transition semigroup

Let X be a Banach space. With the same notations introduced in the chapter 4, for any $T > 0$ and $p \geq 1$ we denote by $\mathcal{I}_p(T, X)$ the space of all adapted processes $u \in L^p(0, T; L^p(\Omega; X))$. $\mathcal{I}_p(T, X)$ is a Banach space, endowed with the norm

$$\|u\|_{\mathcal{I}_p(T,X)}^p = \int_0^T \mathbb{E} |u(t)|_X^p \, dt < \infty.$$

Notice that, if X is a Hilbert space, then $\mathcal{L}_2(T, X)$ is a Hilbert space, as well. Moreover, we denote by $\mathcal{H}_p(T, X)$ the subspace of all processes u in $C([0, T]; L^p(\Omega; X))$ (see section 4.2).

In this section we want to establish some properties of the solution of the following problem

$$du(t) = [Au(t) + F(u(t))]\, dt + Q\, dw(t), \quad u(0) = x, \tag{7.1.1}$$

in order to introduce and describe the associated transition semigroup.

Definition 7.1.1.   *1. The process $u(t; x)$ is said a* mild *solution of the equation (7.1.1) if $u(s; x) \in D(F)$, for any $s > 0$, and*

$$u(t; x) = e^{tA}x + \int_0^t e^{(t-s)A}F(u(s; x))\, ds + w^A(t),$$

where $w^A(t)$ is the process given by (6.1.11).

2. The process $u(t; x)$ is a generalized *solution of the equation (7.1.1) if for any sequence $\{x_n\} \subset E$ converging to x in H, the corresponding sequence of mild solutions $\{u(\cdot; x_n)\}$ converges to $u(\cdot; x)$ in $C([0, T]; H)$, $\mathbb{P}$-a.s. for any $T > 0$.*

We have already seen that under the Hypotheses 6.1 and 6.2 for any $x \in E$ the equation (7.1.1) has a unique mild solution $u(x)$ in $L^p(\Omega; C((0, +\infty); E) \cap L^\infty_{loc}(0, +\infty; E))$, $p \geq 1$, and in the Proposition 6.2.2 the following estimate is established

$$|u(t; x)|_E \leq e^{ct}|x|_E + h(t), \quad \mathbb{P} - a.s. \tag{7.1.2}$$

where h is the process defined by

$$h(t) = c\, e^{ct} \int_0^t \left(1 + \left|w^A(s)\right|_E^{2m+1}\right)\, ds + \sup_{s \in [0,t]} \left|w^A(s)\right|_E. \tag{7.1.3}$$

Moreover, if the Hypothesis 6.3 holds, then for any $t > 0$ we have

$$\sup_{x \in E} |u(t; x)|_E \leq k(t)\, t^{-\frac{1}{2m}}, \quad \mathbb{P} - a.s. \tag{7.1.4}$$

where the random variable $k(t)$ is defined by

$$k(t) = c\left(1 + \sup_{s \in [0,t]} \left|w^A(s)\right|_E^{\frac{\gamma}{2m+1} \vee 1}\right). \tag{7.1.5}$$

Proposition 7.1.2. *Under the Hypotheses 6.1 and 6.2, for any $x, y \in E$ and $t \geq 0$ we have*

$$|u(t; x) - u(t; y)|_H \leq e^{ct}|x - y|_H, \quad \mathbb{P} - a.s. \tag{7.1.6}$$

In particular, for any $x \in H$ the equation (7.1.1) has a unique generalized solution $u(\cdot; x)$ and it holds

$$|u(t; x)|_H \leq e^{ct}|x|_H + h(t), \quad \mathbb{P} - a.s. \tag{7.1.7}$$

where $h(t)$ is the process defined in (7.1.3). Moreover, if the Hypothesis 6.3 is satisfied as well, then for any $t > 0$

$$\sup_{x \in H} |u(t; x)|_H \leq k(t) \, t^{-\frac{1}{2m}}, \quad \mathbb{P} - a.s. \tag{7.1.8}$$

where $k(t)$ is defined as in (7.1.5).

Proof. Let us fix $x, y \in E$. If we define $v(t) = u(t; x) - u(t; y)$, $t \geq 0$, then v is the unique mild solution of the problem

$$\frac{dv}{dt}(t) = Av(t) + F(u(t; x)) - F(u(t; y)), \quad v(0) = x - y. \tag{7.1.9}$$

We can assume that v is a strict solution of the equation (7.1.9). If that is not the case, we can approximate v by means of a more regular sequence in $C^1([0, T]; H) \cap C([0, T]; D(A))$ which converges to v in $C([0, T]; H)$, $\mathbb{P}$-a.s. Indeed for any $\lambda \in \mathbb{N}$ we introduce the problem

$$\frac{dv_\lambda}{dt}(t) = Av_\lambda(t) + F(u(t; x)) - F(u(t; y)), \quad v_\lambda(0) = \lambda(\lambda - I)^{-1}(x - y)$$

which has a unique solution $v_\lambda(t; x, y)$ in $C([0, +\infty); H)$. For any $T > 0$ we have that $F(u(\cdot; x)) - F(u(\cdot; y)) \in C([0, T]; H)$ and $v_\lambda(0) \in D(A)$. Then, as shown in [91, Proposition 4.1.8], v_λ is a strong solution, that is there exists a sequence $\{v_{\lambda,n}\} \subset C^1([0, T]; H) \cap C([0, T]; D(A))$ such that

$$v_{\lambda,n} \to v_\lambda, \quad \frac{dv_{\lambda,n}}{dt} - Av_{\lambda,n} \to F(u(\cdot; x)) - F(u(\cdot; y)) \quad \text{in } C([0, T]; H),$$

as n goes to infinity. Now, if v is a strict solution of (7.1.9), from (6.1.21) it follows

$$\frac{1}{2}\frac{d}{dt}|v(t)|_H^2 = \langle Av(t), v(t)\rangle_H + \langle F(u(t; x)) - F(u(t; y)), v(t)\rangle_H \leq c|v(t)|_H^2,$$

and by the Gronwall lemma this yields (7.1.6). This yields the existence and the uniqueness of a generalized solution for (7.1.1). Indeed, if $\{x_n\} \subset E$ is a sequence converging to x in H, due to (7.1.6) the corresponding sequence of mild solutions $\{u(\cdot; x_n)\}$ is a Cauchy sequence in $C([0, T]; H)$, $\mathbb{P}$-a.s. Hence it admits a limit in $C([0, T]; H)$ as $n \to +\infty$, $\mathbb{P}$-a.s. which is the unique generalized solution $u(\cdot; x)$. Next, for any $x \in H$ we have

$$|u(t; x)|_H \leq |u(t; x) - u(t; 0)|_H + |u(t; 0)|_H \leq |u(t; x) - u(t; 0)|_H + |u(t; 0)|_E,$$

and then (7.1.7) is a consequence of (7.1.2) and (7.1.6). Finally, if $\{x_n\}$ is a sequence in E which converges to x in H as $n \to +\infty$, we have

$$|u(t; x)|_H = \lim_{n \to +\infty} |u(t; x_n)|_H \le \limsup_{n \to +\infty} |u(t; x_n)|_E,$$

so that from (7.1.4) we get (7.1.8). □

Due to the previous proposition, we can introduce the *transition semigroup* associated with the equation (7.1.1). For any $\varphi \in B_b(H)$ and $x \in H$ we define

$$P_t\varphi(x) = \mathbb{E}\varphi(u(t; x)), \quad t \ge 0,$$

where $u(t; x)$ is the unique generalized solution of the equation (7.1.1). From (7.1.6), due to the definition of generalized solution, it easily follows that for any $t \ge 0$ and $x, y \in H$

$$|u(t; x) - u(t; y)|_H \le e^{ct}|x - y|_H, \quad \mathbb{P} - \text{a.s.} \tag{7.1.10}$$

so that P_t is a *Feller* semigroup (that is P_t maps $C_b(H)$ into itself). Moreover, P_t is a contraction semigroup on $C_b(H)$, as

$$\|P_t\varphi\|_0^H = \sup_{x \in H} |\mathbb{E}\varphi(u(t; x))| \le \|\varphi\|_0^H.$$

From (7.1.10) it immediately follows that for any $T > 0$ and $\varphi \in C_b(H)$ the family of functions $\{ P_t\varphi; t \in [0, T] \}$ is equi-uniformly continuous. Moreover, as $u(\cdot; x) \in C([0, T]; H)$, for any $x \in H$ and $T > 0$, it is easy to show that the mapping

$$[0, +\infty) \to \mathbb{R}, \quad t \mapsto P_t\varphi(x),$$

is continuous, for any fixed $\varphi \in C_b(H)$ and $x \in H$. Thus, we can introduce the *infinitesimal generator* of P_t, as for weakly continuous semigroups. Actually, the proofs of the Propositions B.1.3 and B.1.4 can be adapted to the present situation and then, as in the Definition B.1.5 we can define the infinitesimal generator of P_t as the unique closed operator $L : D(L) \subset C_b(H) \to C_b(H)$ such that

$$R(\lambda, L)\varphi(x) = \int_0^{+\infty} e^{-\lambda t} P_t\varphi(x)\, dt, \quad \lambda > 0,$$

for any $\varphi \in C_b(H)$ and $x \in H$. Notice that the Proposition B.2.2 holds in this case, as well, so that for any $\varphi \in D(L)$ and $x \in H$ the mapping $t \mapsto P_t\varphi(x)$ is differentiable and

$$\frac{d}{dt}P_t\varphi(x) = P_t L\varphi(x) = L P_t\varphi(x).$$

In [40] it has been proved that, if $H = L^2(\mathcal{O}; \mathbb{R})$, with $\mathcal{O} \subset \mathbb{R}$ and $Q = I$, the semigroup P_t enjoys the *strong Feller* property, that is

$$\varphi \in B_b(H) \Rightarrow P_t\varphi \in C_b(H), \quad t > 0.$$

Actually, it is proved that $P_t\varphi$ is Lipschitz-continuous, for any $t > 0$, and

$$\|P_t\varphi\|_{0,1}^H \le c\,(t \wedge 1)^{-1/2}\|\varphi\|_0^H.$$

Here we want to show that under the more general Hypotheses 6.1, 6.2 and 6.3 $P_t\varphi \in C_b^1(H)$, for any $t > 0$ and $\varphi \in B_b(H)$, and

$$\|P_t\varphi\|_1^H \le c\,(t \wedge 1)^{-\frac{1+\epsilon}{2}}\|\varphi\|_0^H.$$

In the previous chapter we have proved that for any $x \in E$ the equation (7.1.1) has a unique mild solution $u(\cdot; x)$ which is an E-valued process. Then we have introduced the corresponding transition semigroup. Here we shall denote it by P_t^E, in order to distinguish it from the semigroup P_t in H. In the previous chapter we have proved that P_t^E has a smoothing effect. Namely, if $f \in C^r(\mathbb{R})$ then

$$\varphi \in B_b(E) \Rightarrow P_t^E\varphi(x) \in C_b^r(E), \quad t > 0.$$

Notice that since $B_b(H)$ is continuously embedded in $B_b(E)$, for any $\varphi \in B_b(H)$ and $x \in E$ it holds

$$P_t^E\varphi(x) = P_t\varphi(x), \quad t \ge 0. \tag{7.1.11}$$

7.2 Some approximation results

In the chapter 6 it is proved that if $f \in C^r(\mathbb{R})$, then the unique mild solution of the equation (7.1.1) is r-times differentiable in E in the mean of order p with respect to the initial datum $x \in E$. This means that for any $t \ge 0$ the mapping

$$E \to \mathcal{H}_p(T, E), \quad x \mapsto u(\cdot; x),$$

is r-times Fréchet differentiable. Besides, if $D_x u(t; x)h$ denotes the first order derivative of $u(t; x)$, at the point x and along the direction h, then $D_x u(\cdot; x)h$ is the unique mild solution of the following deterministic problem with random coefficients

$$\frac{dv}{dt}(t) = Av(t) + DF(u(t; x))v(t), \quad v(0) = h. \tag{7.2.1}$$

We are proving now that

$$\sup_{x \in E} |D_x u(t; x)h|_H \le e^{ct}|h|_H, \quad \mathbb{P} - \text{a.s.} \tag{7.2.2}$$

As in the proof of the Proposition 7.1.2 we can assume that $D_x u(\cdot; x)h$ is a strict solution of the equation (7.2.1). Then we have

$$\frac{1}{2}\frac{d}{dt}|D_x u(t; x)h|_H^2$$

$$= \langle AD_x u(t; x)h, D_x u(t; x)h\rangle_H + \langle DF(u(t; x))D_x u(t; x)h, D_x u(t; x)h\rangle_H$$

and from (6.1.20) it follows

$$\frac{1}{2}\frac{d}{dt}\,|D_x u(t;x)h|_H^2 \le c\,|D_x u(t;x)h|_H^2\,,$$

which implies (7.2.2).

Now, recalling how DF is defined and the estimates satisfied by f and $D_\sigma f$, for any $x, y \in E$ we have

$$|DF(x)y|_H^2 = \int_{\mathcal{O}} |D_\sigma f(\xi, x(\xi))y(\xi)|^2\,d\xi \le c \int_{\mathcal{O}} \left(1 + |x(\xi)|^{4m}\right)|y(\xi)|^2\,d\xi,$$

so that

$$|DF(x)y|_H \le c\,|y|_E\left(1 + |x|_E^{2m-1}|x|_H\right).$$

Then, since $D_x u(t;x)h$ is the mild solution of (7.2.1), from (4.1.4) we have

$$|D_x u(t;x)h|_E \le c\,t^{-\frac{d}{4}}|h|_H + c\int_0^t (t-s)^{-\frac{d}{4}}\,|DF(u(s;x))D_x u(s)(x)h|_H\,ds$$

$$\le c\,t^{-\frac{d}{4}}|h|_H + c\int_0^t (t-s)^{-\frac{d}{4}}\left(1 + |u(s;x)|_E^{2m-1}|u(s;x)|_H\right)|D_x u(s;x)h|_E\,ds.$$

By using (7.1.7) and (7.1.4) this yields

$$|D_x u(t;x)h|_E \le c\,t^{-\frac{d}{4}}|h|_H$$

$$+c\int_0^t (t-s)^{-\frac{d}{4}}\left(1 + k(s)^{2m-1}s^{-1+\frac{1}{2m}}\left(e^{cs}|x|_H + h(s)\right)\right)|D_x u(s;x)h|_E\,ds$$

and due to a generalization of the Gronwall lemma, this implies that for any $x, h \in E$

$$|D_x u(t;x)h|_E \le \Phi_{|x|_H}(t)t^{-\frac{d}{4}}|h|_H,\quad \mathbb{P}-\text{a.s.}$$

for a suitable positive process $\Phi_{|x|_H}(t)$ increasing with respect to t, which is $\mathbb{P}$-a.s. finite.

Proposition 7.2.1. *Assume that the Hypotheses 6.1, 6.2 and 6.3 hold. Let $x, h \in H$ and let $\{x_n\}$ and $\{h_n\}$ be any two sequences in E converging respectively to x and h in H. Then the sequence $\{D_x u(\cdot;x_n)h_n\}$ converges in $C([0,T];H)$, $\mathbb{P}$-a.s to a process $v(\cdot;x,h)$ which is called the generalized solution of the first variation equation corresponding to the problem (7.1.1).*

Proof. If we prove that for any $x, y, h, k \in E$ and $t \ge 0$

$$|D_x u(t;x)h - D_x u(t;y)k|_H \le \Phi_{|x|_H}(t)\left(|x-y|_H|k|_H + |h-k|_H\right),\quad \mathbb{P}-\text{a.s.}\quad (7.2.3)$$

then $\{D_x u(\cdot; x_n)h_n\}$ is a Cauchy sequence in $C([0,T]; H)$, $\mathbb{P}$-a.s., and it converges to a process v which is independent of the particular choice of the sequences $\{x_n\}$ and $\{h_n\}$.

If we set

$$z(t) = D_x u(t; x)h - D_x u(t; y)k,$$

from (7.2.1) we have

$$z(t) = e^{tA}(h - k)$$

$$+ \int_0^t e^{(t-s)A} \left(DF(u(s; x))D_x u(s; x)h - DF(u(s; y))D_x u(s; y)k \right) \, ds.$$

Therefore, by using (4.1.5) for any $\theta \in (1, 2]$ it follows

$$|z(t)|_H \leq |h - k|_H + c \int_0^t (t - s)^{-\frac{d(2-\theta)}{4\theta}} |DF(u(s; x))z(s)|_\theta \, ds$$

$$+ c \int_0^t (t - s)^{-\frac{d(2-\theta)}{4\theta}} |(DF(u(s; x)) - DF(u(s; y))) D_x u(s; y)k|_\theta \, ds$$

$$= |h - k|_H + J_1(t) + J_2(t).$$

Now, we estimate the two terms $J_1(t)$ and $J_2(t)$. To this purpose, we first remark that if $x \in E$ and $y \in H$, due to the growth conditions on $D_\sigma f$ for any $1 \leq \theta < 2$ we have

$$|DF(x)y|_\theta^\theta \leq c \int_{\mathcal{O}} \left(1 + |x(\xi)|^{2m}\right)^\theta |y(\xi)|^\theta \, d\xi \leq c|y|_H^\theta \left(1 + \left(\int_{\mathcal{O}} |x(\xi)|^{\frac{4m\theta}{2-\theta}} \, d\xi\right)^{\frac{2-\theta}{2}}\right),$$

so that by easy calculations

$$|DF(x)y|_\theta \leq c|y|_H \left(1 + |x|_E^{2m - \frac{2-\theta}{\theta}} |x|_H^{\frac{2-\theta}{\theta}}\right). \qquad (7.2.4)$$

Hence, if we take $\theta = 1$, we get

$$J_1(t) \leq c \int_0^t (t - s)^{-\frac{d}{4}} \left(1 + |u(s; x)|_E^{2m-1} |u(s; x)|_H\right) |z(s)|_H \, ds$$

$$\leq c \int_0^t (t - s)^{-\frac{d}{4}} \left(1 + k(s)^{2m-1} s^{-1 + \frac{1}{2m}} \left(e^{cs} |x|_H + h(s)\right)\right) |z(s)|_H \, ds,$$

$\mathbb{P}$-a.s., last inequality following from (7.1.4) and (7.1.7).

Next, we remark that for any $x, y \in E$, $z \in H$ and $1 \leq \theta < 2$, we have

$$|(DF(x) - DF(y)) z|_\theta^\theta$$

$$= \int_O \left| \int_0^1 D_\sigma^2 f(\xi, \rho x(\xi) + (1-\rho) y(\xi)) \, d\rho \right|^\theta |x(\xi) - y(\xi)|^\theta |z(\xi)|^\theta \, d\xi$$

$$\leq c \int_O \left(1 + |x(\xi)|^{(2m-1)\theta} + |y(\xi)|^{(2m-1)\theta} \right) |x(\xi) - y(\xi)|^\theta |z(\xi)|^\theta \, d\xi$$

$$\leq c \int_O \left(1 + |x(\xi)|^{(2m-1)\theta} + |y(\xi)|^{(2m-1)\theta} \right) |x(\xi) - y(\xi)|^{2-\theta}$$

$$\times \left(|x(\xi)|^{2(\theta-1)} + |y(\xi)|^{2(\theta-1)} \right) |z(\xi)|^\theta \, d\xi,$$

so that

$$|DF(x)z - DF(y)z|_\theta \leq c \left(1 + |x|_E^{2m - \frac{2-\theta}{\theta}} + |y|_E^{2m - \frac{2-\theta}{\theta}} \right) |x - y|_H^{\frac{2-\theta}{\theta}} |z|_H. \quad (7.2.5)$$

Therefore, if we choose $\theta = 1$, we have

$$J_2(t) \leq c \int_0^t (t-s)^{-\frac{d}{4}} \left(1 + |u(s; x)|_E^{2m-1} + |u(s; y)|_E^{2m-1} \right)$$

$$\times |u(s; x) - u(s; y)|_H |D_x u(s; y)k|_H \, ds,$$

and thanks to (7.1.4), (7.1.6) and (7.2.2) this yields

$$J_2(t) \leq c \int_0^t (t-s)^{-\frac{d}{4}} \left(1 + k(s)^{2m-1} s^{-1 + \frac{1}{2m}} \right) e^{2cs} \, ds |x - y|_H |k|_H, \quad \mathbb{P} - \text{a.s.}$$

Thus, if we define

$$L(t, s) = c (t-s)^{-\frac{d}{4}} \left(1 + k(s)^{2m-1} s^{-1 + \frac{1}{2m}} \left(e^{cs} |x|_H + h(s) \right) \right)$$

$$M(t, s) = c (t-s)^{-\frac{d}{4}} \left(1 + k(s)^{2m-1} s^{-1 + \frac{1}{2m}} \right) e^{2cs},$$

we have that $L(t, s)$ and $M(t, s)$ are both integrable in $[0, t]$ with respect to s, $\mathbb{P}$-a.s. and it holds

$$|z(t)|_H \leq |h - k|_H + |x - y|_H |k|_H \int_0^t M(t, s) \, ds + \int_0^t L(t, s) |z(s)|_H \, ds.$$

By using a modification of the Gronwall lemma, this implies (7.2.3). $\qquad \square$

Before proving next lemma, we recall that, according to (4.1.7) and to the Hypothesis 6.1-3 the operator $\Gamma(t) = Q^{-1}e^{tA}$ is bounded for any $t > 0$ and

$$|\Gamma(t)x|_H \le ct^{-\epsilon/2}|x|_H, \quad x \in H. \tag{7.2.6}$$

Moreover, since $\Gamma(t) = \Gamma(t/2)e^{t/2A}$, from (4.1.5) and (7.2.6) $\Gamma(t)$ maps $L^\theta(\mathcal{O}; \mathbb{R}^n)$ into H, for any $t > 0$ and $1 \le \theta \le 2$, and it holds

$$|\Gamma(t)x|_H \le ct^{-\frac{\epsilon}{2}}|e^{t/2A}x|_H \le ct^{-\frac{\epsilon}{2} - \frac{d(2-\theta)}{4\theta}}|x|_\theta, \quad x \in L^\theta(\mathcal{O}; \mathbb{R}^n). \tag{7.2.7}$$

We recall that in the Proposition 6.4.1 we have proved that if $x, h \in E$ then $D_x u(t; x)h$ belongs to $D(Q^{-1})$ for any $t > 0$ and

$$\mathbb{E} \sup_{x \in E} \int_0^t |Q^{-1}D_x u(s; x)h|_H^2 \, ds \le c(t)t^{1-\epsilon}|h|_H^2, \quad h \in E. \tag{7.2.8}$$

Now, we prove the following estimate.

Lemma 7.2.2. *For any $R > 0$ there exists an increasing process $c_R(t)$ which is finite $\mathbb{P}$-a.s. such that for any $x, y, h, k \in E$, with $x, y \in \{z \in H : |z|_H \le R\}$, it holds*

$$\int_0^t |Q^{-1}(D_x u(s; x)h - D_x u(s; y)k)|_H^2 \, ds \le c_R(t)\left(|x - y|_H^\beta |k|_H + |h - k|_H\right)^2, \tag{7.2.9}$$

$\mathbb{P}$-*a.s., where*

$$\begin{cases} \beta = 1 & \text{if } dm - 2 \le 0 \\[2mm] \beta < \dfrac{2m(1-\epsilon)}{dm - 2} & \text{if } dm - 2 > 0. \end{cases} \tag{7.2.10}$$

Proof. If we define $z(t) = D_x u(t; x)h - D_x u(t; y)k$, we have

$$Q^{-1}z(t) = \Gamma(t)(h - k) + \int_0^t \Gamma(t - s)DF(u(s; x))z(s) \, ds$$

$$+ \int_0^t \Gamma(t - s)\left([DF(u(s; x))] - DF(u(s; y))\right) D_x u(s; y)k \, ds.$$

We remark that if

$$\theta^\star = \begin{cases} 0 & \text{if } dm - 2 \le 0 \\[2mm] \dfrac{2(dm - 2)}{dm - 2 + 2m(1-\epsilon)} & \text{if } dm - 2 > 0, \end{cases}$$

then for any $\theta > \theta^\star$ we have

$$\epsilon + \frac{d(2-\theta)}{2\theta} - \frac{2-\theta}{m\theta} < 1.$$

Hence, by using (7.2.6) and (7.2.7), for any $\theta^\star < \theta < 2$ we get

$$|Q^{-1}z(t)|_H \le c\,t^{-\frac{\epsilon}{2}}|h - k|_H + c\int_0^t (t-s)^{-\frac{\epsilon}{2}-\frac{d(2-\theta)}{4\theta}}\,|DF(u(s;x))z(s)|_\theta\,ds$$

$$+c\int_0^t (t-s)^{-\frac{\epsilon}{2}-\frac{d(2-\theta)}{4\theta}}\,|[DF(u(s;x)) - DF(u(s;y))]\,D_x u(s;y)k|_\theta\,ds$$

$$= c\,t^{-\frac{\epsilon}{2}}|h - k|_H + I_1(t) + I_2(t).$$

By using (7.1.4), (7.1.7), (7.2.3) and (7.2.4) we have

$$I_1(t) \le c\int_0^t (t-s)^{-\frac{\epsilon}{2}-\frac{d(2-\theta)}{4\theta}}\Phi_{|x|_H}(s)$$

$$\times\left(1 + k(s)^{2m-\frac{2-\theta}{\theta}}s^{-1+\frac{2-\theta}{2m\theta}}\left(e^{cs}|x|_H + h(s)\right)^{\frac{2-\theta}{\theta}}\right)ds\,(|x - y|_H|k|_H + |h - k|_H)$$

$$\le c\,\Phi_{|x|_H}(t)t^{-\frac{\epsilon}{2}-\frac{d(2-\theta)}{4\theta}+\frac{2-\theta}{2m\theta}}\,(|x - y|_H|k|_H + |h - k|_H),\quad \mathbb{P}-\text{a.s.}$$

$$(7.2.11)$$

Concerning $I_2(t)$, by using (7.1.4), (7.1.6) and (7.2.2) and according to (7.2.5) we get

$$I_2(t) \le c\int_0^t (t-s)^{-\frac{\epsilon}{2}-\frac{d(2-\theta)}{4\theta}}\left(1 + k(s)^{2m-\frac{2-\theta}{\theta}}s^{-1+\frac{2-\theta}{2\theta m}}\right)e^{\frac{2c}{\theta}s}ds|x - y|_H^{\frac{2-\theta}{\theta}}|k|_H$$

$$\le \Phi(t)t^{-\frac{\epsilon}{2}-\frac{d(2-\theta)}{4\theta}+\frac{2-\theta}{2m\theta}}|x - y|_H^{\frac{2-\theta}{\theta}}|k|_H,\quad \mathbb{P}-\text{a.s.}$$

$$(7.2.12)$$

for some continuous increasing process $\Phi(t)$.

Then, from (7.2.11) and (7.2.12) we get

$$\left|Q^{-1}\left(D_x u(s;x)h - D_x u(s;y)k\right)\right|_H$$

$$\le \Phi_{|x|_H}(t)t^{-\frac{\epsilon}{2}-\frac{d(2-\theta)}{4\theta}+\frac{2-\theta}{2m\theta}}\left(|x - y|_H|k|_H + |x - y|_H^{\frac{2-\theta}{\theta}}|k|_H + |h - k|_H\right)$$

$$\le \Phi_{|x|_H,|y|_H}(t)\,t^{-\frac{\epsilon}{2}-\frac{d(2-\theta)}{4\theta}+\frac{2-\theta}{2m\theta}}\left(|x - y|_H^{\frac{2-\theta}{\theta}}|k|_H + |h - k|_H\right)\quad \mathbb{P}-\text{a.s.}$$

This implies that

$$\int_0^t \left| Q^{-1} \left(D_x u(s; x)h - D_x u(s; y)k \right) \right|_H^2 ds$$

$$\leq \int_0^t \left| \Phi_{|x|_H, |y|_H}(s) \right|^2 s^{-\epsilon - \frac{d(2-\theta)}{2\theta} + \frac{2-\theta}{m\theta}} ds \left(|x - y|_H + |x - y|_H^{\frac{2-\theta}{\theta}} |k|_H + |h - k|_H \right)^2,$$

$\mathbb{P}$-a.s. and hence, by setting

$$c_R(t) = \int_0^t \left| \Phi_{R,R}(s) \right|^2 s^{-\epsilon - \frac{d(2-\theta)}{2\theta} + \frac{2-\theta}{m\theta}} ds,$$

(7.2.9) follows. Finally, since $\theta > \theta^*$ we get (7.2.10).   □

Proposition 7.2.3. *Let $x, h \in H$ and let $\{x_n\}$ and $\{h_n\}$ be any two sequences in E converging to x and h in H. Then, under the Hypotheses 6.1, 6.2 and 6.3 we have*

$$\lim_{n \to +\infty} \mathbb{E} \int_0^t \left| Q^{-1} \left(D_x u(s; x_n)h_n - v(s; x, h) \right) \right|_H^2 ds = 0 \qquad (7.2.13)$$

and it holds

$$\sup_{x \in H} \mathbb{E} \int_0^t \left| Q^{-1} v(s; x, h) \right|_H^2 ds \leq c(t) t^{1-\epsilon} |h|_H^2. \qquad (7.2.14)$$

Proof. Due to (7.2.9) the sequence $\{Q^{-1} D_x u(t; x_n)h_n\}$ is a Cauchy sequence in $L^2(0, T; H)$, $\mathbb{P}$-a.s. for any $T > 0$. Then, since Q^{-1} is closed, from the Proposition 7.2.1 we have that

$$\lim_{n \to +\infty} \int_0^t \left| Q^{-1} \left(D_x u(s; x_n)h_n - v(s; x, h) \right) \right|_H^2 ds = 0, \qquad \mathbb{P} - a.s.$$

Therefore, due to (7.2.8) from the dominated convergence theorem we can conclude that (7.2.13) and (7.2.14) hold true.   □

7.3   Smoothing property of the transition semigroup

As proved in the chapter 6, if $f(\xi, \cdot) \in C^r(\mathbb{R}^n)$, then for any $\varphi \in B_b(E)$ and $t > 0$, $P_t^E \varphi \in C_b^r(E)$. Moreover for any $\varphi \in C_b(E)$ and $x, h \in E$ the following generalization of the Bismut-Elworthy formula holds

$$\left\langle D(P_t^E \varphi)(x), h \right\rangle_E = \frac{1}{t} \mathbb{E} \varphi(u(t; x)) \int_0^t \left\langle Q^{-1} D_x u(s; x)h, dw(s) \right\rangle_H.$$

Now, by using the approximation results proved in the previous section, we can state the following theorem.

Theorem 7.3.1. *Assume the Hypotheses 6.1, 6.2 and 6.3. Then $P_t\varphi \in C_b^1(H)$, for any $\varphi \in B_b(H)$ and $t > 0$. Moreover, if $\varphi \in C_b(H)$, for any $x, h \in H$ it holds*

$$\langle D(P_t\varphi)(x), h\rangle_H = \frac{1}{t}\mathbb{E}\varphi(u(t;x)) \int_0^t \langle Q^{-1}v(s;x,h), dw(s)\rangle_H, \qquad (7.3.1)$$

where $v(t;x,h)$ is the process introduced in the Proposition 7.2.1. In particular, this implies that there exists some $\nu > 0$ such that

$$\|P_t\varphi\|_1^H \leq \nu\,(t \wedge 1)^{-\frac{1+\varepsilon}{2}} \|\varphi\|_0^H. \qquad (7.3.2)$$

Proof. Let us fix $x, h \in H$ and let $\{x_n\}$ and $\{h_n\}$ be two sequences in E, converging respectively to x and h in H, such that $|x_n|_H \leq c\,|x|_H$ and $|h_n|_H \leq c\,|h|_H$, for some constant c independent of n. Since $C_b(H)$ is continuously embedded in $C_b(E)$ and $P_t^E\varphi$ is differentiable in E, for any $n \in \mathbb{N}$ and $\varphi \in C_b(H)$ we have

$$P_t^E\varphi(x_n + h_n) - P_t^E\varphi(x_n)$$

$$= \langle D(P_t^E\varphi)(x_n), h_n\rangle_E + \int_0^1 \langle D(P_t^E\varphi)(x_n + \rho h_h) - D(P_t^E\varphi)(x_n), h_n\rangle_E \, d\rho.$$

Recalling that $u(\cdot; x+h)$ and $u(\cdot; x)$ are the generalized solutions of the equation (7.1.1) corresponding respectively to the initial data $x + h$ and x, we have that $u(\cdot; x_n + h_n)$ converges to $u(\cdot; x+h)$ and $u(\cdot; x_n)$ converges to $u(\cdot; x)$ in $C([0,T]; H)$, $\mathbb{P}$-a.s. as n goes to infinity. Hence, since $\varphi \in C_b(H)$ from the dominated convergence theorem and from (7.1.11) it follows

$$\lim_{n \to +\infty} P_t^E\varphi(x_n + h_n) - P_t^E\varphi(x_n) = P_t\varphi(x + h) - P_t\varphi(x).$$

Besides, by using these arguments and the Proposition 7.2.3 it is easy to check that for any $t > 0$

$$\lim_{n \to +\infty} \langle D(P_t^E\varphi)(x_n), h_n\rangle_E = \frac{1}{t}\mathbb{E}\varphi(u(t;x)) \int_0^t \langle Q^{-1}v(s;x,h), dw(s)\rangle_H.$$

Thus, if we define

$$R_t(x,h) = \lim_{n \to +\infty} \int_0^1 \langle D(P_t^E\varphi)(x_n + \rho h_h) - D(P_t^E\varphi)(x_n), h_n\rangle_E \, d\rho$$

and if we show that

$$\lim_{|h|_H \to 0} \frac{|R_t(x,h)|}{|h|_H} = 0, \qquad (7.3.3)$$

we can conclude that $P_t\varphi$ is differentiable and

$$\langle D(P_t\varphi)(x), h\rangle_H = \frac{1}{t}\mathbb{E}\varphi(u(t;x)) \int_0^t \langle Q^{-1}v(s;x,h), dw(s)\rangle_H.$$

According to (7.2.14), for any $\varphi \in C_b(H)$ we have that

$$|\langle D(P_t\varphi)(x), h\rangle_H| \leq \frac{1}{t}\|\varphi\|_0^H \left(\mathbb{E}\int_0^t |Q^{-1}v(s;x,h)|_H^2 \, ds\right)^{1/2}$$

$$\leq c(t)\, t^{-\frac{1+\epsilon}{2}}\|\varphi\|_0^H |h|_H.$$

Then, since P_t is a contraction semigroup and $P_t\varphi = P_1(P_{t-1}\varphi)$, for any $t > 1$, we get

$$\sup_{x\in H} |\langle D(P_t\varphi)(x), h\rangle_H| \leq \nu\,(t\wedge 1)^{-\frac{1+\epsilon}{2}}\|\varphi\|_0^H |h|_H, \quad h \in H,$$

for some constant $\nu > 0$. In particular, for any $t > 0$ and $x \in H$, the mapping $D(P_t\varphi) : H \to H$ is continuous and bounded, so that $P_t\varphi$ is Lipschitz-continuous and it holds

$$\|P_t\varphi\|_{0,1}^H \leq \nu\,(t\wedge 1)^{-\frac{1+\epsilon}{2}}\|\varphi\|_0^H. \tag{7.3.4}$$

From the semigroup law, recalling (7.1.11) for any $x, y, h \in E$ we have

$$\langle D(P_t^E\varphi)(x) - D(P_t^E\varphi)(y), h\rangle_E$$

$$= \frac{2}{t}\mathbb{E}\left[P_{t/2}\varphi(u(t/2;x)) - P_{t/2}\varphi(u(t/2;y))\right]\int_0^{t/2} \langle Q^{-1}D_x u(s;x)h, dw(s)\rangle_H$$

$$+\frac{2}{t}\mathbb{E}\,P_{t/2}\varphi(u(t/2;y))\int_0^{t/2} \langle Q^{-1}\left(D_x u(s;x)h - D_x u(s;y)h\right), dw(s)\rangle_H$$

and then by easy computations from (7.3.4) we have

$$|\langle D(P_t^E\varphi)(x) - D(P_t^E\varphi)(y), h\rangle_E| \leq \frac{c}{t}(t\wedge 1)^{-\frac{1+\epsilon}{2}}\|\varphi\|_0^H$$

$$\times \left(\mathbb{E}|u(t/2;x) - u(t/2;y)|_H^2\right)^{1/2} \left(\mathbb{E}\int_0^{t/2} |Q^{-1}D_x u(s;x)h|_H^2 \, ds\right)^{1/2}$$

$$+\frac{2}{t}\|\varphi\|_0^H \left(\mathbb{E}\int_0^{t/2} |Q^{-1}\left(D_x u(s;x)h - D_x u(s;y)h\right)|_H^2 \, ds\right)^{1/2}.$$

Due to (7.2.8)) this easily implies that for any $n \in \mathbb{N}$

$$\left| \int_0^1 \langle D(P_t^E \varphi)(x_n + \rho h_h) - D(P_t^E \varphi)(x_n), h_n \rangle_E \, d\rho \right| \leq c \, (t \wedge 1)^{-(1+\epsilon)} \|\varphi\|_0^H \, |h_n|^2$$

$$+ \frac{2}{t} \|\varphi\|_0^H \left(\mathbb{E} \int_0^1 \int_0^{t/2} |Q^{-1} \left(D_x u(s; x_n + \rho h_n) - D_x u(s; x_n) \right) h_n|_H^2 \, ds \, d\rho \right)^{1/2}$$

$$= I_{1,n}(h) + \frac{2}{t} \|\varphi\|_0^H \left(\mathbb{E} \int_0^1 I_{2,n}(t, h, \rho) \, d\rho \right)^{1/2}.$$

Clearly, as $|h_n|_H \leq c|h|_H$,

$$\lim_{|h|_H \to 0} \lim_{n \to +\infty} \frac{I_{1,n}(h)}{|h|_H} = 0.$$

Thus, if we show that

$$\lim_{|h|_H \to 0} \lim_{n \to +\infty} \frac{\mathbb{E} \int_0^1 I_{2,n}(t, h, \rho) d\rho}{|h|_H^2} = 0, \tag{7.3.5}$$

we can conclude that (7.3.3) holds and $P_t \varphi$ is differentiable for any $\varphi \in C_b(H)$. For any $n \in \mathbb{N}$, due to (7.2.9) we have

$$\int_0^{t/2} |Q^{-1} \left(D_x u(s; x_n + \rho h_n) - D_x u(s; x_n) \right) h_n|_H^2 \, ds \leq c(|x|_H, |h|_h)(t)|h_n|_H^{2(\beta+1)}$$

and then

$$\lim_{n \to +\infty} \int_0^1 I_{2,n}(t, h, \rho) d\rho \leq c(|x|_H, |h|_h)(t)|h|_H^{2(\beta+1)}.$$

This implies that

$$\lim_{|h|_H \to 0} \lim_{n \to +\infty} \frac{1}{|h|_H^2} \int_0^1 I_{2,n}(t, h, \rho) d\rho = 0, \qquad \mathbb{P} - \text{a.s.}$$

Moreover, thanks to (7.2.8) for any $n \in \mathbb{N}$

$$\frac{1}{|h|_H^2} \int_0^1 I_{2,n}(t, h, \rho) d\rho \leq c(t) \, (t \wedge 1)^{1-\epsilon}, \qquad \mathbb{P} - \text{a.s.}$$

so that, by using the dominated convergence theorem we have (7.3.5). The continuity of $D(P_t \varphi)$ can be proved in a similar way, recalling that for any $x, h \in H$

$$\langle D(P_t \varphi)(x), h \rangle_H = \lim_{n \to +\infty} \langle D(P_t^E \varphi)(x_n), h_n \rangle_E.$$

Now, in order to prove that $P_t\varphi \in C_b^1(H)$, for any $\varphi \in B_b(H)$, we remark that from (7.3.4) we obtain

$$\text{Var}\,(P_t(x, \cdot) - P_t(y, \cdot)) \leq \nu\,(t \wedge 1)^{-\frac{1+\epsilon}{2}}|x - y|_H,$$

where $P_t(x, \cdot)$ is the law of $u(t; x)$. This implies that for any $\varphi \in B_b(H)$ and $t > 0$ $P_t\varphi \in C_b(H)$. Therefore, since $P_t\varphi = P_{t/2}(P_{t/2}\varphi)$, it follows that $P_t\varphi \in C_b^1(H)$. □

Remark 7.3.2. Notice that P_t is the restriction of P_t^E to $B_b(H)$, that is for any $\varphi \in B_b(H)$ $P_t^E\varphi$ extends to a Borel bounded function defined in the whole space H and $P_t^E\varphi = P_t\varphi$ on E.

Indeed, let $x \in H$ and let $\{x_n\} \subset E$ be a sequence converging to x in H. For any $n \in \mathbb{N}$ and $\varphi \in B_b(H)$ we have that $P_t^E\varphi(x_n) = P_t\varphi(x_n)$. Now, due to the previous theorem $P_t\varphi \in C_b(H)$ for any $t > 0$ and then for any $t > 0$

$$\exists \lim_{n \to +\infty} P_t^E\varphi(x_n) = P_t\varphi(x).$$

Thus, if we define

$$P_t^E\varphi(x) = \lim_{n \to +\infty} P_t^E\varphi(x_n), \quad x \in H,$$

we have that $P_t^E\varphi \in B_b(H)$ and $P_t^E\varphi = P_t\varphi$.

Chapter 8

Ergodicity

In the chapters 4, 6 and 7 we have proved that the semigroup P_t corresponding to the system
$$du(t) = [Au(t) + F(u(t))] \, dt + Q dw(t), \quad u(0) = x \qquad (8.0.1)$$
has a regularizing effect both in E and in H. Here we apply these results to the proof of the existence of a unique invariant measure μ for the semigroup P_t, which is equivalent to all transition probabilities $P_t(x, \cdot)$, $t > 0$ and $x \in H$, and which is concentrated on the space of continuous functions E. Moreover, we want to show that for any $\varphi \in L^2(H, \mu)$

$$\lim_{t \to +\infty} P_t \varphi = \int_H \varphi(y) \, \mu(dy), \quad \text{in } L^2(H; \mu), \qquad (8.0.2)$$

and the following *strong law of large numbers* holds

$$\lim_{t \to +\infty} \frac{1}{t} \int_0^t \varphi(u(s; x)) \, ds = \int_H \varphi(y) \, \mu(dy), \qquad \mathbb{P} - \text{a.s.} \qquad (8.0.3)$$

Concerning the existence of the invariant measure, we prove a stronger result. Actually, we show that there exists an invariant measure for the system (8.0.1) regarded as a system in the space of continuous functions E and as a by-product we get the existence of an invariant measure for P_t which is concentrated on E. Indeed, for any $x \in E$ there exists a unique mild solution $u(\cdot; x)$ for the equation (8.0.1), which is an E-valued adapted process. Then, as in the previous two chapters, for any $\varphi \in B_b(E)$ and $t \geq 0$ we define

$$P_t^E \varphi(x) = \mathbb{E} \varphi(u(t; x)), \quad x \in E.$$

We prove that for any $x_0 \in E$ and $a > 0$ the family of probability measures $P_t^E(x_0, \cdot)$, $t \geq a$, is *tight* on E, and hence from the Krylov-Bogoliubov theorem (see [48] and Theorem 2.1.1) the existence of an invariant measure μ for P_t^E follows. In particular, by setting for any Borel set $\Gamma \subset H$

$$\mu'(\Gamma) = \mu(\Gamma \cap E), \qquad (8.0.4)$$

this implies the existence of an invariant measure μ' for P_t, which is concentrated on E (notice that since E is continuously embedded in H, then μ' is well defined).

As far as the uniqueness and the asymptotic behaviour of the invariant measure μ are concerned, we first prove that, even by weakening the conditions on f given in the chapters 6 and 7, P_t enjoys the *strong Feller* property in H, that is it maps $B_b(H)$ into $C_b(H)$ (notice that in the chapter 7 we have shown that, under stronger assumptions for f, P_t maps $B_b(H)$ into $C_b^1(H)$). Then we show that P_t is *irreducible* on H, that is $P_t(x, A) > 0$, for any non-empty open set $A \subset H$ and for any $t > 0$ and $x \in H$. Thus we can apply the Khas'minskii theorem (see Theorem 2.1.4 for the statement and [80] and [48] for a proof) and we get that all the transition probabilities $P_t(x, \cdot)$, $t > 0$ and $x \in H$, are equivalent. Due to the Doob theorem (see Theorem 2.1.3 and [54] and [48] for a proof), this implies that there exists exactly one invariant measure μ for P_t and (8.0.2) and (8.0.3) hold. It is interesting to notice that, from the uniqueness of the invariant measure on H, the uniqueness of the invariant measure on E follows. Indeed, if μ_1 and μ_2 are two invariant measure for P_t^E, the measures μ_1' and μ_2' defined as in (8.0.4) are invariant for P_t and then they coincide. We will show that this implies that μ_1 and μ_2 coincide, as well.

8.1   Assumptions

Concerning the operators A and Q, we refer for all hypotheses and preliminary results to the chapters 4 and 6 (see the subsection 4.1.1 and the section 6.1). Concerning the reaction term, as in the chapter 6 we assume that for any $k = 1, \ldots, n$

$$f_k(\xi, \sigma, \ldots, \sigma_n) = g_k(\xi, \sigma_k) + h_k(\xi, \sigma_1, \ldots, \sigma_n),$$

where g_k and h_k are two continuous functions. Here we can weaken the conditions on g_k and h_k given in the Hypotheses 6.2 and 6.3.

Hypothesis 8.1.   *1. For any fixed $\xi \in \overline{\mathcal{O}}$, the function $h_k(\xi, \cdot)$ is of class C^1 and the derivative is bounded, uniformly with respect to $\xi \in \overline{\mathcal{O}}$. Moreover the function $D_\sigma h_k : \overline{\mathcal{O}} \times \mathbb{R}^n \to \mathbb{R}$ is continuous.*

   2. For any fixed $\xi \in \overline{\mathcal{O}}$, the function $g_k(\xi, \cdot)$ is of class C^1 and there exists $m \geq 0$ such that

$$\sup_{\xi \in \overline{\mathcal{O}}} \sup_{t \in \mathbb{R}} \left(\frac{|g_k(\xi, t)|}{1 + |t|^{2m+1}} + \frac{|D_t g_k(\xi, t)|}{1 + |t|^{2m}} \right) < \infty.$$

   Moreover, the function $D_t g_k : \overline{\mathcal{O}} \times \mathbb{R} \to \mathbb{R}$ is continuous.

   3. There exists $\beta \in \mathbb{R}$ such that for any $\xi \in \overline{\mathcal{O}}$ and $t \in \mathbb{R}$

$$D_t g_k(\xi, t) \leq \beta I.$$

Thus, the conditions above are analogous to those introduced in the Hypothesis 6.2, with the only difference that $g_k(\xi, \cdot)$ and $h_k(\xi, \cdot)$ are assumed to be only C^1.

Hypothesis 8.2. *There exist $a, \gamma > 0$ and $c \in \mathbb{R}$ such that for any $t, s \in \mathbb{R}$*

$$\sup_{\xi \in \overline{\mathcal{O}}} \left(g_k(\xi, t + s) - g_k(\xi, s) \right) t \leq -a|t|^{2m+2} + c \left(|s|^{\gamma} + 1 \right) |t|.$$

From the definition of δ_h, due to (6.1.2) the hypothesis above implies that if $x, h \in E$, then

$$\langle Ah, \delta_h \rangle_E + \langle F(x + h) - F(x), \delta_h \rangle_E \leq -a|h|_E^{2m+1} + c \left(|x|_E^{\gamma} + 1 \right). \tag{8.1.1}$$

For all notations and preliminary results about invariant measure we refer to the section 2.1.

8.2 Existence

In this section we prove the existence of an invariant measure for the equation (8.0.1). Before proceeding, it is useful to remark that if we are able to prove the existence of an invariant measure μ for P_t^E, this automatically yields the existence of an invariant measure μ' for P_t. Indeed, let μ be a measure on $(E, \mathcal{B}(E))$ such that for any $\varphi \in B_b(E)$ and $t \geq 0$

$$\int_E P_t^E \varphi(x) \, \mu(dx) = \int_E \varphi(x) \, \mu(dx).$$

As E is continuously embedded into H, for any $\Gamma \in \mathcal{B}(H)$ the set $\Gamma \cap E$ belongs to $\mathcal{B}(E)$, so that the measure μ' defined by

$$\mu'(\Gamma) = \mu(\Gamma \cap E), \quad \Gamma \in \mathcal{B}(H), \tag{8.2.1}$$

is well defined. Moreover, for any $\varphi \in B_b(H)$ and $x \in E$ we have $P_t^E \varphi(x) = P_t \varphi(x)$, so that for any $t \geq 0$

$$\int_H P_t \varphi(x) \mu'(dx) = \int_E P_t \varphi(x) \, \mu(dx) = \int_E P_t^E \varphi(x) \, \mu(dx)$$

$$= \int_E \varphi(x) \, \mu(dx) = \int_H \varphi(x) \, \mu'(dx),$$

which means that μ' is invariant for P_t. Notice that μ' is concentrated on E.

Lemma 8.2.1. *Under the Hypotheses 6.1-1 and 6.1-2, for any $p \geq 1$ it holds*

$$\sup_{t \geq 0} \mathbb{E} \, |w^A(t)|_E^p < +\infty. \tag{8.2.2}$$

Proof. As proved in the Lemma 6.1.2, for any $T > 0$ and $p \geq 2$ we have

$$\mathbb{E} \sup_{t \in [0,T]} |w^A(t)|_E^p < \infty.$$

Now, if $t \geq 1$ we have

$$w^A(t) = \int_0^t e^{(t-s)A} Q \, dw(s) = \sum_{k=0}^{[t]-1} \int_k^{k+1} e^{(t-s)A} Q \, dw(s) + \int_{[t]}^t e^{(t-s)A} Q \, dw(s)$$

$$= \sum_{k=0}^{[t]-1} e^{(t-k-1)A} \int_k^{k+1} e^{(k+1-s)A} Q \, dw(s) + \int_{[t]}^t e^{(t-s)A} Q \, dw(s).$$

Then, due to (4.1.2), for any $p \geq 2$ it follows

$$\mathbb{E}|w^A(t)|_E^p \leq c \, e^{-pt+p} \sum_{k=0}^{[t]-1} e^{pk} \, \mathbb{E} \left| \int_k^{k+1} e^{(k+1-s)A} Q \, dw(s) \right|_E^p$$

$$+ c \, \mathbb{E} \left| \int_{[t]}^t e^{(t-s)A} Q \, dw(s) \right|_E^p.$$

Due to the stationarity of $w^A(t)$ it is easy to verify that if $k \in \mathbb{N}$ and $t \geq 0$

$$\mathbb{E} \left| \int_k^{k+1} e^{(k+1-s)A} Q \, dw(s) \right|_E^p + \mathbb{E} \left| \int_{[t]}^t e^{(t-s)A} Q \, dw(s) \right|_E^p \leq c \sup_{s \in [0,1]} \mathbb{E}|w^A(s)|_E^p.$$

Thanks to (6.1.12), this yields

$$\mathbb{E}|w^A(t)|_E^p \leq c \left(e^{-pt+p} \sum_{k=0}^{[t]-1} e^{pk} + 1 \right) = c \left(e^{-pt+p} \left(\frac{e^{p[t]} - 1}{e^p - 1} \right) + 1 \right) \leq \frac{c e^p}{e^p - 1},$$

which implies (8.2.2).   □

Thanks to the Ascoli-Arzelà theorem the embedding of $C^\theta(\overline{\mathcal{O}}; \mathbb{R}^n)$ into E is compact. Hence if we prove that for some $\theta \in (0,1)$ and $a > 0$

$$\sup_{t \geq a} \mathbb{E}|u(t;x)|_{C^\theta(\overline{\mathcal{O}}; \mathbb{R}^n)} \leq c(a,x) < \infty,$$

it follows that the family of probability measures $\{P_t(x, \cdot)\}_{t \geq a}$ is tight on E. According to the Krylov-Bogoliubov theorem, this implies that there exists an invariant measure μ for P_t^E.

Proposition 8.2.2. *Under the Hypotheses 6.1, 8.1 and 8.2, there exists $\theta \in (0,1)$ such that for any $a > 0$ and $x \in E$*

$$\sup_{t \geq a} \mathbb{E} |u(t;x)|_{C^\theta(\overline{\mathcal{O}};\mathbb{R}^n)} < \infty. \tag{8.2.3}$$

Thus there exists an invariant measure both for P_t^E and for P_t.

Proof. We first give an estimate of

$$\sup_{t \geq 0} \mathbb{E} |u(t;x)|_E^p,$$

for any $p \geq 2$ and $x \in E$. To this purpose, we introduce the function $v(t;x) = u(t;x) - w^A(t)$, which is the unique mild solution of the problem

$$\frac{dv}{dt}(t) = Av(t) + F(v(t) + w^A(t)), \quad v(0) = x.$$

We can assume that $v(\cdot;x)$ is a strict solution, otherwise we approximate it as in the proof of the Proposition 6.2.2 by means of a sequence $\{v_n(\cdot;x)\}$ of strict solutions. Thus, if $\delta_t = \delta_{v(t)}$ is the element in $\partial |v(t)|_E$ defined as in (A.1.1), by using (6.1.2), (6.1.19) and (8.1.1) we have

$$\frac{d^-}{dt} |v(t)|_E \leq \langle Av(t), \delta_t \rangle_E + \langle F(v(t) + w^A(t)), \delta_t \rangle_E$$

$$\leq \langle Av(t), \delta_t \rangle_E + \langle F(v(t) + w^A(t)) - F(w^A(t)), \delta_t \rangle_E + \langle F(w^A(t)), \delta_t \rangle_E$$

$$\leq -a |v(t)|_E^{2m+1} + c \left(|w^A(t)|_E^{\gamma \vee (2m+1)} + 1 \right) \leq -a |v(t)|_E + c \left(|w^A(t)|_E^{\gamma \vee (2m+1)} + 1 \right).$$

By comparison it easily follows

$$|v(t;x)|_E \leq e^{-at}|x|_E + c \int_0^t e^{-a(t-s)} \left(|w^A(s)|_E^{\gamma \vee (2m+1)} + 1 \right) ds.$$

Then, due to (8.2.2), by taking the expectation and the supremum over $t \geq 0$ we get

$$\sup_{t \geq 0} \mathbb{E} |u(t;x)|_E^p \leq c \left(\sup_{t \geq 0} \mathbb{E} |v(t;x)|_E^p + \sup_{t \geq 0} \mathbb{E} |w^A(t)|_E^p \right) \leq c \left(|x|_E^p + 1 \right). \tag{8.2.4}$$

Next, we estimate the expectation of the θ-Hölder norm of $u(t;x)$. We recall that if we fix $s \in (0,1)$ and $p \geq 1$ such that $sp > d$, then by the Sobolev embedding theorem $W^{s,p}(\mathcal{O};\mathbb{R}^n)$ is continuously embedded into $C^\theta(\overline{\mathcal{O}};\mathbb{R}^n)$, for any $\theta < s - d/p$. Hence, from (4.1.3) we have that e^{tA} maps $L^p(\mathcal{O};\mathbb{R}^n)$ into $C^\theta(\overline{\mathcal{O}};\mathbb{R}^n)$, for any $p > d$ and $\theta < 1 - d/p$ and

$$|e^{tA}x|_{C^\theta(\overline{\mathcal{O}};\mathbb{R}^n)} \leq c e^{-\frac{\theta p + d}{2p}t} (t \wedge 1)^{-\frac{\theta p + d}{2p}} |x|_p.$$

In particular we have that e^{tA} maps E into $C^\theta(\overline{\mathcal{O}}; \mathbb{R}^n)$ for any $\theta < 1$ and

$$|e^{tA}x|_{C^\theta(\overline{\mathcal{O}};\mathbb{R}^n)} \leq c\, e^{-\frac{\theta}{2}t}(t \wedge 1)^{-\frac{\theta}{2}}|x|_E, \tag{8.2.5}$$

so that for any $\theta \in (0,1)$ we have

$$\left|\int_0^t e^{(t-s)A}F(u(s;x))\,ds\right|_{C^\theta(\overline{\mathcal{O}};\mathbb{R}^n)} \leq c\int_0^t e^{-\frac{\theta}{2}(t-s)}((t-s)\wedge 1)^{-\frac{\theta}{2}}|F(u(s;x))|_E\,ds.$$

Then, since F satisfies the growth condition (6.1.19), we have

$$\left|\int_0^t e^{(t-s)A}F(u(s;x))\,ds\right|_{C^\theta(\overline{\mathcal{O}};\mathbb{R}^n)} \leq c\int_0^t e^{-\frac{\theta}{2}(t-s)}(t-s)^{-\frac{\theta}{2}}\left(1+|u(s;x)|_E^{2m+1}\right)\,ds.$$

Now, as $u(t;x)$ is the unique mild solution of the problem

$$u(t;x) = e^{tA}x + \int_0^t e^{(t-s)A}F(u(s;x))\,ds + w^A(t),$$

by using again (8.2.5), from the Hölder inequality we can conclude that for any $t > 0$ and $p > 2/(2-\theta)$ and for any $\xi, \zeta \in \overline{\mathcal{O}}$ it holds

$$|u(t,\xi;x) - u(t,\zeta;x)|^p \leq c\, e^{-\frac{\theta p}{2}t}t^{-\frac{\theta p}{2}}|x|_E^p\,|\xi-\zeta|^{\theta p}$$

$$+c\left(\int_0^t (t-s)^{-\frac{\theta p}{2(p-1)}}\,ds\right)^{p-1}\int_0^t e^{-\frac{\theta p}{2}(t-s)}\left(1+|u(s;x)|_E^{p(2m+1)}\right)\,ds\,|\xi-\zeta|^{\theta p}$$

$$+c\,|w^A(t,\xi) - w^A(t,\zeta)|^p.$$

Then, as $w^A(t,\xi) - w^A(t,\zeta)$ is gaussian, by taking the expectation we get

$$\mathbb{E}|u(t,\xi;x) - u(t,\zeta;x)|^p \leq c\, e^{-\frac{\theta p}{2}t}t^{-\frac{\theta p}{2}}|x|_H^p\,|\xi-\zeta|^{\theta p}$$

$$+c\,t^{\frac{p(2-\theta)}{2}-1}\int_0^t e^{-\frac{\theta p}{2}(t-s)}\left(1+\mathbb{E}|u(s;x)|_E^{p(2m+1)}\right)\,ds\,|\xi-\zeta|^{\theta p} \tag{8.2.6}$$

$$+c\left(\mathbb{E}|w^A(t,\xi) - w^A(t,\zeta)|^2\right)^{\frac{p}{2}}.$$

We recall that in [47, Theorem 5.2.9] it is proved that for any $\xi, \eta \in \overline{\mathcal{O}}$ and $t \geq 0$

$$\mathbb{E}|w^C(t,\xi) - w^C(t,\eta)|^2 \leq c\,|\xi-\eta|^\rho,$$

where ρ is the constant introduced in the Hypothesis 6.1-2. Moreover, as shown in the proof of the Lemma 6.1.2 it holds

$$w^A(t,\xi) - w^A(t,\eta) = [\psi(w^A)(t)](\xi) - [\psi(w^A)(t)](\eta) + w^C(t,\xi) - w^C(t,\eta),$$

and then we have

$$|w^A(t,\xi) - w^A(t,\eta)|^2 \le c\,|\psi(w^A)(t)|^2_{C^{\rho/2}(\overline{\mathcal{O}};\mathbb{R}^n)}|\xi - \eta|^\rho + c\,|w^C(t,\xi) - w^C(t,\eta)|^2.$$

If we fix any $\delta \in (\rho/2, 1)$, from the Sobolev embedding theorem we have

$$|\psi(w^A)(t)|_{C^{\rho/2}(\overline{\mathcal{O}};\mathbb{R}^n)} \le |\psi(w^A)(t)|_{\delta,\,p},$$

for p sufficiently large and due to the first inequality in (6.1.14) we get

$$|\psi(w^A)(t)|^2_{C^{\rho/2}(\overline{\mathcal{O}};\mathbb{R}^n)} \le c\left(\int_0^t e^{-\frac{t-s}{2}}\left((t-s)\wedge 1\right)^{-\frac{1+\delta}{2}}|w^A(s)|_E\,ds\right)^2.$$

By easy calculations, from (8.2.2) it follows

$$\sup_{t\ge 0}\mathbb{E}|\psi(w^A)(t)|^2_{C^{\rho/2}(\overline{\mathcal{O}};\mathbb{R}^n)} < \infty,$$

so that

$$\mathbb{E}|w^A(t,\xi) - w^A(t,\eta)|^2 \le c\,|\xi - \eta|^\rho.$$

According to (8.2.6), this implies that if we take $\theta = \rho/2$, for any $t > 0$ we have

$$\mathbb{E}|u(t,\xi;x) - u(t,\zeta;x)|^p \le c\left(t^{-\frac{\rho p}{4}}|x|_E^{p(2m+1)} + 1\right)|\xi - \zeta|^{\frac{\rho p}{2}}.$$

Then, for any $\alpha \in (0,1)$ and $t > 0$ we get

$$\mathbb{E}\int_{\mathcal{O}\times\mathcal{O}}\frac{|u(t,\xi;x) - u(t,\zeta;x)|^p}{|\xi - \eta|^{d+\alpha p}}\,d\xi d\eta$$

$$\le c\left(1 + t^{-\frac{\rho p}{4}}|x|_E^{p(2m+1)}\right)\int_{\mathcal{O}\times\mathcal{O}}|\xi - \eta|^{\frac{\rho p}{2}-d-\alpha p}\,d\xi d\eta,$$

which is finite if $\alpha < \rho/2$. This means that $u(t;x) \in W^{\alpha,p}(\mathcal{O};\mathbb{R}^n)$ $\mathbb{P}$-a.s. for any $\alpha < \rho/2$ and $p > 4/(4 - \rho)$ and for any $t > 0$. Moreover if $a > 0$ it holds

$$\sup_{t\ge a}\mathbb{E}|u(t;x)|_{W^{\alpha,\,p}(\mathcal{O};\mathbb{R}^n)} \le c_a\left(1 + |x|_E^{p(2m+1)}\right).$$

By the Sobolev embedding theorem, it follows that if $\alpha p > d$ and $\theta < \alpha - d/p$ then $u(t;x) \in C^\theta(\overline{\mathcal{O}};\mathbb{R}^n)$, for any $t > 0$, and the estimate (8.2.3) holds. $\qquad\square$

8.3   Uniqueness

In the previous section we have proved that there exists an invariant measure for the
semigroup P_t which is concentrated on E. The aim of this section is to prove that
such an invariant measure is unique. We stress that, as a by-product, this yields
the uniqueness of the invariant measure for P_t^E. Indeed, assume that there exists
two invariant measures μ_1 and μ_2 for P_t^E. Then, as shown in the previous section,
the two measures μ_1' and μ_2' defined as in (8.2.1) are invariant for P_t, so that they
coincide. Next lemma proves that this implies that $\mu_1 = \mu_2$.

Lemma 8.3.1. If $\mu_1' = \mu_2'$, then $\mu_1 = \mu_2$.

Proof. It is sufficient to prove that for any $\varphi \in C_b(E)$ it holds

$$\int_E \varphi(x)\,\mu_1(dx) = \int_E \varphi(x)\,\mu_2(dx).$$

For any $\varphi \in C_b(E)$ we define

$$J\varphi(x) = \begin{cases} \varphi(x) & \text{if } x \in E \\ 0 & \text{if } x \in H \setminus E. \end{cases}$$

Our aim is to show $J\varphi \in B_b(H)$. Let $\{\varphi_k\}$ be a sequence in $C_b(H)$ such that $\varphi_k(x)$
converges to $\varphi(x)$, for any $x \in E$, as k goes to infinity (see the Proposition 6.1.5).
We define

$$\widetilde{\varphi}_k(x) = \begin{cases} \varphi_k(x) & \text{if } x \in E \\ 0 & \text{if } x \in H \setminus E. \end{cases}$$

Clearly we have that $\widetilde{\varphi}_k(x)$ converges to $J\varphi(x)$ for any $x \in H$. Then, since

$$\sup_{x \in H} |J\varphi(x)| \leq \|\varphi\|_0^E,$$

if we show that $\widetilde{\varphi}_k \in B_b(H)$, it follows that $J\varphi \in B_b(H)$. For any Borel set $A \subset \mathbb{R}$
we have

$$\widetilde{\varphi}_k^{-1}(A) = \begin{cases} (\varphi_k^{-1}(A) \cap E) \cup H \setminus E & \text{if } 0 \in A \\ \varphi_k^{-1}(A) \cap E & \text{if } 0 \notin A. \end{cases}$$

Then, as φ_k is continuous in H and E is a Borel subset of H, we can conclude that
$\widetilde{\varphi}_k^{-1}(A) \in \mathcal{B}(H)$, so that $\widetilde{\varphi}_k \in B_b(H)$.

Now, if $\varphi \in C_b(E)$ we have

$$\int_E \varphi(x)\,\mu_1(dx) = \int_E J\varphi(x)\,\mu_1(dx) = \int_H I\varphi(x)\,\mu_1'(dx)$$

$$= \int_H I\varphi(x)\,\mu_2'(dx) = \int_E \varphi(x)\,\mu_2(dx),$$

so that $\mu_1 = \mu_2$. $\qquad \Box$

Theorem 8.3.2. *Assume that the Hypotheses 6.1, 8.1 and 8.2 are fulfilled. Then*

1. *the system (8.0.1) admits a unique invariant measure μ which is equivalent to all probability measures $P_t(x, \cdot)$, for $t > 0$ and $x \in H$, and which is concentrated on E;*

2. *for any $x \in H$ and $\varphi \in B_b(H)$ it holds*

$$\lim_{t \to +\infty} \frac{1}{t} \int_0^t \varphi(u(s; x)) \, ds = \int_H \varphi(x) \, \mu(dx), \quad \mathbb{P} - \text{a.s.}$$

3. *for any Borel measure ν on H*

$$\lim_{t \to +\infty} \text{Var} \left(P_t^\star \nu - \mu \right) = 0.$$

In particular, for any $x \in H$ and $\varphi \in B_b(H)$

$$\lim_{t \to +\infty} P_t \varphi(x) = \int_H \varphi(y) \, \mu(dy).$$

The main tools that we are using in order to prove the previous result are given by the Khas'minskii theorem and the Doob theorem and some improvements of it due to Seidler [107] and Stettner [110] (see also chapter 2, section 2.1).

Therefore, if we prove that the semigroup P_t enjoys the *strong Feller* property and is *irreducible*, we can conclude that the Theorem 8.3.2 holds true.

8.3.1 Strong Feller Property

In this subsection we show that even is $f(\xi, \cdot)$ is only C^1 the semigroup P_t enjoys the *strong Feller* property, that is maps $B_b(H)$ into $C_b(H)$, for any $t > 0$.

Proposition 8.3.3. *Under the Hypotheses 6.1, 8.1 and 8.2 we have*

$$\varphi \in B_b(H) \implies P_t \varphi \in C_b^{0,1}(H), \quad t > 0.$$

Proof. If $m = 0$, we are under the hypotheses of the Proposition (4.4.4) and we have that P_t maps $B_b(H)$ into $C_b^1(H)$, for any $t > 0$.

If $m > 0$, in the chapter 6 (see Remark 6.5.2) we have proved that under the Hypotheses 6.1, 8.1 and 8.2, even if $f(\xi, \cdot)$ is only of class C^1, $P_t^E \varphi$ is differentiable in E, for any $\varphi \in B_b(E)$ and $t > 0$. Moreover, if $\varphi \in C_b(E)$ and $x, h \in E$ it holds

$$\left\langle D(P_t^E \varphi)(x), h \right\rangle_E = \frac{1}{t} \mathbb{E} \varphi(u(t; x)) \int_0^t \left\langle Q^{-1} D_x u(s; x) h, dw(s) \right\rangle_H, \tag{8.3.1}$$

where $D_x u(t; x)h$ is the derivative of $u(t; x)$ with respect to x along the direction h. Notice that in order to give a meaning to the formula (8.3.1) we had preliminary to prove that $D_x u(s; x)h \in D(Q^{-1})$, for any $s > 0$, and to this purpose the Hypothesis 6.1-3 was crucial.

Now, let us fix $x, h \in H$ and any two sequences $\{x_n\}$ and $\{h_n\}$ in E, converging respectively to x and h in H. In the chapter 7, Proposition 7.2.3 we have proved that

$$\lim_{n \to +\infty} \int_0^t |Q^{-1} (D_x u(s; x_n)h_n - v(s; x, h))|_H^2 \, ds = 0, \qquad (8.3.2)$$

for a suitable process $v(t; x, h)$ which is independent of the choice of $\{x_n\}$ and $\{h_n\}$ and such that

$$\sup_{x \in H} \mathbb{E} \int_0^t |Q^{-1} v(s; x, h)|_H^2 \, ds \le c(t) t^{1-\epsilon} |h|_H^2. \qquad (8.3.3)$$

Due to the definition of generalized solution, for any $\varphi \in C_b(H)$ we have

$$P_t \varphi(x + h) - P_t \varphi(x) = \lim_{n \to +\infty} P_t \varphi(x_n + h_n) - P_t \varphi(x_n)$$

$$= \lim_{n \to +\infty} P_t^E \varphi(x_n + h_n) - P_t^E \varphi(x_n),$$

and then from (8.3.1) and (8.3.2) and from the estimate (7.2.8) it easily follows

$$P_t \varphi(x + h) - P_t \varphi(x)$$

$$= \lim_{n \to +\infty} \frac{1}{t} \int_0^1 \mathbb{E} \varphi(u(t; x_n + \theta h_n)) \int_0^t \langle Q^{-1} D_x u(s; x_n + \theta h_n)h_n, dw(s) \rangle_H \, d\theta$$

$$= \frac{1}{t} \int_0^1 \mathbb{E} \varphi(u(t; x + \theta h)) \int_0^t \langle Q^{-1} D_x u(s; x + \theta h)h, dw(s) \rangle_H \, d\theta.$$

Hence, by using (8.3.3) we get

$$|P_t \varphi(x + h) - P_t \varphi(x)| \le \|\varphi\|_0^H c(t) t^{-\frac{1-\epsilon}{2}} |h|_H,$$

which means that $P_t \varphi$ is Lipschitz-continuous and for any $x, y \in H$ and $t > 0$

$$\text{Var} \left(P_t(x, \cdot) - P_t(y, \cdot) \right) \le c(t) t^{-\frac{1-\epsilon}{2}} |x - y|_H.$$

This implies that $P_t \varphi$ is Lipschitz-continuous, for any $\varphi \in B_b(H)$. $\qquad \square$

8.3.2   Irreducibility

Consider the deterministic system

$$\frac{dv}{dt}(t) = Av(t) + F(v(t)) + Qz(t), \quad v(0) = x, \qquad (8.3.4)$$

with $z \in L^2(0, T; H)$ and $x \in E$. By arguing as in the proof of the Proposition 6.2.2, we have that the problem (8.3.4) has a unique mild solution $v(t; x, z)$ which belongs to $C((0, +\infty); E) \cap L^\infty_{loc}((0, \infty); E)$. Moreover, for any $x \in H$ there exists a unique generalized solution in $C([0, +\infty); H)$, which we denote by $v(t; x, z)$, as well.

Proposition 8.3.4. *Assume the Hypotheses 6.1 and 8.1. Then, for any $x \in E$ and $y \in \overline{D(A)}^E$ and for any $T, \epsilon > 0$, there exists $z \in L^2(0, T; H)$ such that*

$$|v(T; x, z) - y|_E < \epsilon.$$

Proof. We can assume that $x \in \overline{D(A)}^E$. Indeed, if that is not the case, we first fix any $0 < \delta < T$ and we consider the problem

$$\frac{dv}{dt}(t) = Av(t) + F(v(t)), \quad t \in [0, \delta], \quad v(0) = x.$$

Hence we consider the new problem

$$\frac{dv}{dt}(t) = Av(t) + F(v(t)) + Qz(t), \quad t \in [\delta, T], \quad v(\delta) = v(\delta; x, 0) \in \overline{D(A)}^E,$$

and we find a control $\bar{z} \in L^2(\delta, T; H)$ such that $|v(T; v(\delta; x, 0), z) - y|_E < \epsilon$. Thus, if we set

$$z(t) = \begin{cases} 0 & \text{if } t \in [0, \delta], \\ \bar{z}(t) & \text{if } t \in (\delta, T], \end{cases}$$

we have that $z \in L^2(0, T; H)$ is the good control for the original problem.

Let us fix $\epsilon > 0$. We take $\varphi \in C([0, T]; \overline{D(A)}^E)$ such that $\varphi(0) = x$ and $\varphi(T) = y$ and we define

$$\bar{R} = \sup_{t \in [0,T]} |\varphi(t)|_E.$$

The function F is locally Lipschitz-continuous on E; thus in correspondence of $R = 2\bar{R} + \epsilon$ there exists a constant $L_R > 0$ such that F restricted to $B^E(0; R)$ is Lipschitzcontinuous of Lipschitz constant $L_R > 0$. Now, if we define

$$\psi(t) = \varphi(t) - e^{tA}x - \int_0^t e^{(t-s)A} F(\varphi(s)) \, ds,$$

since $x \in \overline{D(A)}^E$ we easily get that $\psi \in C([0, T]; \overline{D(A)}^E)$ and $\psi(0) = 0$.

Before proceeding, we state a lemma that we are using in what follows, whose proof is postponed.

Lemma 8.3.5. *For any $\psi \in C([0, T]; \overline{D(A)}^E)$, with $\psi(0) = 0$, and for any $\epsilon > 0$, there exists $z \in L^2(0, T; H)$ such that*

$$\sup_{t \in [0,T]} \left| \psi(t) - \int_0^t e^{(t-s)A} Qz(s) \, ds \right|_E < \epsilon.$$

By using the previous lemma, we see that there exists $z \in L^2(0, T; H)$ such that

$$\sup_{t \in [0,T]} \left| \psi(t) - \int_0^t e^{(t-s)A} Qz(s) \, ds \right|_E < \epsilon e^{-ML_R T},$$

where

$$M_T = \sup_{t \in [0,T]} \|e^{tA}\|_{\mathcal{L}(E)}.$$

If $v(t; x, z)$ is the mild solution of

$$\frac{dv}{dt}(t) = Av(t) + F(v(t)) + Qz(t), \quad v(0) = x,$$

we have

$$\varphi(t) - v(t; x, z)$$

$$= \psi(t) - \int_0^t e^{(t-s)A} Qz(s) \, ds + \int_0^t e^{(t-s)A} \left[F(\varphi(s)) - F(v(s; x, z)) \right] \, ds.$$

Therefore, if

$$\sup_{s \in [0,t]} |v(s; x, z)|_E \leq R$$

for some $t \in [0, T]$, it follows that

$$|\varphi(t) - v(t; x, z)|_E \leq \left| \psi(t) - \int_0^t e^{(t-s)A} Qz(s) \, ds \right|_E$$

$$+ M_T \int_0^t |F(\varphi(s)) - F(v(s; x, z))|_E \, ds$$

$$< \epsilon e^{-M_T L_R T} + M_T L_R \int_0^t |\varphi(s) - v(s; x, z)|_E \, ds.$$

By the Gronwall lemma, this yields

$$\sup_{s \in [0,t]} |v(s; x, z)|_E \leq R \Longrightarrow |\varphi(t) - v(t; x, z)|_E < \epsilon. \qquad (8.3.5)$$

Next, let us introduce the set

$$I_\epsilon = \{ t \in [0, T] \ : \ |\varphi(t) - v(t; x, z)|_E < \epsilon \}.$$

As $0 \in I_\epsilon$, we have that I_ϵ is non-empty and then we can define $T_\epsilon = \sup I_\epsilon$. If we show that $T_\epsilon = T$ it follows that $|\varphi(T) - v(T; x, z)|_E < \epsilon$, that is $|y - v(T; x, z)|_E < \epsilon$. Assume that $T_\epsilon < T$. For any $t \in [0, T_\epsilon)$ we have

$$|v(t; x, z)|_E \leq |v(t; x, z) - \varphi(t)|_E + |\varphi(t)|_E < \epsilon + \bar{R} = R - \bar{R}.$$

Then, since the mapping $[0, T] \to \mathbb{R}$, $t \mapsto |v(t; x, z)|_E$, is continuous, there exists $\delta > 0$ such that

$$t \in [0, T_\epsilon + \delta] \Longrightarrow |v(t; x, z)|_E \leq R \Longrightarrow |\varphi(t) - v(t; x, z)|_E < \epsilon,$$

last implication following from (8.3.5). This means that T_ϵ does not coincide with $\sup I_\epsilon$, which is a contradiction. $\qquad \square$

Proof of the Lemma 8.3.5 - The Banach space $D(A)$, endowed with the graph norm

$$|x|_{D(A)} = |x|_E + |Ax|_E,$$

is densely embedded into $\overline{D(A)}^E$, then, by using the technique of the Bernstein polynomials, it is possible to prove that $C^1([0, T]; D(A))$ is dense into $C([0, T]; \overline{D(A)}^E)$ (see [91, Proposition 0.1.2 and Corollary 0.1.3]). Thus it is sufficient to prove the lemma for any $\varphi \in C^1([0, T]; D(A))$.

Take $\varphi \in C^1([0, T]; D(A))$ such that $\varphi(0) = 0$ and define $u(t) = \varphi'(t) - A\varphi(t)$. We have $u \in C([0, T]; \overline{D(A)}^E)$ and

$$\varphi(t) = \int_0^t e^{(t-s)A} u(s) \, ds.$$

Now, for any $k \in \mathbb{N}$ and $t \geq 0$ we define

$$z^k(t) = \sum_{i=1}^k \lambda_i^{-1} \langle u(t), e_i \rangle_H e_i.$$

Clearly $z^k \in L^2(0, T; H)$ (actually it belongs to $C([0, T]; E)$) and $Qz^k(t) = P_k u(t)$, so that for any $t \in [0, T]$

$$\lim_{k \to +\infty} \left| Qz^k(t) - u(t) \right|_H^2 = \lim_{k \to +\infty} \sum_{i=k+1}^\infty |\langle u(t), e_i \rangle_H|^2 = 0. \qquad (8.3.6)$$

Moreover we have

$$\sup_{t \in [0, T]} \left| Qz^k(t) - u(t) \right|_H \leq \sup_{t \in [0, T]} |u(t)|_H \leq c |u|_{C([0, T]; E)}. \qquad (8.3.7)$$

Therefore, since for any $k \in \mathbb{N}$ we have

$$\left| \varphi(t) - \int_0^t e^{(t-s)A} Qz^k(t) \, ds \right|_E = \left| \int_0^t e^{(t-s)A} \left(u(s) - Qz^k(s) \right) ds \right|_E$$

$$\leq c \int_0^t (t-s)^{-\frac{d}{4}} \left| u(s) - Qz^k(s) \right|_H \, ds,$$

due to (8.3.6) and (8.3.7), by using the dominated convergence theorem we can take the limit for k going to infinity and we get

$$\lim_{k \to +\infty} \left| \varphi(t) - \int_0^t e^{(t-s)A} Q z^k(t)\, ds \right|_E = 0.$$

$\square$

We can conclude with the proof of the irreducibility of the semigroup.

Proposition 8.3.6. *Assume the Hypotheses 6.1 and 8.1 and fix $x, y \in H$. Then for any $t, \epsilon > 0$ we have*

$$P_t\left(x, B^H(y; \epsilon)\right) > 0,$$

where $B^H(y; \epsilon) = \{z \in H\, ;\, |z - y|_H < \epsilon\}$.

Proof. We want to prove that for any $t, \epsilon > 0$

$$\mathbb{P}\left(u(t; x) \in B^H(y; \epsilon)\right) = \mathbb{P}\left(|u(t; x) - y|_H < \epsilon\right) > 0.$$

Notice that we can assume that $x \in E$ and $y \in \overline{D(A)}^E$. Indeed, thanks to (7.1.6) and to the definition of generalized solution, for any $x \in H$ and $\epsilon > 0$ there exists $x_\epsilon \in E$ such that

$$\mathbb{P}\left(\sup_{t \in [0,T]} |u(t; x) - u(t; x_\epsilon)|_H < \frac{\epsilon}{3} \right) = 1.$$

Moreover, since $\overline{D(A)}^E$ is dense in H, there exists $y_\epsilon \in \overline{D(A)}^E$ such that $|y - y_\epsilon|_H < \epsilon/3$. Thus, as

$$|u(t; x) - y|_H \leq |u(t; x) - u(t; x_\epsilon)|_H + |u(t; x_\epsilon) - y_\epsilon|_H + |y_\epsilon - y|_H,$$

we have that

$$\mathbb{P}\left(|u(t; x) - y|_H < \epsilon\right) \geq \mathbb{P}\left(|u(t; x_\epsilon) - y_\epsilon|_H < \frac{\epsilon}{3}\right).$$

From the previous proposition we know there exists a control z_ϵ in $L^2(0, t; H)$ such that

$$|v(t; x_\epsilon, z_\epsilon) - y_\epsilon|_E < \frac{\epsilon}{6}.$$

Then if we are able to prove that in correspondence of such a control

$$\mathbb{P}\left(|u(t; x_\epsilon) - v(t; x_\epsilon, z_\epsilon)|_E < \frac{\epsilon}{6}\right) > 0, \tag{8.3.8}$$

our statement is proved.

As in the proof of the Proposition 8.3.4, we define

$$\overline{R} = \sup_{s \in [0,t]} |v(s; x_\epsilon, z_\epsilon)|_E,$$

where $v(s; x_\epsilon, z_\epsilon)$ is the solution of the controlled deterministic system (8.3.4) and we take L_R as the Lipschitz constant of F restricted to the ball $B^E(0; R)$, with $R = 2\overline{R} + \epsilon/6$. We introduce the set

$$\Omega_{R,\epsilon} = \left\{ \sup_{s \in [0,t]} \left| w^A(s) - \int_0^t e^{(t-s)A} Q z(s)\, ds \right|_E < \frac{\epsilon}{6} e^{-L_R t} \right\}$$

and we claim that

$$\omega \in \Omega_{R,\epsilon} \implies \sup_{s \in [0,t]} |u(s; x_\epsilon)(\omega) - v(s; x_\epsilon, z)|_E < \frac{\epsilon}{6}. \tag{8.3.9}$$

By arguing as in the proof of Lemma 8.3.5, it is possible to prove that the law of w^A is full on $C([0,t]; \overline{D(A)}^E)$, so that the probability of $\Omega_{R,\epsilon}$ is strictly positive. Therefore from (8.3.9) we get that (8.3.8) holds true and the statement of the proposition follows. Finally, the proof of the implication (8.3.9) is obtained by using arguments which are almost identical to those used in the proof of Proposition 8.3.4. □

Chapter 9

Hamilton-Jacobi-Bellman equations in Hilbert spaces

We are here concerned with the study of the following class of infinite dimensional Hamilton-Jacobi-Bellman problems

$$
\begin{cases}
\dfrac{\partial y}{\partial t}(t,x) + \mathcal{L}(x,D)y(t,x) - K(Dy(t,x)) + g(x) = 0 \\[2mm]
y(0,x) = \varphi(x),
\end{cases}
\tag{9.0.1}
$$

and

$$
\lambda\varphi(x) - \mathcal{L}(x,D)\varphi(x) + K(D\varphi(x)) = g(x), \qquad \lambda > 0, \tag{9.0.2}
$$

where $\mathcal{L}$ is the diffusion operator associated with the system (6.0.1), that is

$$
\mathcal{L}(x,D)\psi(x) = \frac{1}{2}\mathrm{Tr}\left[QQ^{*}D^{2}\psi(x)\right] + \langle Ax + F(x), D\psi(x)\rangle_{H}.
$$

Here we assume that $K : H \to \mathbb{R}$ is Fréchet differentiable and locally Lipschitz-continuous together with its derivative and $g, \varphi : H \to \mathbb{R}$ are uniformly continuous and bounded. Our aim is showing that the problems (9.0.1) and (9.0.2) admit regular solutions.

In the section 9.1 we investigate the controlled state equation

$$
du(t) = [Au(t) + F(u(t)) + z(t)]\,dt + Q\,dw(t), \qquad u(s) = 0. \tag{9.0.3}
$$

In its uncontrolled version it has been widely studied in the chapter 6. Here we prove some approximation results. Actually, we show that if $u_{\alpha}(t,s;x,z)$ is the solution of the problem

$$
du(t) = [Au(t) + F_{\alpha}(u(t)) + z(t)]\,dt + Q\,dw(t), \qquad u(s) = x, \tag{9.0.4}
$$

where F_α are the Yosida approximations of F introduced in the subsection 6.1.2, then for any $p \geq 1$

$$\lim_{\alpha \to 0} \mathbb{E} \sup_{t \in [s,T]} |u_\alpha(t, s; x, z) - u(t, s; x, z)|_E^p = 0.$$

Moreover, if $u_{\alpha,n}(t, s; x, z)$ is the solution of the finite dimensional problem

$$du(t) = [A_n u(t) + F_{\alpha,n}(u(t)) + z_n(t)] \, dt + Q_n \, dw(t), \qquad u(s) = x \qquad (9.0.5)$$

(here A_n, $F_{\alpha,n}$, z_n and Q_n are finite dimensional Galerkin approximations), we show that

$$\lim_{n \to +\infty} \mathbb{E} \sup_{t \in [s,T]} |u_{\alpha,n}(t, s; x, z) - u_\alpha(t, s; x, z)|_H^2 = 0.$$

In the section 9.2 we prove that the solution of the first variation equation corresponding to the problem (9.0.3) can be approximated by the derivative $Du_\alpha(t, s; x, z)$ of $u_\alpha(t, s; x, z)$. Analogously, we show that $Du_\alpha(t, s; x, z)$ can be approximated by the derivative of $u_{\alpha,n}(t, s; x, z)$.

In the section 9.3 we introduce the transition semigroups P_t^α and $P_t^{\alpha,n}$ associated respectively with the systems (9.0.4) and (9.0.5) and we prove that they approximate in the right way P_t and its derivative. The advantage of considering P_t^α and $P_t^{\alpha,n}$ lies in the fact that P_t^α has a stronger regularizing effect than P_t and $P_t^{\alpha,n}$ allows us to use Itô's formula. Unfortunately we can not approximate P_t directly by P_t^n.

In the section 9.4 the parabolic problem (9.0.1) is studied. As K is only locally Lipschitz-continuous, we first prove that for any $\varphi, g \in C_b^{0,1}(H)$ there exists a local solution in the space of continuous functions $y : [0, T] \times H \to \mathbb{R}$ such that $y(t, \cdot) \in C_b^1(H)$ for any $t \in (0, T]$ and the mapping

$$(0, T] \times H \to H, \qquad (t, x) \mapsto t^{\frac{1+\epsilon}{2}} |Dy(t, x)|_H,$$

is bounded and measurable. Then, by using some a-priori estimates and the approximation results proved in the previous sections, we show that such solution is global. This second step is more delicate and the proof is quite technical.

In the last section we consider the elliptic problem (9.0.2). First we prove that, if L is the weak generator of P_t, then $D(L) \subset C_b^1(H)$. Then we show that when K is Lipschitz-continuous, the problem

$$\lambda \varphi(x) - N(\varphi)(x) = \lambda \varphi(x) - L\varphi(x) + K(D\varphi(x)) = g(x) \qquad (9.0.6)$$

has a unique solution $\varphi \in D(L)$, for any $g \in C_b(H)$ and λ large enough. Then, by using suitable approximation results, we show that the operator N is m-dissipative and $D(\overline{N}) = D(L)$. This provides the existence of a solution for (9.0.6), for any $\lambda > 0$. Finally, we consider the case of K beeing only locally Lipschitz-continuous. This case is much more delicate and the results we get are weaker. In fact, we can only show that for any $g \in C_b^1(H)$ and λ sufficiently large the problem (9.0.6) has a unique solution $\varphi \in D(L) \subseteq C_b^1(H)$. This is possible due to a suitable a-priori estimate. Moreover, we show that the operator $\bar{N}$ is m-dissipative and then the equation

$$\lambda \varphi - \bar{N}(\varphi) = g$$

has a unique solution in $D(\bar{N})$, for any $g \in C_b(H)$ and $\lambda > 0$. Unfortunately, at present we are not able to characterize the domain of $\bar{N}$, and then we can not prove any further regularity of the solution.

9.1   The state equation

In this section we are concerned with the controlled version of the stochastic reaction-diffusion system (6.0.1), that is

$$\begin{cases} \dfrac{\partial u_k}{\partial t}(t,\xi) = A_k\, u_k(t,\xi) + f_k(\xi, u_1(t,\xi), \ldots, u_n(t,\xi)) + z_k(t,\xi) + Q_k \dfrac{\partial^2 w_k}{\partial t \partial \xi}(t,\xi) \\[2mm] u_k(s,\xi) = x_k(\xi), \quad 0 \le s < t \le T, \quad \xi \in \overline{\mathcal{O}} \\[2mm] \mathcal{B}_k\, u_k(t,\xi) = 0, \qquad \xi \in \partial \mathcal{O}, \qquad k = 1, \ldots, n. \end{cases}$$

By using the notations introduced in the chapters 4 and 6 it can be rewritten in the following abstract form

$$du(t) = [Au(t) + F(u(t)) + z(t)]\, dt + Q\, dw(t), \qquad u(s) = x, \tag{9.1.1}$$

for $0 \le s \le t \le T$.

Definition 9.1.1.   1. *Let us fix an adapted process $z \in L^2(\Omega; L^p(0,T;H))$, with $p > 4/(4-d)$, and $x \in E$. An E-valued adapted process $u(t) = u(t,s;x,z)$ is a mild solution for the problem (9.1.1) if*

$$u(t) = e^{(t-s)A} x + \int_s^t e^{(t-r)A}\left(F(u(r)) + z(r)\right) dr + w^A(t,s),$$

where the process $w^A(t,s)$ is the solution of the problem (6.1.10) starting from 0 at time s.

2. *Let us fix an adapted process $z \in L^2(\Omega; L^2(0,T;H))$ and $x \in H$. A H-valued process $u(t,s;x,z)$ is a generalized solution for the problem (9.1.1) if for any sequences $\{x_n\} \subset E$ converging to x in H and $\{z_n\} \subset L^2(\Omega; L^2(0,T;E))$ converging to z in $L^2(\Omega; L^2(0,T;H))$, the corresponding sequence of mild solutions $\{u(\cdot,s;x_n,z_n)\}$ converges to $u(\cdot,s;x,z)$ in $C([s,T];H)$, $\mathbb{P}$-a.s.*

By proceeding as for the uncontrolled system (see Propositions 6.2.2 and 7.1.2 and also [47, Theorem 7.13]) we have

Theorem 9.1.2. *Assume the Hypotheses 6.1 and 6.2 and fix $0 \le s \le T$.*

1. *For any $z \in L^2(\Omega; L^p(s, T; H))$, with $p > 4/(4 - d)$, and $x \in E$, the problem (9.1.1) admits a unique mild solution $u(\cdot, s; x, z) \in L^2(\Omega; C((s, T]; E) \cap L^\infty(s, T; E))$ such that*

$$|u(t, s; x, z)|_E \le c_T \left(|x|_E + |z|^{2m+1}_{L^p(s,t;H)} + \sup_{r \in [s,t]} |w^A(r, s)|^{2m+1}_E \right), \qquad \mathbb{P} - a.s.$$
(9.1.2)

2. *For any $z \in L^2(\Omega; L^2(s, T; H))$ and $x \in H$, the problem (9.1.1) admits a unique generalized solution $u(\cdot, s; x, z) \in L^2(\Omega; C([s, T]; H))$ such that*

$$|u(t, s; x, z)|_H \le c_T \left(|x|_H + |z|^{2m+1}_{L^2(s,T;H)} + \sup_{r \in [s,t]} |w^A(r, s)|^{2m+1}_E \right). \qquad \mathbb{P} - a.s.$$
(9.1.3)

3. *The unique generalized solution $u(\cdot, s; x, z)$ belongs to $L^{p_*}(s, T; L^{p_*}(\mathcal{O}; \mathbb{R}^n))$, $\mathbb{P}$-a.s., where $p_* = 2m + 2$, and*

$$u(t, s; x, z) = e^{(t-s)A}x + \int_s^t e^{(t-r)A} \left(F(u(r, s; x, z)) + z(r) \right) dr + w^A(t, s).$$

4. *For any $x_1, x_2 \in H$ and $z_1, z_2 \in L^2(\Omega; L^2(s, T; H))$*

$$|u(t, s; x_1, z_1) - u(t, s; x_2, z_2)|_H \le c_T \left(|x_1 - x_2|_H + |z_1 - z_2|_{L^2(s,T;H)} \right).$$
(9.1.4)

For any $\alpha > 0$, as in the chapter 6 we consider the approximating problem

$$du(t) = [Au(t) + F_\alpha(u(t)) + z(t)] \, dt + Q dw(t), \qquad u(s) = x, \qquad (9.1.5)$$

with $s \le t \le T$. Clearly an existence theorem analogous to the Theorem 9.1.2 holds for (9.1.5). Actually, for each $x \in E$ and $z \in L^2(\Omega; L^p(s, T; H))$, with $p > 4/(4 - d)$, there exists a unique mild solution $u_\alpha(\cdot, s; x, z)$ in $L^2(\Omega; C((s, T]; E) \cap L^\infty(s, T; E))$ and for each $x \in H$ and $z \in L^2(\Omega; L^2(s, T; H))$ there exists a unique generalized solution $u_\alpha(\cdot, s; x, z)$ which belongs to $L^2(\Omega; C([s, T]; H))$. Moreover estimates analogous to (9.1.2) and (9.1.3) hold, uniformly with respect to $\alpha > 0$.

Lemma 9.1.3. *Under the Hypotheses 6.1 and 6.2, if $z \in L^p(\Omega; L^\infty(s, T; H))$, with p sufficiently large, and $x \in E$, then for any fixed $q \ge 1$*

$$\lim_{\alpha \to 0} \sup_{t \in [s,T]} \mathbb{E}|u_\alpha(t, s; x, z) - u(t, s; x, z)|^q_E = 0, \qquad (9.1.6)$$

uniformly with respect to x in bounded subsets of E and z in the set

$$\mathcal{M}_R = \left\{ z \in L^2(\Omega; L^\infty(s, T; H)) \; ; \; \sup_{t \in [s,T]} |z(t)|_H \leq R, \; \mathbb{P} - a.s. \right\},$$

for any $R > 0$.

Proof. If we set $v_\alpha(t) = u_\alpha(t) - u(t)$, we have that v_α is the unique mild solution of the problem

$$\frac{dv}{dt}(t) = Av(t) + F_\alpha(u_\alpha(t)) - F(u(t)), \qquad v(s) = 0.$$

Thus, by using some properties of the subdifferential of the norm in E introduced in the appendix A, if $\delta_{v_\alpha(t)}$ is the element of $\partial |v_\alpha(t)|_E$ defined in (A.1.1), we have

$$\frac{d^-}{dt} |v_\alpha(t)|_E \leq \left\langle Av_\alpha(t), \delta_{v_\alpha(t)} \right\rangle_E + \left\langle F_\alpha(u_\alpha(t)) - F(u(t)), \delta_{v_\alpha(t)} \right\rangle_E.$$

From (6.1.2) and (6.1.33) this easily implies that

$$\frac{d^-}{dt} |v_\alpha(t)|_E \leq c |v_\alpha(t)|_E + |F_\alpha(u(t)) - F(u(t))|_E.$$

so that, due to the Gronwall lemma and (6.1.29), we have

$$|v_\alpha(t)|_E \leq \alpha c \int_s^t e^{c(t-r)} \left(1 + |u(r)|_E^{4m+1}\right) \, dr.$$

This implies (9.1.6), as from (6.1.12) and (9.1.2) for any $q \geq 1$ we have

$$\sup_{z \in \mathcal{M}_R} \mathbb{E} |u(\cdot, s; x, z)|_{C([s,T];E)}^q < \infty.$$

$\square$

For $n \in \mathbb{N}$ we define

$$A_n = P_n A P_n, \qquad C_n = C P_n, \qquad G_n = P_n G P_n, \qquad Q_n = Q P_n,$$

where P_n is the projection of H onto the finite dimensional space H_n generated by the eigenfunctions $\{e_1, \ldots, e_n\}$. By using a factorization argument in the Lemma 5.3.3 we have proved that for any $p \geq 1$

$$\lim_{n \to +\infty} \mathbb{E} |w^A(\cdot, s) - w^{A,n}(\cdot, s)|_{C([s,T];H)}^p = 0,$$

where $w^{A,n}(t, s)$ is the solution of the finite dimensional problem

$$dv(t) = A_n v(t) \, dt + Q_n \, dw(t), \qquad v(s) = 0.$$

Now, for any $n \in \mathbb{N}$ and $\alpha > 0$ we define

$$F_{\alpha,n}(x) = P_n F_\alpha(P_n x), \qquad x \in H.$$

It is immediate to check that for any $x, y \in H$ it holds

$$\langle F_{\alpha,n}(x) - F_{\alpha,n}(y), x - y \rangle_H \leq c |x - y|_H^2, \qquad (9.1.7)$$

for a constant c independent both of n and α. Moreover

$$|F_{\alpha,n}(x) - F_{\alpha,n}(y)|_H \leq c_\alpha |x - y|_H, \qquad (9.1.8)$$

for some constant c_α independent of n. In correspondence with each $n \in \mathbb{N}$, $\alpha > 0$ and $0 \leq s \leq T$, we consider the approximating problem

$$du(t) = [A_n u(t) + F_{\alpha,n}(u(t)) + z_n(t)]\, dt + Q_n dw(t), \qquad u(s) = P_n x, \qquad (9.1.9)$$

where $z_n(t) = P_n z(t)$ and z is an adapted process in $L^2(\Omega; L^2(s, T; H))$. Such a problem is a finite dimensional problem with Lipschitz coefficients. Thus, for any $x \in H$ there exists a unique strong solution $u_{\alpha,n}(\cdot, s; x, z) \in L^2(\Omega; C([s, T]; H))$.

Lemma 9.1.4. *Let z be an adapted process in $L^2(\Omega; L^2(s, T; H))$. If $u_{\alpha,n}(\cdot, s; x, z)$ is the unique solution of the approximating problem (9.1.9), it holds*

$$\lim_{n \to +\infty} \mathbb{E} \sup_{t \in [s,T]} |u_{\alpha,n}(t, s; x, z) - u_\alpha(t, s; x, z)|_H^2 = 0, \qquad (9.1.10)$$

uniformly for x in bounded subsets of H.

Proof. Step 1: For each $n, k \in \mathbb{N}$, we consider the problem

$$du(t) = [A_n u(t) + F_{\alpha,n}(u(t)) + z_{n \wedge k}(t)]\, dt + Q_{k \wedge n}\, dw(t), \qquad u(0) = P_n x.$$

If we denote by $u_{\alpha,n}^k(t)$ its solution, we have

$$\lim_{k \to +\infty} \sup_{n \in \mathbb{N}} |u_{\alpha,n}^k(\cdot, s; x, z) - u_{\alpha,n}(\cdot, s; x, z)|_{L^2(\Omega; C([s,T];H))} = 0. \qquad (9.1.11)$$

Actually, let us define

$$w_{n,k}^A(t, s) = \int_s^t e^{(t-r)A_n}\, (Q_n - Q_{n \wedge k})\, dw(r),$$

and $v_{\alpha,n}^k(t) = u_{\alpha,n}(t) - u_{\alpha,n}^k(t) - w_{n,k}^A(t, s)$. We have that $v_{\alpha,n}^k(t)$ is the unique solution of the problem

$$\frac{dv}{dt}(t) = A_n v(t) + F_{\alpha,n}(u_{\alpha,n}(t)) - F_{\alpha,n}(u_{\alpha,n}^k(t)) + z_n(t) - z_{n \wedge k}(t), \qquad v(s) = 0.$$

Therefore, by multiplying each side by $v_{\alpha,n}^k(t)$ and by integrating with respect to $\xi \in \mathcal{O}$, we get

$$\frac{1}{2}\frac{d}{dt}|v_{\alpha,n}^k(t)|_H^2 + |(-A)^{1/2}v_{\alpha,n}^k(t)|_H^2 =$$

$$+\left\langle F_\alpha(u_{\alpha,n}(t)) - F_\alpha(u_{\alpha,n}^k(t)), v_{\alpha,n}^k(t)\right\rangle_H + \left\langle z_n(t) - z_{n\wedge k}(t), v_{\alpha,n}^k(t)\right\rangle_H.$$

Thus, as F_α is Lipschitz-continuous, by using the Young inequality and recalling that $u_{\alpha,n}^k(t) - u_{\alpha,n}(t) = v_{\alpha,n}^k(t) + w_{n,k}^A(t,s)$, we have

$$\frac{d}{dt}|v_{\alpha,n}^k(t)|_H^2 \le c_\alpha |v_{\alpha,n}^k(t)|_H^2 + c_\alpha \left(|w_{n,k}^A(t,s)|_H^2 + |z_n(t) - z_{n\wedge k}(t)|_H^2\right).$$

Thanks to the Gronwall lemma, this yields

$$|u_{\alpha,n}^k(t) - u_{\alpha,n}(t)|_H^2 \le 2|v_{\alpha,n}^k(t)|_H^2 + 2|w_{n,k}^A(t,s)|_H^2$$

$$\le c_{\alpha,T}\left(\sup_{t\in[s,T]}|w_{n,k}^A(t,s)|_H^2 + \int_s^T |z_n(t) - z_{n\wedge k}(t)|_{L^2(s,T;H)}^2\,dt\right)$$

and, by taking first the supremum for $t \in [s,T]$ and then the expectation, we have

$$\mathbb{E}\sup_{t\in[s,T]}|u_{\alpha,n}^k(t) - u_{\alpha,n}(t)|_H^2 \le c_{\alpha,T}\mathbb{E}\left(\sup_{t\in[s,T]}|w_{n,k}^A(t,s)|_H^2 + |z_n - z_{n\wedge k}|_{L^2(s,T;H)}^2\right).$$

Thanks to a factorization argument, for any $p \ge 1$ we have

$$\lim_{k\to+\infty}\sup_{n\in\mathbb{N}}\mathbb{E}|w_{n,k}^A(\cdot,s)|_{C([s,T];H)}^p = 0$$

and, since $z \in L^2(\Omega; L^2(s,T;H))$, we have

$$\lim_{k\to+\infty}\sup_{n\in\mathbb{N}}\mathbb{E}|z_n - z_{n\wedge k}|_{L^2(s,T;H)}^2 = 0.$$

Thus we can conclude that (9.1.11) holds true.

   Step 2: For any $k \in \mathbb{N}$, we consider the problem

$$du(t) = [Au(t) + F_\alpha(u(t)) + z_k(t)]\,dt + Q_k\,dw(t), \qquad u(s) = x.$$

It is immediate to check that $w^{A,k} \in L^p(\Omega, C((s,T];D((-A)^\delta)))$, for any $\delta \in \mathbb{R}$ and $p \ge 1$. Hence, by straightforward computations, thanks to (6.1.5) we have that such a problem admits a unique mild solution $u_\alpha^k(\cdot,s;x,z)$ such that

$$|u_\alpha^k(t)|_H^2 + \int_0^t |(-C)^{1/2}u_\alpha^k(s)|_H^2\,ds$$

$$\le c_{\alpha,T}\left(|x|_H^2 + \sup_{t\in[s,T]}|w^{A,k}(t,s)|_{D((-A)^{1/2})}^2 + |z|_{L^2(s,T;H)}^2\right). \tag{9.1.12}$$

Moreover, for any fixed $k \in \mathbb{N}$ we have

$$\lim_{n \to +\infty} \left| u_{\alpha,n}^k(\cdot, s; x, z) - u_\alpha^k(\cdot, s; x, z) \right|_{L^2(\Omega; C([s,T];H))} = 0. \tag{9.1.13}$$

Actually, if we define $v_{\alpha,n}^k(t) = u_{\alpha,n}^k(t) - u_\alpha^k(t)$, for any $n > k$, we have that $v_{\alpha,n}^k$ is the unique mild solution of the problem

$$\frac{dv}{dt}(t) = Cv(t) + G_n u_{\alpha,n}^k(t) - G u_\alpha^k(t) + F_{\alpha,n}(u_{\alpha,n}^k(t)) - F_\alpha(u_\alpha^k(t)), \quad v(s) = P_n x - x.$$

Thus, we have

$$\frac{1}{2}\frac{d}{dt}|v_{\alpha,n}^k(t)|_H^2 + |(-C)^{1/2} v_{\alpha,n}^k(t)|_H^2 = \left\langle G_n u_{\alpha,n}^k(t) - G u_\alpha^k(t), v_{\alpha,n}^k(t) \right\rangle_H$$

$$+ \left\langle F_{\alpha,n}(u_{\alpha,n}^k(t)) - F_{\alpha,n}(u_\alpha^k(t)), v_{\alpha,n}^k(t) \right\rangle_H + \left\langle F_{\alpha,n}(u_\alpha^k(t)) - F_\alpha(u_\alpha^k(t)), v_{\alpha,n}^k(t) \right\rangle_H.$$

Due to (6.1.6), as $(-C)^{1/2}$ commutes with P_n, by using the Young inequality we have

$$\left\langle G_n u_{\alpha,n}^k(t) - G u_\alpha^k(t), v_{\alpha,n}^k(t) \right\rangle_H =$$

$$\left\langle v_{\alpha,n}^k(t), G^* P_n v_{\alpha,n}^k(t) \right\rangle_H + \left\langle (P_n - I) G u_\alpha^k(t), v_{\alpha,n}^k(t) \right\rangle_H$$

$$\leq \frac{1}{2}|(-C)^{1/2} v_{\alpha,n}^k(t)|_H^2 + c|v_{\alpha,n}^k(t)|_H^2 + |(P_n - I)G u_\alpha^k(t)|_H^2.$$

Therefore, thanks to (9.1.8), by using once more the Young inequality we get

$$\frac{d}{dt}|v_{\alpha,n}^k(t)|_H^2 \leq c_\alpha |v_{\alpha,n}^k(t)|_H^2 + c|F_{\alpha,n}(u_\alpha^k(t)) - F_\alpha(u_\alpha^k(t))|_H^2 + |(P_n - I)G u_\alpha^k(t)|_H^2$$

and from the Gronwall lemma this yields

$$\mathbb{E} \sup_{t \in [s,T]} |u_{\alpha,n}^k(t) - u_\alpha^k(t)|_H^2 \leq |P_n x - x|_H^2$$

$$+ c_{\alpha,T} \int_s^T \mathbb{E}\left(|F_{\alpha,n}(u_\alpha^k(r)) - F_\alpha(u_\alpha^k(r))|_H^2 + |(P_n - I)G u_\alpha^k(r)|_H^2 \right) dr.$$

By some calculations it is possible to show that

$$|F_{\alpha,n}(u_\alpha^k(r)) - F_\alpha(u_\alpha^k(r))|_H^2$$

$$\leq c_\alpha |P_n u_\alpha^k(t) - u_\alpha^k|_H^2 + 2|P_n F_\alpha(u_\alpha^k(t)) - F_\alpha(u_\alpha^k(t))|_H^2$$

and then, due to (9.1.12), we can apply the dominated convergence theorem and we get

$$\lim_{n \to +\infty} \int_s^T \mathbb{E}\left(|F_{\alpha,n}(u_\alpha^k(r)) - F_\alpha(u_\alpha^k(r))|_H^2 + |(P_n - I)Gu_\alpha^k(r)|_H^2 \right) dr = 0,$$

so that (9.1.13) follows.

Step 3: Finally, we have that

$$\lim_{k \to +\infty} \left| u_\alpha^k(\cdot, s; x, z) - u_\alpha(\cdot, s; x, z) \right|_{L^2(\Omega; C([s,T]; H))} = 0. \qquad (9.1.14)$$

Indeed, if we define $v_\alpha^k(t) = u_\alpha(t) - u_\alpha^k(t) - w^A(t, s) + w^{A,k}(t, s)$, we have that $v_\alpha^k(t)$ is the unique solution of the problem

$$\frac{dv}{dt}(t) = Av(t) + F_\alpha(u_\alpha(t)) - F_\alpha(u_\alpha^k(t)) + z(t) - z_k(t), \qquad v(s) = 0.$$

Thus, by multiplying each side by $v_\alpha^k(t)$, we have

$$\frac{1}{2}\frac{d}{dt}|v_\alpha^k(t)|_H^2 + |(-A)^{1/2}v_\alpha^k(t)|_H^2$$

$$= \left\langle F_\alpha(u_\alpha(t)) - F_\alpha(u_\alpha^k(t)), v_\alpha^k(t) \right\rangle_H + \left\langle z(t) - z_k(t), v_\alpha^k(t) \right\rangle_H,$$

so that, as F_α is Lipschitz-continuous, we easily get

$$\frac{1}{2}\frac{d}{dt}|v_\alpha^k(t)|_H^2 \le c_\alpha |v_\alpha^k(t)|_H^2 + c_\alpha |w^A(t, s) - w^{A,k}(t, s)|_H^2 + c|z(t) - z_k(t)|_H^2.$$

By applying the Gronwall lemma, and by taking first the supremum over $t \in [s, T]$ and then the expectation, we get

$$\mathbb{E} \sup_{t \in [s,T]} |v_\alpha^k(t)|_H^2 \le c_{\alpha,T} \int_s^T \mathbb{E}\left(|w^A(t, s) - w^{A,k}(t, s)|_H^2 + |z(t) - z_k(t)|_H^2 \right) dt$$

and this immediately implies (9.1.14).

Step 4: Due to (9.1.11) and (9.1.14), for any $\epsilon > 0$ there exists $k_\epsilon \in \mathbb{N}$ such that for any $n \in \mathbb{N}$ it holds

$$\mathbb{E} \sup_{t \in [s,T]} \left(|u_{\alpha,n}^{k_\epsilon}(t) - u_{\alpha,n}(t)|_H^2 + |u_\alpha^{k_\epsilon}(t) - u_\alpha(t)|_H^2 \right) < \epsilon/2.$$

Besides, due to (9.1.13) there exists $n_\epsilon \in \mathbb{N}$ such that

$$\mathbb{E} \sup_{t \in [s,T]} |u_{\alpha,n}^{k_\epsilon}(t) - u_\alpha^{k_\epsilon}(t)|_H^2 < \epsilon/2,$$

for any $n \ge n_\epsilon$, and then (9.1.10) follows. $\qquad \square$

9.2   The first variation equation

Here and in what follows, we shall denote by $u(\cdot;x)$, $u_\alpha(\cdot;x)$ and $u_{\alpha,n}(\cdot;x)$ the mild solutions of the problems (9.1.1) and (9.1.5) and (9.1.9) respectively, when $z = 0$ and $s = 0$.

In the present section we study the first variation equation associated with the problem (9.1.1)

$$\frac{dv}{dt}(t) = Av(t) + DF(u(t;x))v(t), \qquad v(0) = h. \qquad (9.2.1)$$

In the Theorem 6.3.3 we have proved that for any $x, h \in E$ there exists a unique mild solution for the problem (9.2.1) and for any $p \geq 1$ such a solution is given by the Fréchet derivative of the mapping

$$E \to \mathcal{H}_p(T, E), \qquad x \mapsto u(\cdot;x),$$

along the direction h. Moreover, in the Proposition 7.2.1 we have proved that if $x, h \in H$, then the problem (9.2.1) admits a unique generalized solution $v(\cdot;x,h)$.

We have already seen that, since F_α is Lipschitz-continuous, we can apply the Theorem 4.2.4 and we have that the solution $u_\alpha(\cdot;x)$ is twice differentiable with respect to $x \in H$ in the mean of order p. Moreover, the first derivative $Du_\alpha(\cdot;x)h$ is the unique solution of the first variation equation corresponding to the problem (9.1.5)

$$\frac{dv}{dt}(t) = Av(t) + DF_\alpha(u_\alpha(t))v(t), \qquad v(0) = h.$$

We have the following approximation result.

Lemma 9.2.1. *Under the Hypotheses 6.1, 6.2 and 6.3, for any $x \in E$*

$$\lim_{\alpha \to 0} \mathbb{E} \sup_{|h|_H \leq 1} |Du_\alpha(\cdot;x)h - Du(\cdot;x)h|^2_{L^\infty(0,T;H) \cap L^2(0,T;D((-A)^{1/2}))} = 0, \qquad (9.2.2)$$

uniformly for x in bounded subsets of E.

Proof. By arguing as in the proof of the Proposition 6.4.1, for any $h \in E$ it holds

$$\sup_{x \in E} \left(|Du(t;x)h|^2_H + \int_0^t |(-A)^{1/2}Du(s;x)h|^2_H \, ds \right) \leq c_T |h|^2_H, \qquad \mathbb{P} - \text{a.s.} \qquad (9.2.3)$$

If we define $v_\alpha(t) = Du_\alpha(t;x)h - Du(t;x)h$, we have that $v_\alpha(t)$ is the unique solution of the problem

$$\frac{dv}{dt}(t) = Av(t) + DF_\alpha(u_\alpha(t;x))Du_\alpha(t;x)h - DF(u(t;x))Du(t;x)h, \qquad v(0) = 0.$$

Thus we have

$$\frac{1}{2}\frac{d}{dt}|v_\alpha(t)|_H^2 + |(-A)^{1/2}v_\alpha(t)|_H^2$$

$$= \langle DF_\alpha(u_\alpha(t))v_\alpha(t), v_\alpha(t)\rangle_H + \langle (DF_\alpha(u_\alpha(t)) - DF(u(t))) Du(t;x)h, v_\alpha(t)\rangle_H$$

$$\leq c|v_\alpha(t)|_H^2 + c_T |DF_\alpha(u_\alpha(t)) - DF(u(t))|_E^2 |h|_H^2,$$

last inequality following from (6.1.34), (9.2.3) and the Young inequality. Since F_α verifies (6.1.32), for any $x, y \in E$ we have

$$|DF_\alpha(x) - DF(y)|_E \leq |DF_\alpha(x) - DF_\alpha(y)|_E + |DF_\alpha(y) - DF(y)|_E$$

$$\leq c\left(1 + |x|_E^{2m-1} + |y|_E^{2m-1}\right)|x - y|_E + |DF_\alpha(y) - DF(y)|_E.$$

Therefore, thanks to the Gronwall lemma and the above inequality, we have

$$|v_\alpha(t)|_H^2 \leq c_T \int_0^t \left(1 + |u_\alpha(s)|_E^{4m-2} + |u(s)|_E^{4m-2}\right)|u_\alpha(s) - u(s)|_E^2 \, ds |h|_H^2$$

$$+ c_T \int_0^t |DF_\alpha(u(s)) - DF(u(s))|_E^2 \, ds |h|_H^2.$$

We have seen that

$$\sup_{\alpha > 0} |u_\alpha(t;x)|_E \leq c_T \left(|x|_E + \sup_{t \in [0,T]} |w^A(t)|_E^{2m+1}\right), \qquad \mathbb{P} - a.s. \qquad (9.2.4)$$

for some constant c_T independent of α, and hence, by using (6.1.35), (9.1.2) and (9.1.6), we have

$$\lim_{\alpha \to 0} \mathbb{E} \sup_{|h|_H \leq 1} |v_\alpha(t)|_H^2 = 0.$$

This immediately yields

$$\lim_{\alpha \to 0} \mathbb{E} \sup_{|h|_H \leq 1} \int_0^t |(-A)^{1/2}v_\alpha(s)|_H^2 \, ds = 0$$

and (9.2.2) holds true. $\qquad \square$

We have already seen that due to the Hypothesis 6.1-3, and the closed graph theorem, the operator $\Gamma_\epsilon = Q^{-1}(-A)^{-\epsilon/2}$ is bounded in H. Thus, for any $x \in D((-A)^{1/2})$ by interpolation we get

$$|Q^{-1}x|_H \leq \|\Gamma_\epsilon\| \, |(-A)^{1/2}x|_H^\epsilon \, |x|_H^{1-\epsilon}. \qquad (9.2.5)$$

Thanks to (9.2.2) this yields

$$\lim_{\alpha \to 0} \mathbb{E} \sup_{|h|_H \leq 1} \int_0^t |Q^{-1}(Du_\alpha(s;x)h - Du(s;x)h)|_H^2 \, ds = 0, \qquad (9.2.6)$$

uniformly for x in bounded sets of E.

For each $n \in \mathbb{N}$ and $\alpha > 0$, the solution of (9.1.9) is twice differentiable in the mean of order p, for any $p \geq 1$, with respect to $x \in H$. In the next lemma we show that we can approximate in a suitable sense $Du_\alpha(t;x)h$ by means of $Du_{\alpha,n}(t;x)h$.

Lemma 9.2.2. *Assume that the Hypotheses 6.1, 6.2 and 6.3 hold. Then*

$$\lim_{n \to +\infty} \mathbb{E} \sup_{|h|_H \leq 1} |Du_{\alpha,n}(\cdot;x)h - Du_\alpha(\cdot;x)P_n h|_{L^\infty(0,T;H) \cap L^2(0,T;D((-A)^{1/2}))}^2 = 0,$$
$$\qquad (9.2.7)$$

uniformly for x in bounded subsets of H.

Proof. If we set $v_{\alpha,n}(t) = Du_{\alpha,n}(t;x)h - Du_\alpha(t;x)P_n h$, we have that $v_{\alpha,n}(t)$ is the unique solution of the problem

$$\frac{dv}{dt}(t) = Cv(t) + G_n Du_{\alpha,n}(t)h - GDu_\alpha(t)P_n h$$

$$+DF_{\alpha,n}(u_{\alpha,n}(t;x))Du_{\alpha,n}(t)h - DF_\alpha(u_\alpha(t;x))Du_\alpha(t)P_n h, \qquad v(0) = 0.$$

Multiplying each side by $v_{\alpha,n}(t)$, we have

$$\frac{1}{2}\frac{d}{dt}|v_{\alpha,n}(t)|_H^2 + |(-C)^{1/2}v_{\alpha,n}(t)|_H^2$$

$$= \langle G_n Du_{\alpha,n}(t)h - GDu_\alpha(t)P_n h, v_{\alpha,n}(t) \rangle_H + \langle DF_{\alpha,n}(u_{\alpha,n}(t;x))v_{\alpha,n}(t), v_{\alpha,n}(t) \rangle_H$$

$$+ \langle (DF_{\alpha,n}(u_{\alpha,n}(t;x)) - DF_\alpha(u_\alpha(t;x))) Du_\alpha(t)P_n h, v_{\alpha,n}(t) \rangle_H.$$
$$\qquad (9.2.8)$$

Thanks to (6.1.5) and (6.1.6), as $(-C)^{1/2}$ commutes with P_n, from the Young inequality we have

$$\langle G_n Du_{\alpha,n}(t)h - GDu_\alpha(t)P_n h, v_{\alpha,n}(t) \rangle_H = \langle v_{\alpha,n}(t), G^* P_n v_{\alpha,n}(t) \rangle_H$$

$$+ \langle GDu_\alpha(t;x)P_n h, (P_n - I)v_{\alpha,n}(t) \rangle_H \leq \frac{1}{4}|(-C)^{1/2}v_{\alpha,n}(t)|_H^2$$

$$+c\,|v_{\alpha,n}(t)|_H^2 + c\,\|P_n - I\|_{\mathcal{L}(D((-C)^{1/2});H)}^2 |Du_\alpha(t;x)P_n h|_{D((-A)^{1/2})}^2.$$

Moreover, by using again (6.1.5) and the Sobolev embedding theorem, we have

$$\langle (DF_{\alpha,n}(u_{\alpha,n}(t;x)) - DF_\alpha(u_\alpha(t;x)))\, Du_\alpha(t)P_n h, v_{\alpha,n}(t)\rangle_H$$

$$= \langle (DF_\alpha(u_{\alpha,n}(t;x)) - DF_\alpha(u_\alpha(t;x)))\, Du_\alpha(t)P_n h, P_n v_{\alpha,n}(t)\rangle_H$$

$$+ \langle DF_\alpha(u_\alpha(t;x))Du_\alpha(t)P_n h, (P_n - I)v_{\alpha,n}(t)\rangle_H \le \frac{1}{4}|(-C)^{1/2}v_{\alpha,n}(t)|_H^2$$

$$+ c_\alpha |Du_\alpha(t;x)P_n h|^2_{D((-A)^{1/2})} \left(|u_{\alpha,n}(t) - u_\alpha(t)|_H^2 + \|P_n - I\|^2_{\mathcal{L}(D((-C)^{1/2});H)}\right).$$

Therefore, from (9.1.8) and (9.2.8) we conclude that

$$\frac{d}{dt}|v_{\alpha,n}(t)|_H^2 + |(-C)^{1/2}v_{\alpha,n}(t)|_H^2 \le c_\alpha |v_{\alpha,n}(t)|_H^2$$

$$+ c_{\alpha,T}\left(\|P_n - I\|^2_{\mathcal{L}(D((-C)^{1/2});H)} + |u_{\alpha,n} - u_\alpha|^2_{C([0,T];H)}\right)|Du_\alpha(t;x)P_n h|^2_{D((-A)^{1/2})}.$$

As shown in the Remark 6.4.2, for each $h \in H$ we easily have

$$\sup_{x \in H}\int_0^t |Du_\alpha(s;x)h|^2_{D((-A)^{1/2})}\, ds \le c_T |h|_H^2, \qquad \mathbb{P}-\text{a.s.}$$

for some constant c_T independent of α and then, by using the Gronwall lemma it follows

$$\sup_{t \in [0,T]} |v_{\alpha,n}(t)|_H^2 \le c_{\alpha,T}\left(\|P_n - I\|^2_{\mathcal{L}(D((-C)^{1/2});H)} + |u_{\alpha,n} - u_\alpha|^2_{C([0,T];H)}\right)|h|_H^2.$$

Thus, as

$$\lim_{n \to +\infty}\|P_n - I\|_{\mathcal{L}(D((-C)^{1/2});H)} = 0,$$

from the Lemma 9.1.4 we also get

$$\lim_{n \to +\infty}\mathbb{E}\sup_{|h|_H \le 1}\sup_{t \in [0,T]}|Du_{\alpha,n}(t;x)h - Du_\alpha(t;x)P_n h|_H^2 = 0.$$

Thanks to (9.2.8), from the limit above we get

$$\lim_{n \to +\infty}\mathbb{E}\sup_{|h|_H \le 1}\int_0^t |(-A)^{1/2}(Du_{\alpha,n}(s;x)h - Du_\alpha(s;x)P_n h)|_H^2\, ds = 0,$$

so that (9.2.7) follows. $\qquad\square$

Due to the interpolation inequality (9.2.5), the previous lemma yields

$$\lim_{n \to +\infty}\mathbb{E}\sup_{|h|_H \le 1}\int_0^t \left|Q^{-1}(Du_{\alpha,n}(s;x)h - Du_\alpha(s;x)P_n h)\right|_H^2\, ds = 0,$$

9.3   The approximating transition semigroups

As in the section 6.5, for any $\alpha > 0$ we denote by P_t^α the transition semigroup corresponding to the approximating problem (9.1.5), with $z = 0$. We have seen that the semigroup P_t^α maps $B_b(H)$ into $C_b^2(H)$, for any $t > 0$, and if $\varphi \in C_b(H)$ it holds

$$\langle D(P_t^\alpha \varphi)(x), h \rangle_H = \frac{1}{t} \mathbb{E} \varphi(u_\alpha(t; x)) \int_0^t \langle Q^{-1} D_x u_\alpha(s; x) h, dw(s) \rangle_H,$$

for all $x, h \in H$. Moreover, for $0 \leq i \leq j \leq 2$

$$\|P_t^\alpha \varphi\|_j^H \leq c_\alpha \, (t \wedge 1)^{-\frac{(j-i)(1+\epsilon)}{2}} \|\varphi\|_i^H. \tag{9.3.1}$$

Thanks to (6.1.33), we have also seen that

$$\sup_{x \in H} \left(|Du_\alpha(t; x) h|_H^2 + \int_0^t |(-A)^{1/2} Du_\alpha(s; x) h|_H^2 \, ds \right) \leq c_T |h|_H^2, \quad \mathbb{P} - \text{a.s.}, \tag{9.3.2}$$

for some constant c_T independent of α. Thus, for $i = 0, 1$ we immediately have

$$\|P_t^\alpha \varphi\|_1^H \leq c \, (t \wedge 1)^{-\frac{(1-i)(1+\epsilon)}{2}} \|\varphi\|_i^H, \tag{9.3.3}$$

and here the constant c is independent of α.

As shown in section 6.5 (see also Lemma 9.1.3), if $\varphi \in C_b(H)$ we have

$$\lim_{\alpha \to 0} \sup_{t \in [0,T]} \sup_{|x|_E \leq R} |P_t^\alpha \varphi(x) - P_t \varphi(x)| = 0, \tag{9.3.4}$$

for any $R, T > 0$. Moreover, from the Lemma 9.2.1, we have

$$\lim_{\alpha \to 0} \sup_{|x|_E \leq R} |D(P_t^\alpha \varphi)(x) - D(P_t \varphi)(x)|_H = 0, \quad t > 0. \tag{9.3.5}$$

Actually, for each $\alpha > 0$ we have

$$\langle D(P_t^\alpha \varphi)(x), h \rangle_H = \frac{1}{t} \mathbb{E} \varphi(u_\alpha(t; x)) \int_0^t \langle Q^{-1} Du_\alpha(s; x) h, dw(s) \rangle_H$$

and then by easy calculations we have

$$|\langle D(P_t^\alpha \varphi)(x) - D(P_t \varphi)(x), h \rangle_H|$$

$$\leq \frac{\|\varphi\|_1^H}{t} \left(\mathbb{E} |u_\alpha(t, x) - u(t; x)|_H^2 \right)^{1/2} \left(\mathbb{E} \int_0^t |Q^{-1} Du_\alpha(s; x) h|_H^2 \, ds \right)^{1/2}$$

$$+ \frac{\|\varphi\|_0^H}{t} \left(\mathbb{E} \int_0^t |Q^{-1} (Du_\alpha(s; x) h - Du(s; x) h)|_H^2 \, ds \right)^{1/2}.$$

Thus, (9.3.5) follows from (9.1.6) and (9.2.6).

In correspondence of each $n \in \mathbb{N}$, we can introduce the transition semigroup $P_t^{\alpha,n}$ associated with the system (9.1.9). The semigroup $P_t^{\alpha,n}$ fulfills all the regularizing properties described above for P_t^{α}, so that for $0 \leq i \leq j \leq 2$

$$\|P_t^{\alpha,n}\|_j^H \leq c_\alpha (t \wedge 1)^{-\frac{(j-i)(1+\epsilon)}{2}} \|\varphi\|_i^H, \qquad t > 0. \tag{9.3.6}$$

Note that here the constant c_α does not depend on $n \in \mathbb{N}$. Moreover, due to (9.1.7) it is not difficult to check that for $i = 0, 1$

$$\|P_t^{\alpha,n}\|_1^H \leq c (t \wedge 1)^{-\frac{(1-i)(1+\epsilon)}{2}} \|\varphi\|_i^H, \qquad t > 0, \tag{9.3.7}$$

for a constant c which does not depend of n and α. In the next theorem we prove that it is possible to approximate $P_t^{\alpha}\varphi$ and its first derivative by means of $P_t^{\alpha,n}$ and its first derivative, respectively.

Proposition 9.3.1. *Under the Hypotheses 6.1, 6.2 and 6.3, for any $\varphi \in C_b(H)$ and $R, T > 0$ we have*

$$\lim_{n \to +\infty} \sup_{t \in [0,T]} \sup_{|x|_H \leq R} |P_t^{\alpha,n}\varphi(x) - P_t^{\alpha}\varphi(x)| = 0. \tag{9.3.8}$$

Furthermore, for any $\delta \in (0, T)$

$$\lim_{n \to +\infty} \sup_{t \in [\delta,T]} \sup_{|x|_H \leq R} |D(P_t^{\alpha,n}\varphi)(x) - D(P_t^{\alpha}\varphi)(x)|_H = 0. \tag{9.3.9}$$

Proof. The limit (9.3.8) follows directly from the Lemma 9.1.4. As far as the limit (9.3.9) is concerned, we have

$$\langle D(P_t^{\alpha}\varphi)(x) - D(P_t^{\alpha,n}\varphi)(x), P_n h\rangle_H$$

$$= \frac{1}{t}\mathbb{E}\left(\varphi(u_\alpha(t;x)) - \varphi(u_{\alpha,n}(t;x))\right) \int_0^t \langle Q^{-1}Du_\alpha(s;x)P_n h, dw(s)\rangle_H$$

$$+ \frac{1}{t}\mathbb{E}\varphi(u_{\alpha,n}(t;x)) \int_0^t \langle Q^{-1}(Du_\alpha(s;x)P_n h - D_x u_{\alpha,n}(s;x)P_n h), dw(s)\rangle_H.$$

Thus, if $\varphi \in C_b^1(H)$ we get

$$\left|\langle D(P_t^{\alpha}\varphi)(x) - D(P_t^{\alpha,n}\varphi)(x), P_n h\rangle_H\right|$$

$$\leq \frac{2}{t}\|\varphi\|_1^H \left(\mathbb{E}|u_\alpha(t;x) - u_{\alpha,n}(t;x)|^2\right)^{1/2} \left(\mathbb{E}\int_0^t |Q^{-1}Du_\alpha(s;x)P_n h|_H^2 \, ds\right)^{1/2}$$

$$+ \frac{2}{t}\|\varphi\|_0^H \left(\mathbb{E}\int_0^t |Q^{-1}(Du_\alpha(s;x)P_n h - D_x u_n(s;x)P_n h)|_H^2 \, ds\right)^{1/2}.$$

By taking the supremum over $|h|_H \leq 1$, due to (9.1.10), (9.2.6) and (9.3.2), it follows that

$$\lim_{n \to +\infty} |P_n D(P_t^\alpha \varphi)(x) - D(P_t^{\alpha,n} \varphi)(x)|_H = 0.$$

Thus, as

$$\lim_{n \to +\infty} |P_n D(P_t^\alpha \varphi)(x) - D(P_t^\alpha \varphi)(x)|_H = 0,$$

(9.3.9) follows for any $\varphi \in C_b^1(H)$. Now, if $\varphi \in C_b(H)$, since $C_b^1(H)$ is dense in $C_b(H)$, due to (9.3.3) and (9.3.7) there exists $\psi \in C_b^1(H)$ such that

$$\sup_{t \in [\delta,T]} \left(\|D(P_t^{\alpha,n} \varphi) - D(P_t^{\alpha,n} \psi)\|_0^H + \|D(P_t^\alpha \varphi) - D(P_t^\alpha \psi)\|_0^H \right) \leq \epsilon/2,$$

for any $\alpha > 0$ and $n \in \mathbb{N}$. Hence, as (9.3.9) holds for ψ, we can conclude that (9.3.9) holds for any $\varphi \in C_b(H)$. $\square$

As shown in the Proposition 4.3.1, P_t^α and $P_t^{\alpha,n}$ are weakly continuous semigroups in $C_b(H)$. Thus we can introduce their weak generators L_α and $L_{\alpha,n}$ (see appendix B for the definition and main properties). Moreover in the section 7.1 we have introduced the infinitesimal generator L of the semigroup P_t. We have seen that if $\varphi \in D(L)$ and $x \in H$ the mapping

$$[0, +\infty) \to \mathbb{R}, \qquad t \mapsto P_t \varphi(x),$$

is differentiable and

$$\frac{d}{dt} P_t \varphi(x) = L(P_t \varphi)(x) = P_t(L\varphi)(x).$$

As the same holds for L_α and $L_{\alpha,n}$, we are proving that if $\varphi \in C_b^2(H)$ then $P_s^{\alpha,n} \varphi \in D(L_{\alpha,n})$, for any $\alpha > 0$, $n \in \mathbb{N}$ and $s \geq 0$, and

$$L_{\alpha,n}(P_s^{\alpha,n} \varphi) = \mathcal{L}_{\alpha,n}(P_s^{\alpha,n} \varphi), \tag{9.3.10}$$

where the differential operator $\mathcal{L}_{\alpha,n}$ is defined by

$$\mathcal{L}_{\alpha,n} \varphi(x) = \frac{1}{2} \mathrm{Tr} \left[Q_n^2 D^2 \varphi(x) \right] + \langle A_n x + F_{\alpha,n}(x), D\varphi(x) \rangle_H, \qquad x \in H. \tag{9.3.11}$$

Actually, if we define $\psi = \lambda P_s^{\alpha,n} \varphi - \mathcal{L}_{\alpha,n}(P_s^{\alpha,n} \varphi)$, for some $\lambda > 0$, we have that

$$R(\lambda, L_{\alpha,n}) \psi(x) = \int_0^{+\infty} e^{-\lambda t} \left(\lambda P_{t+s}^{\alpha,n} \varphi(x) - P_t^{\alpha,n} \mathcal{L}_{\alpha,n}(P_s^{\alpha,n} \varphi)(x) \right) dt.$$

It is not difficult to prove that in general, if φ is twice differentiable, then

$$P_t^{\alpha,n}(\mathcal{L}_{\alpha,n} \varphi)(x) = \mathcal{L}_{\alpha,n}(P_t^{\alpha,n} \varphi)(x).$$

Thus, as $P_s^{\alpha,n} \varphi \in C_b^2(H)$, from Itô's formula we have

$$P_t^{\alpha,n} \mathcal{L}_{\alpha,n}(P_s^{\alpha,n} \varphi)(x) = \mathcal{L}_{\alpha,n} \left(P_{t+s}^{\alpha,n} \varphi\right)(x) = \frac{d}{dt}(P_{t+s}^{\alpha,n} \varphi(x)).$$

This allows us to conclude that

$$R(\lambda, \mathcal{L}_{\alpha,n}) \psi(x) = -\int_0^{+\infty} \frac{d}{dt}(e^{-\lambda t} P_{t+s}^{\alpha,n} \varphi(x))\, dt = P_s^{\alpha,n} \varphi(x),$$

so that $P_s^{\alpha,n} \varphi \in D(\mathcal{L}_{\alpha,n})$ and (9.3.10) holds.

9.4   The parabolic Hamilton-Jacobi-Bellman equation

We are here concerned with the infinite dimensional Cauchy problem

$$\frac{\partial y}{\partial t}(t, x) = \mathcal{L}y(t, x) - K(Dy(t, x)) + g(x), \qquad y(0, x) = \varphi(x), \qquad (9.4.1)$$

where $\mathcal{L}$ is the differential operator defined by

$$\mathcal{L}\psi(x) = \frac{1}{2}\mathrm{Tr}\left[QQ^* D^2\psi(x)\right] + \langle Ax + F(x), D\psi(x)\rangle_H, \qquad x \in D(A) \cap D(F).$$

In addition to the Hypotheses 6.1, 6.2 and 6.3, the following condition shall be assumed.

Hypothesis 9.1. *The hamiltonian $K : H \to \mathbb{R}$ is Fréchet differentiable and locally Lipschitz-continuous together with its derivative. Moreover, $K(0) = 0$.*

Notice that the requirement $K(0) = 0$ is not restrictive, as we can substitute g by $g - K(0)$.

The problem (9.4.1) can be rewritten in the *mild* form

$$y(t, x) = P_t\varphi(x) - \int_0^t P_{t-s}\left(K(Dy(s, \cdot))\right)(x)\, ds + \int_0^t P_{t-s} g(x)\, ds. \qquad (9.4.2)$$

As we noticed before, the semigroup P_t is not strongly continuous in $C_b(H)$, in general. Nevertheless, the mapping

$$[0, +\infty) \to \mathbb{R}, \qquad t \mapsto P_t\varphi(x),$$

is continuous, for any fixed $\varphi \in C_b(H)$ and $x \in H$. Thus, the integrals in the formula (9.4.2) have a meaning only for fixed $x \in H$.

We define $\mathcal{V}_T^1$ as the space of bounded continuous functions $y : [0, T] \times H \to \mathbb{R}$, such that $y(t, \cdot) \in C_b^1(H)$, for all $t \in (0, T]$, and the mapping

$$(0, T] \times H \to H, \qquad (t, x) \mapsto Dy(t, x)$$

is bounded and measurable. We endow $\mathcal{V}_T^1$ with the norm

$$\|y\|_{\mathcal{V}_T^1} = \sup_{t \in [0,T]} \|y(t,\cdot)\|_0^H + \sup_{t \in (0,T]} \|Dy(t,\cdot)\|_0^H.$$

Moreover, we define $\mathcal{Z}_T^1$ as the space of bounded continuous functions $y : [0,T] \times H \to \mathbb{R}$, such that $y(t,\cdot) \in C_b^1(H)$, for all $t \in (0,T]$, and the mapping

$$(0,T] \times H \to H, \qquad (t,x) \mapsto t^{\frac{1+\varepsilon}{2}} Dy(t,x)$$

is bounded and measurable. It is easy to check that $\mathcal{Z}_T^1$, endowed with the norm

$$\|y\|_{\mathcal{Z}_T^1} = \sup_{t \in [0,T]} \|y(t,\cdot)\|_0^H + \sup_{t \in (0,T]} t^{\frac{1+\varepsilon}{2}} \|Dy(t,\cdot)\|_0^H,$$

is a Banach space.

Finally, we say that a function $y \in \mathcal{V}_T^1$ belongs to the space $\mathcal{Z}_T^2$ if $y(0,\cdot) \in C_b^1(H)$, the function $y(t,\cdot)$ is in $C_b^2(H)$ for any $t > 0$ and the mapping

$$(0,T] \times H \to \mathcal{L}(H), \qquad (t,x) \mapsto t^{\frac{1+\varepsilon}{2}} D^2y(t;x)$$

is bounded and measurable. $\mathcal{Z}_T^2$, endowed with the norm

$$\|y\|_{\mathcal{Z}_T^2} = \sup_{t \in [0,T]} \|y(t,\cdot)\|_1^H + \sup_{t \in (0,T]} t^{\frac{1+\varepsilon}{2}} \|D^2y(t,\cdot)\|_0^H,$$

is a Banach space.

A proof of the following lemma, in the case $F = 0$, can be found in [71] (see the Lemmata 4.8 and 4.12). Such a proof is quite long and completely adapts to our case where $F \neq 0$, thus we do not repeat it.

Lemma 9.4.1. *For any function* $\psi : [0,T] \times H \to \mathbb{R}$ *bounded and measurable, we define*

$$\lambda_\psi(t,x) = \int_0^t P_{t-s}\psi(s,\cdot)(x)\,ds, \qquad (t,x) \in [0,T] \times H.$$

Then $\lambda_\psi : [0,T] \times H \to \mathbb{R}$ *is continuous and bounded,* $\lambda_\psi(t,\cdot) \in C_b^1(H)$, *for any* $t \geq 0$, *and*

$$\sup_{t \in [0,T]} \|\lambda_\psi(t,\cdot)\|_1^H < \infty.$$

It is immediate to check that the Lemma 9.4.1 adapts to the approximating semigroups P_t^α and $P_t^{\alpha,n}$.

For each $\alpha > 0$, we consider the approximating problem

$$\frac{\partial y}{\partial t}(t,x) = \mathcal{L}_\alpha y(t,x) - K(Dy(t,x)) + g(x), \qquad y(0,x) = \varphi(x),$$

where

$$\mathcal{L}_\alpha \psi(x) = \frac{1}{2} \mathrm{Tr}\left[QQ^* D^2 \psi(x)\right] + \langle Ax + F_\alpha(x), D\psi(x)\rangle_H.$$

In mild form it can be rewritten as

$$y(t,x) = P_t^\alpha \varphi(x) - \int_0^t P_{t-s}^\alpha \left(K(Dy(s,\cdot))\right)(x)\, ds + \int_0^t P_{t-s}^\alpha g(x)\, ds. \qquad (9.4.3)$$

Theorem 9.4.2. *Assume that the Hypotheses 6.1, 6.2, 6.3 and 9.1 hold and fix $T > 0$. Then, for any $\varphi, g \in C_b^{0,1}(H)$ the equation (9.4.2) admits a unique mild solution $y(t,x)$ in $\mathcal{V}_T^1$.*

If $y_\alpha(t,x)$ denotes the unique mild solution of the problem (9.4.3), we have

$$\lim_{\alpha \to 0} |y_\alpha(t,x) - y(t,x)| + |Dy_\alpha(t,x) - Dy(t,x)|_H = 0, \qquad (9.4.4)$$

uniformly for t in compact subsets of $(0,T]$ and x in bounded subsets of E. Moreover, if $\varphi, g \in C_b^1(H)$, then the limit (9.4.4) is uniform for $t \in [0,T]$.

We first prove some preliminary results.

Lemma 9.4.3. *Fix $\varphi, g \in C_b^1(H)$ and $R \geq 2\nu \|\varphi\|_1^H$, where ν is the constant introduced in (7.3.2). Then the problem (9.4.2) admits a unique local solution y in $[0, \tau_R]$, for some constant*

$$\tau_R = \tau_R\left(\|\varphi\|_1^H, \|g\|_0^H, \sup_{x \in B_R^H} |DK(x)|\right).$$

Proof. For any $\tau > 0$, we define $\Lambda_R(\tau)$ as the set of all continuous and bounded functions $y : [0, \tau] \times H \to \mathbb{R}$ such that $y(t, \cdot) \in C_b^1(H)$, for all $t \in [0, \tau]$, the mapping

$$[0, \tau] \times H \to H, \qquad (t, x) \mapsto Dy(t, x)$$

is bounded and measurable and

$$\sup_{t \in [0,\tau]} \|y(t, \cdot)\|_1^H \leq R.$$

We claim that for some τ_R sufficiently small, the operator Γ defined by

$$\Gamma(v)(t, x) = P_t \varphi(x) - \int_0^t P_{t-s}\left(K(Dv(s, \cdot))\right)(x)\, ds + \int_0^t P_{t-s} g(x)\, ds$$

maps $\Lambda_R(\tau_R)$ into itself as a contraction. According to the Lemma 9.4.1, $\Gamma(v)(t, x)$ is well defined for any x and t. Due to (7.3.2) we have

$$\|P_t \varphi\|_1^H \leq \nu \|\varphi\|_1^H \leq \frac{R}{2}.$$

Moreover, if we set

$$\Gamma_1(v)(t,x) = -\int_0^t P_{t-s}\left(K(Dv(s,\cdot))\right)(x)\,ds + \int_0^t P_{t-s}g(x)\,ds,$$

for any $v \in \Lambda_R(\tau)$ we have

$$\|\Gamma_1(v)(t,\cdot)\|_0^H \le \int_0^t \|K(Dv(s,\cdot))\|_0^H\,ds + t\|g\|_0^H \le \tau \left(\sup_{x \in B_R^H} |K(x)| + \|g\|_0^H\right).$$

Concerning the derivative, due to the estimate (7.3.2) it holds

$$\|D(\Gamma_1(v))(t,\cdot)\|_0^H \le \nu \int_0^t (t-s)^{-\frac{1+\epsilon}{2}}\left(\|K(Dv(s,\cdot))\|_0^H + \|g\|_0^H\right)\,ds$$

$$\le \nu \tau^{\frac{1-\epsilon}{2}}\left(\sup_{x \in B_R^H} |K(x)| + \|g\|_0^H\right).$$

This implies that

$$\sup_{t \in [0,\tau]} \|\Gamma(v)(t,\cdot)\|_1^H \le \frac{R}{2} + c\left(\tau + \tau^{\frac{1-\epsilon}{2}}\right)\left(\sup_{x \in B_R^H} |K(x)| + \|g\|_0^H\right),$$

so that it is possible to find $\bar{\tau}_R$ sufficiently small such that

$$\sup_{t \in [0,\bar{\tau}_R]} \|\Gamma(v)(t,\cdot)\|_1^H \le R.$$

In an completely analogous way it is possible to show that Γ is a contraction on $\Lambda_R(\tau_R)$, for some $\tau_R \le \bar{\tau}_R$. This allows us to conclude that there exists a unique fixed point u for Γ in $\Lambda_R(\tau_R)$, which is the unique solution of (9.4.2) in $[0,\tau_R]$. $\quad\square$

Remark 9.4.4. In an identical way it is possible to prove that for each $\alpha > 0$ the mapping

$$\Gamma_\alpha(v)(t,x) = P_t^\alpha \varphi(x) - \int_0^t P_{t-s}^\alpha\left(K(Dv(s,\cdot))\right)(x)\,ds + \int_0^t P_{t-s}^\alpha g(x)\,ds$$

is a contraction in $\Lambda_R(\tau_R)$, where τ_R is the same as in the Lemma 9.4.3. This implies that if $\varphi, g \in C_b^1(H)$ there exists a unique solution $y_\alpha(t,x)$ in $\Lambda_R(\tau_R)$ for the problem (9.4.3). Moreover, it is useful to remark that thanks to (9.3.3) the contraction constant of Γ_α in $\Lambda_R(\tau_R)$ can be taken the same for all $\alpha > 0$.

Lemma 9.4.5. *If $y(t,x)$ and $y_\alpha(t,x)$ are respectively the solutions of the problems (9.4.2) and (9.4.3), with $\varphi, g \in C_b^1(H)$, we have*

$$\lim_{\alpha \to 0} |y_\alpha(t,x) - y(t,x)| + |Dy_\alpha(t,x) - Dy(t,x)|_H = 0, \qquad (9.4.5)$$

uniformly for t in $[0,\tau_R]$ and x in bounded subsets of E.

Proof. In order to prove the existence of the solutions $y(t, x)$ and $y_\alpha(t, x)$ for the problems (9.4.2) and (9.4.3), in the Lemma 9.4.3 and in the Remark 9.4.4 we have applied a contraction theorem and we have shown that for each $\epsilon > 0$ there exists $k_\epsilon \in \mathbb{N}$ such that

$$\sup_{t \in [0, \tau_R]} \left(\|y_\alpha(t, \cdot) - \Gamma_\alpha^{k_\epsilon}(0)(t, \cdot)\|_1^H + \|y(t, \cdot) - \Gamma^{k_\epsilon}(0)(t, \cdot)\|_1^H \right) \leq \epsilon/2, \qquad (9.4.6)$$

for each $\alpha > 0$.

Now, by using an induction argument we prove that for each $k \in \mathbb{N}$

$$\lim_{\alpha \to 0} \left| \Gamma_\alpha^k(0)(t, x) - \Gamma^k(0)(t, x) \right| + \left| D\Gamma_\alpha^k(0)(t, x) - D\Gamma^k(0)(t, x) \right|_H = 0, \qquad (9.4.7)$$

uniformly for $t \in [0, \tau_R]$ and x in bounded subsets of E. For $k = 1$, (9.4.7) follows directly from (9.3.4) and (9.3.5). Now, assume that (9.4.7) holds for some $k \geq 1$. We have

$$\Gamma_\alpha^{k+1}(0)(t, x) - \Gamma^{k+1}(0)(t, x) = \Gamma_\alpha(\Gamma_\alpha^k(0))(t, x) - \Gamma(\Gamma^k(0))(t, x)$$

$$= P_t^\alpha \varphi(x) - P_t \varphi(x) + \int_0^t \left(P_{t-s}^\alpha g(x) - P_{t-s} g(x) \right) \, ds$$

$$- \int_0^t \left(P_{t-s}^\alpha \left[K(D(\Gamma_\alpha^k(0))(s, \cdot)) \right](x) - P_{t-s} \left[K(D(\Gamma^k(0))(s, \cdot)) \right](x) \right) ds.$$

Since $\Gamma_\alpha^k(0)$ and $\Gamma^k(0)$ belong to $\Lambda_R(\tau_R)$ and (9.4.7) holds for k, by using (9.3.4) and the boundedness of K on bounded subsets of H, from the dominated convergence theorem it follows

$$\lim_{\alpha \to 0} \int_0^t \left(P_{t-s}^\alpha \left[K(D(\Gamma_\alpha^k(0))(s, \cdot)) \right](x) - P_{t-s} \left[K(D(\Gamma^k(0))(s, \cdot)) \right](x) \right) ds = 0,$$

uniformly on bounded sets of $[0, \tau_R] \times E$. By using (9.3.4) once more, we have

$$\lim_{\alpha \to 0} P_t^\alpha \varphi(x) - P_t \varphi(x) + \int_0^t \left(P_{t-s}^\alpha g(x) - P_{t-s} g(x) \right) \, ds = 0,$$

uniformly on bounded sets of $[0, \tau_R] \times E$, so that we get

$$\lim_{\alpha \to 0} \Gamma_\alpha^{k+1}(0)(t, x) - \Gamma^{k+1}(0)(t, x) = 0.$$

The second part of the limit (9.4.7) for $k + 1$ follows by analogous arguments. By induction we can conclude that (9.4.7) holds for any $k \in \mathbb{N}$.

Now, from (9.4.6) we have

$$|y_\alpha(t,x) - y(t,x)| + |Dy_\alpha(t,x) - Dy(t,x)|_H \le$$

$$\sup_{t \in [0,\tau_R]} \|y_\alpha(t,\cdot) - \Gamma_\alpha^{k_\epsilon}(0)(t,\cdot)\|_1^H + |\Gamma_\alpha^{k_\epsilon}(0)(t,x) - \Gamma^{k_\epsilon}(0)(t,x)|$$

$$+|D\Gamma_\alpha^{k_\epsilon}(0)(t,x) - D\Gamma^{k_\epsilon}(0)(t,x)|_H + \sup_{t \in [0,\tau_R]} \|y(t,\cdot) - \Gamma^{k_\epsilon}(0)(t,\cdot)\|_1^H$$

$$\le \epsilon/2 + |\Gamma_\alpha^{k_\epsilon}(0)(t,x) - \Gamma^{k_\epsilon}(0)(t,x)| + \left|D\Gamma_\alpha^{k_\epsilon}(0)(t,x) - D\Gamma^{k_\epsilon}(0)(t,x)\right|_H$$

and due to (9.4.7) we can conclude that (9.4.5) holds. □

9.4.1 An a priori estimate

We prove here an *a-priori* estimate for the C^1 norm of the solution y_α of the approximating problem (9.4.3). As in [72] we represent y_α and Dy_α by means of the transition semigroups associated with suitable stochastic problems. This allows us to have a maximum principle both for y_α and Dy_α.

Lemma 9.4.6. *Let $U : [0,T] \times H \to H$ be a bounded and measurable mapping such that $U(t,\cdot)$ is Lipschitz-continuous for any $t > 0$ and*

$$\sup_{t \in [\epsilon,T]} \|U(t,\cdot)\|_{0,1}^H < \infty,$$

for any $\epsilon > 0$. Then, for any $\alpha > 0$ the stochastic problem

$$du(t) = [Au(t) + F_\alpha(u(t)) + U(T-t,u(t))]\,dt + Q\,dw(t), \qquad u(r) = x,$$

admits a unique solution $u_\alpha(t;r,x) \in L^2(\Omega; C([r,T);H)) \cap L^\infty(r,T;H))$.

Proof. For any $\epsilon > 0$ the function $U(T-t,\cdot)$ is Lipschitz-continuous, uniformly for $t \in [0, T-\epsilon]$, and then there exists a unique solution $u_\alpha(t)$ in the interval $[0, T-\epsilon]$, for any $\epsilon > 0$. If we define $v_\alpha(t) = u_\alpha(t) - w^A(t,r)$, we have that $v_\alpha(t)$ is the unique solution of the problem

$$\frac{dv}{dt}(t) = Av(t) + F_\alpha(u_\alpha(t)) + U(T-t,u_\alpha(t)), \qquad v(r) = x. \qquad (9.4.8)$$

Thus, by multiplying each side of (9.4.8) by $v_\alpha(t)$ we get

$$\frac{1}{2}\frac{d}{dt}|v_\alpha(t)|_H^2 + |(-A)^{1/2}v_\alpha(t)|_H^2 = \left\langle F_\alpha(v_\alpha(t) + w^A(t,r)) - F_\alpha(w^A(t,r)), v_\alpha(t) \right\rangle_H$$

$$+ \left\langle F_\alpha(w^A(t,r)), v_\alpha(t) \right\rangle_H + \left\langle U(T-t,u_\alpha(t)), v_\alpha(t) \right\rangle_H.$$

Due to the Lipschitz-continuity of F_α and the boundedness of U, this implies that

$$\frac{1}{2}\frac{d}{dt}|v_\alpha(t)|_H^2 \le c_\alpha\,|v_\alpha(t)|_H^2 + c_\alpha\left(|w^A(t,r)|_H^2 + 1\right).$$

Therefore, by integrating with respect to t, by taking the supremum for $t \in [r, T-\epsilon]$ and finally by taking the expectation, due to the Gronwall lemma we have

$$\mathbb{E}\sup_{t\in[r,T-\epsilon]}|v_\alpha(t)|_H^2 \le c_{\alpha,T}\left(|x|_H^2 + \mathbb{E}\sup_{t\in[r,T]}|w^A(t,r)|_H^2\right).$$

Thanks to the regularity of $w^A(t,r)$, this allows us to conclude that

$$\mathbb{E}\sup_{t\in[r,T-\epsilon]}|u_\alpha(t)|_H^2 \le 2\sup_{t\in[r,T-\epsilon]}\left(|v_\alpha(t)|_H^2 + |w^A(t,r))|_H^2\right) \le c_{\alpha,T}(|x|_H^2 + 1).$$

As the constant $c_{\alpha,T}$ does not depend on ϵ, by a uniqueness argument we have that the solution $u_\alpha(t)$ is defined for any $t \in [r, T)$ and $u_\alpha(t; r, x) \in L^2(\Omega; C([r, T); H)) \cap L^\infty(r, T; H))$. $\square$

Next, for each $\alpha > 0$ and $n \in \mathbb{N}$, we introduce the approximating Hamilton-Jacobi-Bellman equation

$$\frac{\partial y}{\partial t}(t, x) = \mathcal{L}_{\alpha,n}y(t, x) - K_n(Dy(t, x)) + g_n(x), \qquad y(0, x) = \varphi_n(x),\qquad (9.4.9)$$

where $K_n(x) = K(P_n x)$, $g_n(x) = g(P_n x)$ and $\varphi_n(x) = \varphi(P_n x)$, for any $x \in H$ and $n \in \mathbb{N}$, and

$$\mathcal{L}_{\alpha,n}\psi(x) = \frac{1}{2}\mathrm{Tr}\left[Q_n^2 D^2\psi(x)\right] + \langle A_n x + F_{\alpha,n}(x), D\psi(x)\rangle_H.$$

By arguing as for the problem (9.4.3) (see the Remark 9.4.4), if $\varphi, g \in C_b^{0,1}(H)$ the problem (9.4.9) admits a unique local solution $y_{\alpha,n}$ in $\Lambda_R(\tau_R)$. The solution $y_{\alpha,n}$ is the unique fixed point for the functional

$$\Gamma_{\alpha,n}(v)(t, x) = P_t^{\alpha,n}\varphi(x) - \int_0^t P_{t-s}^{\alpha,n}K(Dv(s,\cdot))(x)\,ds + \int_0^t P_{t-s}^{\alpha,n}g(x)\,ds.$$

Due to (9.3.7) the contraction constant of $\Gamma_{\alpha,n}$ is independent of n and α and we can proceed as in the proof of the Lemma 9.4.5 and thanks to the Proposition 9.3.1 we conclude that for any $\alpha > 0$

$$\lim_{n\to+\infty}|y_{\alpha,n}(t, x) - y_\alpha(t, x)| + |Dy_{\alpha,n}(t, x) - Dy_\alpha(t, x)|_H = 0,\qquad (9.4.10)$$

uniformly for $t \in [0, \tau_R]$ and x in bounded sets of H.

According to (9.3.1) it is possible to show that both y_α and $y_{\alpha,n}$ have a stronger regularity.

Lemma 9.4.7. *If* $\varphi, g \in C_b^1(H)$, *then the solutions* y_α *and* $y_{\alpha,n}$ *of the problems* *(9.4.2) and (9.4.3) belong to* $\mathcal{Z}_{\tau_*}^2$, *for some* $\tau_* = \tau_*(\alpha) \leq T$ *which can be taken independent of* n.

Proof. We have to show that there exists some τ sufficiently small such that the mappings Γ_α and $\Gamma_{\alpha,n}$ are contractions in $\mathcal{Z}_\tau^2$. We prove this fact for Γ_α, as the proof for $\Gamma_{\alpha,n}$ is identical.

Thanks to (9.3.1) we have that the mapping

$$\gamma_1 : [0,T] \times H \to \mathbb{R}, \qquad (t,x) \mapsto P_t^\alpha \varphi(x) + \int_0^t P_{t-s}^\alpha g(x)\, ds,$$

belongs to $\mathcal{Z}_\tau^2$, for any $\tau \leq T$, and

$$\|\gamma_1\|_{\mathcal{Z}_\tau^2} \leq c_\alpha \left(\|\varphi\|_0^H + (\tau + \tau^{\frac{1+\epsilon}{2}}) \|g\|_0^H \right), \tag{9.4.11}$$

where the constant c_α is possibly different from the one in (9.3.1). Now, for any $v, w \in \mathcal{Z}_T^2$ we consider the mapping

$$\gamma_2 : [0,T] \times H \to \mathbb{R}, \qquad (t,x) \mapsto \int_0^t P_{t-s}^\alpha \left(K(Dv(s,\cdot)) - K(Dw(s,\cdot)) \right)(x)\, ds.$$

If we denote $\psi(s,x) = K(Dv(s,x)) - K(Dw(s,x))$, for $s \in (0,t)$ we have

$$D^2 P_{t-s}^\alpha \left[K(Dv(s,\cdot)) - K(Dw(s,\cdot)) \right](x)(h,k)$$

$$= \frac{1}{t-s} \mathbb{E} \langle D_x \psi(s, u_\alpha(t-s;x)), k \rangle_H \int_0^{t-s} \langle Q^{-1} D_x u_\alpha(r;x)h, dw(r) \rangle_H$$

$$+ \frac{1}{t-s} \mathbb{E} \psi(s, u_\alpha(t-s;x)) \int_0^{t-s} \langle Q^{-1} D_x^2 u_\alpha(r;x)(h,k), dw(r) \rangle_H$$

Hence, if we set

$$R_\tau = \|v\|_{\mathcal{Z}_\tau^2} \vee \|w\|_{\mathcal{Z}_\tau^2}, \qquad M_\tau = \sup_{|y|_H \leq R_\tau} |DK(y)|_H$$

we have

$$|\psi(s,x)| = |K(Dv(s,x)) - K(Dw(s,x))|$$

$$\leq M_\tau |Dv(s,x) - Dw(s,x)|_H \leq M_\tau \|v - w\|_{\mathcal{Z}_\tau^2}.$$

Moreover, as

$$\langle D_x \psi(s,x), k \rangle_H = \langle DK(Dv(s,x)) - DK(Dw(s,x)), D^2(v-w)(s,x)k \rangle_H,$$

we have

$$\|D_x \psi(s,\cdot)\|_0^H \leq c M_\tau (s \wedge 1)^{-\frac{1+\epsilon}{2}} \|v - w\|_{\mathcal{Z}_\tau^2}.$$

This implies that

$$\|D^2 P_{t-s}^\alpha \psi(s,\cdot)\|_0^H \le c_\alpha\, M_\tau \left((t-s)\wedge 1\right)^{-\frac{1+\epsilon}{2}}\left(1+(s\wedge 1)^{-\frac{1+\epsilon}{2}}\right)\|v-w\|_{\mathcal{Z}_\tau^2},$$

so that

$$\|D^2\gamma_2(t,\cdot)\|_0^H \le c_\alpha\, M_\tau\, t^{\frac{1-\epsilon}{2}}\|v-w\|_{\mathcal{Z}_\tau^2}.$$

Hence, as

$$\|\gamma_2(t,\cdot)\|_1^H \le c\, M_\tau\, t\,\|v-w\|_{\mathcal{Z}_\tau^2},$$

we can conclude that Γ_α maps $\mathcal{Z}_\tau^2$ into itself for any $\tau \le T$ and

$$\|\Gamma_\alpha(v)-\Gamma_\alpha(w)\|_{\mathcal{Z}_\tau^2} \le c_\alpha\, M_\tau\left(\tau+\tau^{\frac{1-\epsilon}{2}}\right)\|v-w\|_{\mathcal{Z}_\tau^2}.$$

Furthermore, due to (9.4.11)

$$\|\Gamma_\alpha(v)\|_{\mathcal{Z}_\tau^2} \le c_\alpha\,\|\varphi\|_1^H + c_\alpha\left(\tau+\tau^{\frac{1-\epsilon}{2}}\right)\left(\|g\|_1^H + M_\tau\|v\|_{\mathcal{Z}_\tau^2}\right).$$

Therefore, if we fix $R \ge 2c_\alpha\,\|\varphi\|_1^H$ and $\tau_* = \tau_*(\alpha)$ such that

$$c_\alpha\left(\tau+\tau^{\frac{1-\epsilon}{2}}\right)\left(\|g\|_1^H + M_\tau R\right) \le \frac{1}{2},$$

we conclude that Γ_α maps the ball of radius R in $\mathcal{Z}_{\tau_*}^2$ into itself as a contraction and then admits a unique fixed point.

Finally, the independence of $\tau_*(\alpha)$ from $n \in \mathbb{N}$ follows from (9.3.6). $\qquad\square$

Proposition 9.4.8. *Let us fix* $\varphi, g \in C_b^1(H)$ *and assume that* y_α *is the unique solution of the problem (9.4.3) in* $\mathcal{Z}_{\tau_*}^2$, *with* $\tau_* = \tau_*(\alpha) \le T$. *Then, under the Hypotheses 6.1, 6.2, and 9.1 we have*

$$\sup_{\alpha>0}\left(\sup_{t\in[0,\tau_*]}\|y_\alpha(t,\cdot)\|_1^H\right) \le c\,(1+T)e^{cT}\left(\|\varphi\|_1^H + \|g\|_1^H\right).$$

Proof. If we define

$$U_{\alpha,n}(t,x) = \int_0^1 DK_n(\lambda Dy_{\alpha,n}(t,x))\,d\lambda, \qquad (9.4.12)$$

the problem (9.4.9) can be rewritten as

$$\begin{cases} \dfrac{\partial y}{\partial t}(t,x) = \dfrac{1}{2}\mathrm{Tr}\left[Q_n^2 D^2 y(t,x)\right] + \langle A_n x + F_{\alpha,n}(x) + U_{\alpha,n}(t,x), Dy(t,x)\rangle + g_n(x) \\[2mm] y(0,x) = \varphi_n(x). \end{cases}$$

Since $\varphi, g \in C_b^1(H)$, as proved in the Lemma 9.4.7 the solution $y_{\alpha,n} \in \mathcal{Z}_{\tau_*}^2$, for some $\tau_* = \tau_*(\alpha) \le T$ independent of $n \in \mathbb{N}$. Then $U_{\alpha,n} : [0,\tau_*] \times H \to H_n$ is continuous and bounded. Moreover, since DK is locally Lipschitz-continuous, if we define $M_{\alpha,n}$ as the Lipschitz constant of DK in the ball $\{ x \in H \; ; \; |x|_H \leq \|y_{\alpha,n}\|_{Z^2_{\tau_*}} \}$, for any $x, y \in H$ and $t > 0$, we have that

$$|U_{\alpha,n}(t,x) - U_{\alpha,n}(t,y)|_H \leq \int_0^1 |DK_n(\lambda Dy_{\alpha,n}(t,x)) - DK_n(\lambda Dy_{\alpha,n}(t,y))|_H \, d\lambda$$

$$\leq M_{\alpha,n} |Dy_{\alpha,n}(t,x) - Dy_{\alpha,n}(t,y)|_H \leq c M_{\alpha,n} \sup_{z \in H} \|D^2 y_{\alpha,n}(t,z)\|_{\mathcal{L}(H)} |x - y|_H$$

$$\leq c M_{\alpha,n} t^{-\frac{1+\epsilon}{2}} \|y_{\alpha,n}\|_{Z^2_{\tau_*}} |x - y|_H.$$

This means that the function $U_{\alpha,n}$ fulfills the hypotheses of the Lemma 9.4.6, so that for each $0 \leq r < \tau_*$ the stochastic problem

$$du(t) = [A_n u(t) + F_{\alpha,n}(u(t)) + U_{\alpha,n}(t, u(t))] \, dt + Q_n dw(t), \quad u(r) = P_n x, \quad (9.4.13)$$

admits a unique strong solution $u_{\alpha,n}(t; r, x) \in L^2(\Omega; C([r, \tau_*); H)) \cap L^\infty(r, \tau_*; H))$.

If we denote by $R^{\alpha,n}_{s,t}$ the corresponding transition semigroup, that is

$$R^{\alpha,n}_{s,t} \varphi(x) = \mathbb{E}\varphi(u_{\alpha,n}(t; s, x)), \quad 0 \leq s \leq t \leq \tau_*,$$

for $\varphi \in B_b(H)$ and $x \in H$, we have

$$y_{\alpha,n}(t,x) = R^{\alpha,n}_{\tau_*-t,\tau_*} \varphi(x) + \int_0^t R^{\alpha,n}_{\tau_*-t,\tau_*-s} g(x) \, ds. \quad (9.4.14)$$

Indeed, since $y_{\alpha,n}(t)$ is a strong solution and $y_{\alpha,n}(t,x)$ is twice differentiable, we can apply Itô's formula to the function $s \mapsto y_{\alpha,n}(\tau_* - s, u_{\alpha,n}(s; \tau_* - t, x))$ and by integrating with respect to $s \in [\tau_* - t, \tau_*]$ and taking the expectation, we get

$$y_{\alpha,n}(t,x) = \mathbb{E}\varphi(u_{\alpha,n}(\tau_*; \tau_* - t, x)) + \int_0^t \mathbb{E}g(u_{\alpha,n}(\tau_* - s; \tau_* - t, x)) \, ds,$$

which coincides with (9.4.14). As an immediate consequence this yields

$$\sup_{t \in [0,\tau_*]} \|y_{\alpha,n}(t, \cdot)\|_0^H \leq \|\varphi\|_0^H + T\|g\|_0^H. \quad (9.4.15)$$

The proof of the analogous estimate for the derivative of $y_{\alpha,n}(t,x)$ is more complicate, but is based on similar arguments.

The problem (9.4.9) can be rewritten as

$$\begin{cases} \dfrac{\partial y}{\partial t}(t,x) = \dfrac{1}{2} \displaystyle\sum_{k=1}^n \lambda_k^2 D_k^2 y(t,x) + \sum_{k,h=1}^n a_{kh} x_h D_k y(t,x) - K_n(Dy(t,x)) \\[2mm] \qquad\qquad + \displaystyle\sum_{k=1}^n \langle F_{\alpha,n}(x), e_k \rangle_H D_k y(t,x) + g_n(x), \\[2mm] y(0,x) = \varphi_n(x), \end{cases} \quad (9.4.16)$$

where for each $k, h \in \mathbb{N}$, we denote $D_k y = \langle Dy, e_k \rangle$ and $a_{kh} = \langle Ae_k, e_h \rangle$. By differentiating each side of (9.4.16) with respect to x_j and setting $v_j = D_j y$, we get

$$\frac{\partial v_j}{\partial t} = \frac{1}{2} \sum_{k=1}^{n} \lambda_k^2 D_k^2 v_j + \sum_{k=1}^{n} a_{kh} x_h D_k v_j + \sum_{k=1}^{n} a_{kj} v_k + \sum_{k=1}^{n} \langle F_{\alpha,n}(x), e_k \rangle_H D_k v_j$$

$$+ \sum_{k=1}^{n} \langle DF_{\alpha,n}(x)e_j, e_k \rangle_H v_k - \sum_{k=1}^{n} \langle DK_n(Dy_{\alpha,n}), e_k \rangle_H D_k v_j + \langle Dg_n(x), e_j \rangle_H .$$

Multiplying each side by v_j and summing up on j, we obtain

$$\frac{1}{2} \left(\frac{\partial}{\partial t} \sum_{j=1}^{n} v_j^2 \right) = \frac{1}{2} \sum_{k,j=1}^{n} \lambda_k^2 v_j D_k^2 v_j + \sum_{k,h,j=1}^{n} a_{kh} x_h v_j D_k v_j + \sum_{k,j=1}^{n} a_{kj} v_k v_j$$

$$+ \sum_{k,j=1}^{n} \langle F_{\alpha,n}(x), e_k \rangle v_j D_k v_j + \sum_{k,j=1}^{n} \langle DF_{\alpha,n}(x)v_j e_j, v_k e_k \rangle_H$$

$$- \sum_{k,j=1}^{n} \langle DK_n(Dy_{\alpha,n}), e_k \rangle_H v_j D_k v_j + \sum_{j=1}^{n} \langle Dg_n(x), v_j e_j \rangle_H .$$

Now, if we define $2z_{\alpha,n}(t,x) = |Dy_{\alpha,n}(t,x)|_H^2$, it holds

$$\sum_{k,j=1}^{n} \lambda_k^2 v_j D_k^2 v_j = \mathrm{Tr} \left[Q_n^2 D^2 z_{\alpha,n} \right] - \sum_{k,j=1}^{n} \lambda_k^2 (D_k v_j)^2 ,$$

$$\sum_{k,h,j=1}^{n} a_{kh} x_h v_j D_k v_j + \sum_{k,j=1}^{n} a_{kj} v_k v_j = \langle A_n x, Dz_{\alpha,n} \rangle_H + \langle A_n Dy_{\alpha,n}, Dy_{\alpha,n} \rangle_H ,$$

and

$$\sum_{k,j=1}^{n} \langle F_{\alpha,n}(x), e_k \rangle_H v_j D_k v_j = \langle F_{\alpha,n}(x), Dz_{\alpha,n} \rangle_H ,$$

$$\sum_{k,j=1}^{n} \langle DK_n(Dy_{\alpha,n}), e_k \rangle_H v_j D_k v_j = \langle DK_n(Dy_{\alpha,n}), Dz_{\alpha,n} \rangle_H .$$

Moreover, we have

$$\sum_{k,j=1}^{n} \langle DF_{\alpha,n}(x)v_j e_j, v_k e_k \rangle_H = \langle DF_{\alpha,n}(x)Dy_{\alpha,n}, Dy_{\alpha,n} \rangle_H ,$$

$$\sum_{j=1}^{n} \langle Dg_n, v_j e_j \rangle_H = \langle Dg_n, Dy_{\alpha,n} \rangle_H .$$

Thus, by substituting and by taking into account of (4.1.6) and (9.1.7), we can conclude that

$$\frac{\partial z_{\alpha,n}}{\partial t}(t,x) \leq \mathcal{M}_{\alpha,n} z_{\alpha,n}(t,x) + c\, z_{\alpha,n}(t,x) + |Dg_n|_H^2,$$

where the differential operator $\mathcal{M}_{\alpha,n}$ is defined by

$$\mathcal{M}_{\alpha,n}\psi(x) = \frac{1}{2}\mathrm{Tr}\,[Q_n^2 D^2\psi(x)] + \langle A_n x + F_{\alpha,n}(x) - DK_n(Dy_{\alpha,n}(t,x)), D\psi(x)\rangle_H .$$

Now, we define
$$V_{\alpha,n}(t,x) = -DK_n(Dy_{\alpha,n}(t,x)).$$

By arguing as we did above for the function $U_{\alpha,n}(t,x)$ defined in (9.4.12), we have that $V_{\alpha,n} : [0,\tau_*] \times H \to H$ satisfies the hypotheses of the Lemma 9.4.6, so that the stochastic problem

$$du(t) = [A_n u(t) + F_{\alpha,n}(u(t)) + V_{\alpha,n}(t,u(t))]\, dt + Q_n dw(t), \qquad u(r) = P_n x, \quad (9.4.17)$$

admits a unique strong solution $y_{\alpha,n}(\cdot;r,x)$, for any $0 \leq r \leq \tau_*$. If we denote by $S_{s,t}^{\alpha,n}$ the transition semigroup associated with the equation (9.4.17), by arguing as before for the semigroup associated with the problem (9.4.13), we have that the solution of the problem

$$\frac{\partial v}{\partial t}(t,x) = \mathcal{M}_{\alpha,n} v(t,x) + c\, v(t,x) + |Dg_n(x)|_H^2, \qquad v(0,x) = |D\varphi_n(x)|_H^2,$$

is given by

$$v_{\alpha,n}(t,x) = e^{ct} S_{\tau_*-t,\tau_*}^{\alpha,n}\left(|D\varphi_n(\cdot)|_H^2\right)(x) + \int_0^t e^{c(t-s)} S_{\tau_*-t,\tau_*-s}^{\alpha,n}\left(|Dg_n(\cdot)|_H^2\right)(x)\, ds.$$

This yields

$$\sup_{t\in[0,\tau_*]} \|v_{\alpha,n}(t,\cdot)\|_0^H \leq c(1+T)\, e^{cT}\left(\|\varphi\|_1^H + \|g\|_1^H\right)^2,$$

so that by a comparison argument we conclude

$$\sup_{t\in[0,\tau_*]}\left(\|Dy_{\alpha,n}(t,\cdot)\|_0^H\right)^2 \leq 2 \sup_{t\in[0,T]} \|z_{\alpha,n}(t,\cdot)\|_0^H \leq c(1+T)\, e^{cT}\left(\|\varphi\|_1^H + \|g\|_1^H\right)^2.$$

Thanks to (9.4.15) this implies that

$$\sup_{t\in[0,\tau_*]} \|y_{\alpha,n}(t,\cdot)\|_1^H \leq c(1+\sqrt{T})\, e^{cT}\left(\|\varphi\|_1^H + \|g\|_1^H\right)$$

and hence, due to (9.4.10), we can conclude. $\square$

9.4.2 Proof of the Theorem 9.4.2

We fix $T > 0$ and $\varphi, g \in C_b^1(H)$ and we define

$$R = 2c(1+T)e^{cT}\left((1+2\nu)\|\varphi\|_1^H + \|g\|_1^H\right),$$

where c is the constant introduced in the Proposition 9.4.8. Due to the Lemmata 9.4.3 and 9.4.5, the problem (9.4.2) admits a mild solution $y(t, x)$ which is defined for $t \in [0, \tau_R]$ and which fulfills (9.4.4), uniformly with respect to $t \in [0, \tau_R]$ and x in bounded sets of E.

In the Proposition 9.4.8 we have shown that there exists $\tau_* = \tau_*(\alpha) \leq T$ such that

$$\sup_{t \in [0, \tau_*]} \|y_\alpha(t, \cdot)\|_1^H \leq c(1+T)e^{cT}\left(\|\varphi\|_1^H + \|g\|_1^H\right).$$

According to the Lemma 9.4.5, this implies that for any $t \in [0, \tau_*]$ and $x \in E$

$$|y(t, x)| + |Dy(t, x)|_H \leq c(1+T)e^{cT}\left(\|\varphi\|_1^H + \|g\|_1^H\right)$$

and, since $y(t, \cdot) \in C_b^1(H)$ for $t \in [0, \tau_*]$, this yields

$$\sup_{t \in [0, \tau_*]} \|y(t, \cdot)\|_1^H \leq c(1+T)e^{cT}\left(\|\varphi\|_1^H + \|g\|_1^H\right). \tag{9.4.18}$$

In particular, due to the definition of R we have that

$$\|y(\tau_*, \cdot)\|_1^H \leq \frac{R}{2}.$$

This allows us to repeat all these arguments in the intervals $[\tau_*, 2\tau_*]$, $[2\tau_*, 3\tau_*]$ and so on, up to time T, and hence to get a global solution.

Now, assume that $\varphi, g \in C_b^{0,1}(H)$. It is possible to find two bounded sequences $\{\varphi_k\}$ and $\{g_k\}$ in $C_b^1(H)$ converging respectively to φ and g in $C_b(H)$. In correspondence with each k, there exists a unique global solution $y_k(t, x)$ to the problem

$$y_k(t, x) = P_t\varphi_k(x) - \int_0^t P_{t-s}\left(K(Dy_k(s, \cdot))\right)(x)\,ds + \int_0^t P_{t-s}g_k(x)\,ds.$$

Our aim is to show that $\{y_k\}$ is a Cauchy sequence in $\mathcal{Z}_T^1$ and that the limit y fulfills the equation (9.4.2).

For each $k, h \in \mathbb{N}$ we have

$$y_k(t, x) - y_h(t, x) = P_t\left(\varphi_k - \varphi_h\right)(x)$$

$$- \int_0^t P_{t-s}\left[K(Dy_k(s, \cdot)) - K(Dy_h(s, \cdot))\right](x)\,ds + \int_0^t P_{t-s}\left(g_k - g_h\right)(x)\,ds.$$

Due to (9.4.18) we easily have

$$\sup_{k \in \mathbb{N}} \sup_{t \in [0,T]} \|Dy_k(t,\cdot)\|_0^H \le c(1+T)e^{cT} \sup_{k \in \mathbb{N}} (\|\varphi_k\|_1^H + \|g_k\|_1^H) = c_T. \qquad (9.4.19)$$

If M is the Lipschitz constant of K in $B_{c_T}^H$, we have

$$\|y_k(t,\cdot) - y_h(t,\cdot)\|_0^H \le \|\varphi_k - \varphi_h\|_0^H$$

$$+ M \int_0^t \|Dy_k(s,\cdot) - Dy_h(s,\cdot)\|_0^H \, ds + t \|g_k - g_h\|_0. \qquad (9.4.20)$$

Moreover, we have

$$Dy_k(t,x) - Dy_h(t,x) = DP_t(\varphi_k - \varphi_h)(x)$$

$$- \int_0^t DP_{t-s} \left[K(Dy_k(s,\cdot)) - K(Dy_h(s,\cdot)) \right](x) \, ds + \int_0^t DP_{t-s}(g_k - g_h)(x) \, ds,$$

so that, thanks to (7.3.2), we get

$$\|Dy_k(t,\cdot) - Dy_h(t,\cdot)\|_0^H \le c t^{-\frac{1+\epsilon}{2}} \|\varphi_k - \varphi_h\|_0^H$$

$$+ cM \int_0^t (t-s)^{-\frac{1+\epsilon}{2}} \|Dy_k(s,\cdot) - Dy_h(s,\cdot)\|_0^H \, ds + c \int_0^t (t-s)^{-\frac{1+\epsilon}{2}} \, ds \|g_k - g_h\|_0^H.$$

This implies that

$$t^{\frac{1+\epsilon}{2}} \|Dy_k(t,\cdot) - Dy_h(t,\cdot)\|_0^H \le c \|\varphi_k - \varphi_h\|_0^H$$

$$+ cM t^{\frac{1+\epsilon}{2}} \int_0^t (t-s)^{-\frac{1+\epsilon}{2}} \|Dy_k(s,\cdot) - Dy_h(s,\cdot)\|_0^H \, ds + c t \|g_k - g_h\|_0^H. \qquad (9.4.21)$$

By combining together (9.4.20) and (9.4.21), we have

$$\|y_k(t,\cdot) - y_h(t,\cdot)\|_0^H + t^{\frac{1+\epsilon}{2}} \|Dy_k(t,\cdot) - Dy_h(t,\cdot)\|_0^H$$

$$\le c \left(\|\varphi_k - \varphi_h\|_0^H + T\|g_k - g_h\|_0^H \right) + M(1 + cT^{\frac{1+\epsilon}{2}}) \int_0^t s^{-\frac{1+\epsilon}{2}} \left((t-s)^{-\frac{1+\epsilon}{2}} + 1 \right)$$

$$\left(\|y_k(s,\cdot) - y_h(s,\cdot)\|_0^H + s^{\frac{1+\epsilon}{2}} \|Dy_k(s,\cdot) - Dy_h(s,\cdot)\|_0^H \right) ds.$$

Thus, from a generalization of the Gronwall lemma, we can conclude that

$$\|y_k - y_h\|_{Z_T^1} \le c_T \left(\|\varphi_k - \varphi_h\|_0^H + T\|g_k - g_h\|_0^H \right), \qquad (9.4.22)$$

for some constant c_T independent of k and h. This implies that $\{y_k\}$ is a Cauchy sequence in $\mathcal{Z}_T^1$ and hence it converges to a limit $y \in \mathcal{Z}_T^1$. Moreover, from (9.4.19) we have that

$$\sup_{t \in [0,T]} \|Dy(t,\cdot)\|_0^H < +\infty,$$

so that $y \in \mathcal{V}_T^1$.

Now, we show that y is the mild solution of the problem (9.4.2). Actually, for any $s > 0$ and $x \in H$

$$\lim_{k \to +\infty} K(Dy_k(s,x)) = K(Dy(s,x)).$$

Due to (9.4.19) we can apply the dominated convergence theorem and we get

$$\lim_{k \to +\infty} \int_0^t P_{t-s} K(Dy_k(s,\cdot))(x)\,ds = \int_0^t P_{t-s} K(Dy(s,\cdot))(x)\,ds.$$

Therefore, since

$$\lim_{k \to +\infty} P_t \varphi_k(x) = P_t \varphi(x)$$

and

$$\lim_{k \to +\infty} \int_0^t P_{t-s} g_k(x)\,ds = \int_0^t P_{t-s} g(x)\,ds,$$

we conclude that y is a solution of (9.4.2).

Finally, uniqueness follows from the Gronwall lemma and the local Lipschitz-continuity of K. Indeed, if y_1 and y_2 are two solutions in $\mathcal{V}_T^1$ we have

$$y_1(t,x) - y_2(t,x) = -\int_0^t P_{t-s} \left[K(Dy_1(s,\cdot)) - K(Dy_2(s,\cdot)) \right](x)\,ds$$

and then, if M is the Lipschitz constant of K in $B_{c_T}^H$, where

$$c_T = \|y_1\|_{\mathcal{V}_T^1} + \|y_2\|_{\mathcal{V}_T^1},$$

we have

$$\|y_1 - y_2\|_{\mathcal{V}_T^1} \le M \int_0^t \left(1 + (t-s)^{-\frac{1+\epsilon}{2}} \right) ds \|y_1 - y_2\|_{\mathcal{V}_T^1}.$$

This implies that $y_1 = y_2$.
$\square$

9.5 The elliptic Hamilton-Jacobi-Bellman equation

We are here concerned with the stationary Hamilton-Jacobi-Bellman equation

$$\lambda \varphi(x) - L\varphi(x) + K(D\varphi(x)) = g(x), \qquad x \in H, \qquad (9.5.1)$$

where L is the weak generator of P_t. Our aim is to study the existence and uniqueness of solutions for it. To this purpose we first prove a regularity result for the elements of $D(L)$.

Lemma 9.5.1. *Assume the Hypotheses 6.1, 6.2 and 6.3. Then $D(L) \subset C_b^1(H)$ and for any $\lambda > 0$ and $g \in C_b(H)$ it holds*

$$\|R(\lambda, L)g\|_1^H \leq \rho(\lambda)\|g\|_0^H, \tag{9.5.2}$$

where $\rho(\lambda) = c\left(\lambda^{\frac{\epsilon-1}{2}} + \lambda^{-1}\right)$.

Proof. We recall that if $\varphi \in C_b(H)$, then $P_t\varphi \in C_b^1(H)$, for any $t > 0$. Thus for any $x, h \in H$ and $\lambda > 0$ we have

$$R(\lambda, L)g(x+h) - R(\lambda, L)g(x) = \int_0^{+\infty} e^{-\lambda t}(P_t g(x+h) - P_t g(x)) \, dt$$

$$= \int_0^{+\infty} e^{-\lambda t} \langle D(P_t g)(x), h \rangle_H \, dt + E(x, h),$$

where

$$E(x, h) = \int_0^{+\infty} e^{-\lambda t} \int_0^1 \langle D(P_t g)(x + \theta h) - D(P_t g)(x), h \rangle_H \, d\theta \, dt.$$

Due to (7.3.2) we have

$$\left| \int_0^{+\infty} e^{-\lambda t} \langle D(P_t g)(x), h \rangle_H \, dt \right|$$

$$\leq c \int_0^{+\infty} e^{-\lambda t}(t \wedge 1)^{-\frac{1+\epsilon}{2}} \, dt \, |h|_H \, \|g\|_0 = c\left(\lambda^{\frac{\epsilon-1}{2}} + \lambda^{-1}\right) |h|_H \, \|g\|_0.$$

Moreover, as $D(P_t g)$ is continuous in H, by the dominated convergence theorem we easily have

$$\lim_{|h|_H \to 0} \frac{|E(x, h)|}{|h|_H} = 0.$$

This implies that $R(\lambda, L)g \in C_b^1(H)$ and for any $x, h \in H$

$$\langle D(R(\lambda, L)g)(x), h \rangle_H = \int_0^{+\infty} e^{-\lambda t} \langle D(P_t g)(x), h \rangle_H \, dt. \tag{9.5.3}$$

Furthermore, the estimate (9.5.2) holds true. □

Remark 9.5.2. Notice that due to (9.3.3) and (9.3.7) we can repeat the arguments used in the previous proposition and we can show that both $D(L_\alpha)$ and $D(L_{\alpha,n})$ are contained in $C_b^1(H)$ and for any $g \in C_b(H)$ a formula analogous to (9.5.3) holds for the derivatives of $R(\lambda, L_\alpha)g$ and $R(\lambda, L_{\alpha,n})g$. In particular it holds

$$\|R(\lambda, L_\alpha)g\|_1^H + \|R(\lambda, L_{\alpha,n})g\|_1^H \leq \rho(\lambda)\|g\|_0^H. \tag{9.5.4}$$

Moreover, as

$$\|P_t^\alpha \varphi\|_j^H + \|P_t^{\alpha,n} \varphi\|_j^H \le c_\alpha \, (t \wedge 1)^{-\frac{(j-i)(1+\epsilon)}{2}} \|\varphi\|_i^H, \quad i \le j \le 2,$$

for a constant c_α independent of $n \in \mathbb{N}$, by interpolating for any $\theta_1, \theta_2 \in [0,1]$ we have

$$\|P_t^\alpha \varphi\|_{1+\theta_1}^H + \|P_t^{\alpha,n} \varphi\|_{1+\theta_1}^H \le c_\alpha \, (t \wedge 1)^{-\frac{(\theta_1 - \theta_2 + 1)(1+\epsilon)}{2}} \|\varphi\|_{\theta_2}^H.$$

By proceeding as in the proof of the previous lemma, this implies that if $\varphi \in C_b^{\theta_2}(H)$ then both $R(\lambda, L_\alpha)\varphi$ and $R(\lambda, L_{\alpha,n})\varphi$ are in $C_b^{1+\theta_1}(H)$, for any $\theta_1 < \theta_2 + (1-\epsilon)/(1+\epsilon)$ and

$$\|R(\lambda, L_\alpha)\varphi\|_{1+\theta_1}^H + \|R(\lambda, L_{\alpha,n})\varphi\|_{1+\theta_1}^H \le c_\alpha \left(\lambda^{\frac{(\theta_1 - \theta_2 + 1)(\epsilon+1)}{2} - 1} + \lambda^{-1} \right) \|g\|_{\theta_2}^H. \quad (9.5.5)$$

In particular, we have that $D(L_\alpha)$ and $D(L_{\alpha,n})$ are contained in $C_b^{1+\theta}(H)$, for any $\theta < (1-\epsilon)/(1+\epsilon)$.

9.5.1 Lipschitz hamiltonian K

In the proof of the existence and uniqueness of solutions for the problem (9.5.1) we proceed in several steps. First we assume the Lipschitz-continuity of the hamiltonian K.

Hypothesis 9.2. *The mapping $K : H \to \mathbb{R}$ is Fréchet differentiable and Lipschitz-continuous together with its derivative. Moreover, $K(0) = 0$.*

Notice that the condition $K(0) = 0$ is not restrictive, as we can substitute g by $g - K(0)$.

Thanks to the Lemma 9.5.1, we can prove the following result.

Proposition 9.5.3. *Under the Hypotheses 6.1, 6.2, 6.3 and 9.2, there exists $\lambda_0 > 0$ such that the equation (9.5.1) admits a unique solution $\varphi(\lambda, g) \in C_b^1(H)$, for any $\lambda > \lambda_0$ and $g \in C_b(H)$.*

Proof. The equation (9.5.1) is equivalent to the equation

$$\varphi = R(\lambda, L) \, (g - K(D\varphi)) = \Gamma(\lambda, g)(\varphi).$$

Due to the Lemma 9.5.1, if $\varphi \in C_b^1(H)$ and $g \in C_b(H)$, then $\Gamma(\lambda, g)(\varphi) \in C_b^1(H)$. Thus if we show that for some $\lambda_0 > 0$ the mapping $\Gamma(\lambda, g)$ is a contraction in $C_b^1(H)$ for any $\lambda > \lambda_0$, our conclusion follows.

As K is Lipschitz-continuous, for any $\varphi_1, \varphi_2 \in C_b^1(H)$ we have

$$\|R(\lambda, L)(K(D\varphi_1) - K(D\varphi_2))\|_1^H \le c\,\rho(\lambda) \|D\varphi_1 - D\varphi_2\|_0^H \le c\,\rho(\lambda) \|\varphi_1 - \varphi_2\|_1^H.$$

Thus, if we choose λ_0 such that $c\rho(\lambda_0) = 1$, we have that $\Gamma(\lambda, g)$ is a contraction in $C_b^1(H)$, for any $\lambda > \lambda_0$. This implies that it admits a unique fixed point $\varphi \in C_b^1(H)$ which is the unique solution of (9.5.1). $\qquad\square$

Remark 9.5.4. By using (9.5.4) it is possible to prove that there exists $\lambda_0 > 0$ sufficiently large such that the mappings

$$\Gamma_\alpha(\lambda, g)(\varphi) = R(\lambda, L_\alpha)(g - K(D\varphi)), \qquad \alpha > 0,$$

are contractions in $C_b^1(H)$, for any $\lambda > \lambda_0$ and for any $g \in C_b(H)$, so that the approximating Hamilton-Jacobi-Bellman equations

$$\lambda\varphi - L_\alpha\varphi + K(D\varphi) = g, \tag{9.5.6}$$

admit a unique solution $\varphi_\alpha(\lambda, g) \in C_b^1(H)$. Moreover, as the constant c in the right hand side of (9.5.4) does not depend of $\alpha > 0$, the constant λ_0 does not depend on α, as well.

Lemma 9.5.5. *Under the Hypotheses 6.1, 6.2, 6.3 and 9.2, if $\lambda > 0$ and $g \in C_b(H)$ we have*

$$\lim_{\alpha \to 0} \sup_{|x|_E \le R} \left| D^j \left(\Gamma_\alpha^k(\lambda, g)(0) - \Gamma^k(\lambda, g)(0) \right)(x) \right|_{\mathcal{L}^j(H)} = 0, \qquad j = 0, 1, \tag{9.5.7}$$

for any $k \in \mathbb{N}$ and $R > 0$.

Proof. We proceed by induction. For $k = 1$ the limit (9.5.7) is trivially verified. Next, assume that (9.5.7) holds for some $k \ge 1$. We show that this implies that (9.5.7) holds for $k + 1$. We have

$$D^j \left(\Gamma_\alpha^{k+1}(\lambda, g)(0) - \Gamma^{k+1}(\lambda, g)(0) \right)$$

$$= D^j \left[R(\lambda, L_\alpha) \left(g - K \left(D(\Gamma_\alpha^k(\lambda, g)(0)) \right) \right) - R(\lambda, L) \left(g - K \left(D(\Gamma^k(\lambda, g)(0)) \right) \right) \right].$$

We remark that in general, if $f \in C_b(H)$ and $\{f_\alpha\}$ is any bounded generalized sequence of $C_b(H)$ such that for any $R > 0$

$$\lim_{\alpha \to 0} \sup_{|x|_E \le R} |f_\alpha(x) - f(x)| = 0, \tag{9.5.8}$$

then for any $R > 0$ and $j = 0, 1$ we have

$$\lim_{\alpha \to 0} \sup_{|x|_E \le R} |D^j(R(\lambda, L_\alpha)(f_\alpha - f))(x)|_H = 0. \tag{9.5.9}$$

Indeed, as the formula (9.5.3) holds also for the derivative of $R(\lambda, L_\alpha)$, for any $x \in H$ we have

$$D^j(R(\lambda, L_\alpha)(f_\alpha - f))(x) = \int_0^{+\infty} e^{-\lambda t} D^j \left(P_t^\alpha(f_\alpha - f) \right)(x) \, dt.$$

In the previous section we have seen that there exists an increasing continuous function $c(t)$ independent of $\alpha > 0$ and $n \in \mathbb{N}$ such that

$$|u_\alpha(t; x, z)|_E \leq c(t) \left(|x|_E + |z|_{L^p(0,+\infty;H)}^{2m+1} + \sup_{s \in [0,t]} |w^A(s)|_E^{2m+1} \right), \qquad \mathbb{P} - \text{a.s.}$$

and

$$|u_{\alpha,n}(t; x, z)|_H \leq c(t) \left(|x|_H + |z|_{L^2(0,+\infty;H)}^{2m+1} + \sup_{s \in [0,t]} |w^A(s)|_E^{2m+1} \right), \qquad \mathbb{P} - \text{a.s.}$$

$$(9.5.10)$$

Whence, for any $R > 0$ there exists a process $c_R(t)$ which is finite $\mathbb{P}$-a.s. such that

$$|x|_E \leq R \Rightarrow |u_\alpha(t; x)(\omega)| \leq c_R(t, \omega), \qquad \mathbb{P} - \text{a.s.}$$

Therefore, by (9.5.8) this yields

$$\lim_{\alpha \to 0} \sup_{|x|_E \leq R} |(f_\alpha - f)(u_\alpha(t; x))| = 0, \qquad \mathbb{P} - \text{a.s.} \qquad (9.5.11)$$

and by applying the dominated convergence theorem we obtain (9.5.9) for $j = 0$. As proved in the Proposition 4.4.3, for any $t > 0$ we have

$$\langle D(P_t^\alpha(f_\alpha - f))(x), h \rangle_H = \frac{1}{t} \mathbb{E}(f_\alpha - f)(u_\alpha(t; x)) \int_0^t \langle Q^{-1} Du_\alpha(s; x)h, dw(s) \rangle_H,$$

where $Du_\alpha(t; x)h$ is the derivative of $u_\alpha(t; x)$ along the direction $h \in H$. Hence, thanks to (9.3.2) we easily get

$$|D(P_t^\alpha(f_\alpha - f))(x)|_H \leq c \, (t \wedge 1)^{-\frac{1+\epsilon}{2}} \left(\mathbb{E}|(f_\alpha - f)(u_\alpha(t; x))|^2 \right)^{1/2}$$

and thanks to (9.5.11) this implies (9.5.9) for $j = 1$.

Thus, since from the inductive hypothesis and the Lipschitz-continuity of K, the sequence $\{K(D(\Gamma_\alpha^k(\lambda, g)(0)))\}$ and $K(D(\Gamma^k(\lambda, g)(0)))$ fulfill (9.5.8), we can conclude that for any $R > 0$ and $j = 0, 1$

$$\lim_{\alpha \to 0} \sup_{|x|_E \leq R} \left| D^j \left[R(\lambda, L_\alpha) \left(K(D(\Gamma_\alpha^k(\lambda, g)(0))) - K(D(\Gamma^k(\lambda, g)(0))) \right) \right] (x) \right| = 0.$$

$$(9.5.12)$$

Now, if $f \in C_b^1(H)$, for any $x \in H$ we have

$$D^j \left[(R(\lambda, L_\alpha) - R(\lambda, L)) \, (g - K(Df)) \right] (x)$$

$$= \int_0^{+\infty} e^{-\lambda t} D^j \left[(P_t^\alpha - P_t)(g - K(Df)) \right] (x) \, dt.$$

Then, by using the estimates (7.3.2) and (9.3.1) and the limit (9.3.5), we get

$$\lim_{\alpha \to 0} \sup_{|x|_E \leq R} \left| D^j \left[(R(\lambda, L_\alpha) - R(\lambda, L)) \left(g - K(Df) \right) \right] (x) \right|_{\mathcal{L}^j(H)} = 0,$$

for any $R > 0$. As $\Gamma^k(\lambda, L)(0) \in C_b^1(H)$, this implies that

$$\lim_{\alpha \to 0} \sup_{|x|_E \leq R} \left| D^j \left[(R(\lambda, L_\alpha) - R(\lambda, L)) \left(g - K(\Gamma^k(\lambda, g)(0)) \right) \right] (x) \right|_H = 0$$

and recalling (9.5.12) we can conclude that

$$\lim_{\alpha \to 0} \sup_{|x|_E \leq R} \left| D^j \left(\Gamma_\alpha^{k+1}(\lambda, g)(0) - \Gamma^{k+1}(\lambda, g)(0) \right) (x) \right| = 0.$$

By induction this yields (9.5.7), for any $k \in \mathbb{N}$. $\square$

In the next proposition we show that the solution $\varphi(\lambda, g)$ of the problem (9.5.1) can be approximated by the solutions $\varphi_\alpha(\lambda, g)$ of the problems (9.5.6).

Proposition 9.5.6. *Assume the Hypotheses 6.1, 6.2, 6.3 and 9.2. Then if $\lambda_0 > 0$ is the constant introduced in the Proposition 9.5.3, for any $\lambda > \lambda_0$ and $g \in C_b(H)$ it holds*

$$\lim_{\alpha \to 0} \sup_{|x|_E \leq R} \left| D^j \left(\varphi(\lambda, g) - \varphi_\alpha(\lambda, g) \right) (x) \right|_{\mathcal{L}^j(H)} = 0, \qquad j = 0, 1, \qquad (9.5.13)$$

for any $R > 0$.

Proof. Let us fix λ_0 as in the Proposition 9.5.3. We have seen that $\varphi = \varphi(\lambda, g)$ and $\varphi_\alpha = \varphi_\alpha(\lambda, g)$ are respectively the unique fixed points of the mappings $\Gamma(\lambda, g)$ and $\Gamma_\alpha(\lambda, g)$. Since for any $\lambda > \lambda_0$ and $g \in C_b(H)$ the contraction constants of $\Gamma_\alpha(\lambda, g)$ are the same for all $\alpha > 0$, for any $\epsilon > 0$ there exists $k_\epsilon \in \mathbb{N}$ such that

$$\left\| \Gamma^{k_\epsilon}(\lambda, g)(0) - \varphi \right\|_1^H + \sup_{\alpha > 0} \left\| \Gamma_\alpha^{k_\epsilon}(\lambda, g)(0) - \varphi_\alpha \right\|_1^H \leq \epsilon.$$

Thus, for $j = 0, 1$ and $x \in H$ we have

$$\left| D^j (\varphi - \varphi_\alpha)(x) \right|_{\mathcal{L}^j(H)} \leq \epsilon + \left| D^j \left(\Gamma^{k_\epsilon}(\lambda, g)(0) - \Gamma_\alpha^{k_\epsilon}(\lambda, g)(0) \right) (x) \right|_{\mathcal{L}^j(H)},$$

and from (9.5.7) we obtain (9.5.13). $\square$

Assume that for some $\lambda > 0$ the problem (9.5.1) has a solution $\varphi(\lambda, g)$. Then as an immediate consequence of the previous proposition we have

$$\lim_{\alpha \to 0} \sup_{|x|_E \leq R} \left| D^j \left[\varphi(\lambda, g) - \varphi_\alpha(\lambda + \lambda_0, g + \lambda_0 \, \varphi(\lambda, g)) \right] (x) \right|_{\mathcal{L}^j(H)} = 0, \qquad j = 0, 1.$$
$$(9.5.14)$$

Actually, as $\varphi(\lambda, g) = \varphi(\lambda + \lambda_0, g + \lambda_0 \, \varphi(\lambda, g))$, we can proceed by using (9.5.13).

Remark 9.5.7. For any $\alpha > 0$ and $n \in \mathbb{N}$, consider the problem

$$\lambda\varphi - L_{\alpha,n}\varphi + K_n(D\varphi) = g_n, \qquad (9.5.15)$$

where $K_n(x) = K(P_n x)$ and $g_n(x) = g(P_n x)$, for $x \in H$. By proceeding as for the problems (9.5.1) and (9.5.6), it is possible to show that there exists λ_0 large enough such that for any $g \in C_b(H)$ and $\lambda > \lambda_0$ there exists a unique solution $\varphi_{\alpha,n}(\lambda, g) \in C_b^1(H)$. Such a solution is given by the unique fixed point of the mapping

$$\Gamma_{\alpha,n}(\lambda, g)(\varphi) = R(\lambda, L_{\alpha,n})(g_n - K_n(D\varphi)).$$

By using arguments analogous to those used in the Lemma 9.5.5, due to the estimates (9.3.1) and (9.5.10), and due to the limits (9.3.8) and (9.3.9), we have that there exists $\lambda_0 > 0$ such that for $\lambda > \lambda_0$ and $g \in C_b(H)$ it holds

$$\lim_{n \to +\infty} \sup_{|x|_H \leq R} \left| D^j \left(\Gamma_{\alpha,n}^k(\lambda, g)(0) - \Gamma_\alpha^k(\lambda, g)(0) \right)(x) \right|_{\mathcal{L}^j(H)} = 0, \qquad j = 0, 1,$$

for any $\alpha > 0$, $k \in \mathbb{N}$ and $R > 0$. Thus, by proceeding as in the proof of the Proposition 9.5.6, it is possible to verify that there exists $\lambda_0 > 0$ such that if $\lambda > \lambda_0$ then

$$\lim_{n \to +\infty} \sup_{|x|_H \leq R} \left| D^j \left(\varphi_\alpha(\lambda, g) - \varphi_{\alpha,n}(\lambda, g) \right)(x) \right|_{\mathcal{L}^j(H)} = 0. \qquad (9.5.16)$$

In the next proposition we show that if the datum $g \in C_b^1(H)$ then the approximating problems (9.5.6) and (9.5.15) have a solution of class C^2.

Lemma 9.5.8. *Under the Hypotheses 6.1, 6.2, 6.3 and 9.2, if $g \in C_b^1(H)$ and $\lambda > 0$ then the solutions $\varphi_\alpha(\lambda, g)$ and $\varphi_{\alpha,n}(\lambda, g)$ of the problems (9.5.6) and (9.5.15) belong to $C_b^2(H)$. Moreover, for any $R > 0$ and $\lambda > 0$*

$$\sup_{\|g\|_1^H \leq R} \left(\|\varphi_\alpha(\lambda, g)\|_2^H + \|\varphi_{\alpha,n}(\lambda, g)\|_2^H \right) = c_\alpha < \infty. \qquad (9.5.17)$$

Proof. We prove the lemma only for the solution $\varphi_\alpha(\lambda, g)$ of the problem (9.5.6), as the proof for $\varphi_{\alpha,n}(\lambda, g)$ is identical.

As shown in the Remark 9.5.2, $D(L_\alpha) \subset C_b^{1+\theta}(H)$, for any $\theta < (1 - \epsilon)/(1 + \epsilon)$. Thus, if $\varphi_\alpha(\lambda, g)$ is the solution of the problem (9.5.6), we have that $\varphi_\alpha(\lambda, g) \in C_b^{1+\theta_0}(H)$, for some $0 < \theta_0 < (1 - \epsilon)/(1 + \epsilon)$. As we have

$$\varphi_\alpha(\lambda, g) = R(\lambda, L_\alpha) \left(g - K(D\varphi_\alpha(\lambda, g)) \right),$$

by using again the Remark 9.5.2 it follows that $\varphi_\alpha(\lambda, g) \in C_b^{1+2\theta_0}(H)$. Therefore, by repeating this argument a finite number of steps we get that $\varphi_\alpha(\lambda, g) \in C_b^2(H)$.

The estimate (9.5.17) follows as above, by applying (9.5.5) a finite number of times. $\qquad \square$

Due to (9.3.10), the previous lemma implies that if $g \in C_b^1(H)$ then $\varphi_{\alpha,n} = \varphi_{\alpha,n}(\lambda, g)$ is a strict solution of the problem (9.5.15), that is

$$\lambda \varphi_{\alpha,n} - \mathcal{L}_{\alpha,n} \varphi_{\alpha,n} + K_n(D\varphi) = g_n,$$

where $\mathcal{L}_{\alpha,n}$ is the differential operator introduced in (9.3.11).

Now, for any $\varphi \in D(L)$ we define

$$N(\varphi) = L\varphi - K(D\varphi). \tag{9.5.18}$$

In the same way, for any $\alpha > 0$ and $n \in \mathbb{N}$ we define $N_\alpha(\varphi) = L_\alpha \varphi - K(D\varphi)$ and $N_{\alpha,n}(\varphi) = L_{\alpha,n}\varphi - K_n(D\varphi)$.

We recall that a linear operator $A : D(A) \subset X \to X$ is said *m-dissipative* if it is dissipative and $\text{Range}(I - A) = X$.

Theorem 9.5.9. *Under the Hypotheses 6.1, 6.2, 6.3 and 9.2, the operator N defined by (9.5.18) is m-dissipative. Thus for any $\lambda > 0$ and for any $g \in C_b(H)$ there exists a unique solution $\varphi(\lambda, g) \in D(L)$ for the problem (9.5.1).*

Thanks to the Proposition 9.5.3, there exists $\lambda_0 > 0$ such that $R(\lambda, L)(C_b(H)) = C_b(H)$, for any $\lambda > \lambda_0$. Thus, in order to show that N is m-dissipative, it suffices to show that N is dissipative. To this purpose, we first give the following preliminary result.

Lemma 9.5.10. *Assume that the assumptions 6.1, 6.2, 6.3 and 9.2 hold. Then, there exists $\lambda_0 > 0$ such that for any $\lambda > \lambda_0$ and $\varphi_1, \varphi_2 \in D(L_\alpha)$*

$$\|\varphi_1 - \varphi_2\|_0^H \leq \frac{1}{\lambda} \|\lambda(\varphi_1 - \varphi_2) - (N_\alpha(\varphi_1) - N_\alpha(\varphi_2))\|_0^H.$$

Proof. We set $g_1 = \lambda \varphi_1 - N_\alpha(\varphi_1)$ and $g_2 = \lambda \varphi_2 - N_\alpha(\varphi_2)$ and for any $n \in \mathbb{N}$ we set $g_{1,n}(x) = g_1(P_n x)$ and $g_{2,n}(x) = g_2(P_n x)$, $x \in H$. Then for any λ large enough there exist $\varphi_{1,n}$ and $\varphi_{2,n}$ in $D(L_{\alpha,n})$ such that

$$\lambda \varphi_{1,n} - N_{\alpha,n}(\varphi_{1,n}) = g_{1,n}, \qquad \lambda \varphi_{2,n} - N_{\alpha,n}(\varphi_{2,n}) = g_{2,n}.$$

If we show that

$$\|\varphi_{1,n} - \varphi_{2,n}\|_0^H \leq \frac{1}{\lambda} \|g_{1,n} - g_{2,n}\|_0^H, \tag{9.5.19}$$

we are done. Actually, this implies that for any $x \in H$

$$|\varphi_{1,n}(x) - \varphi_{2,n}(x)| \leq \frac{1}{\lambda} \|g_{1,n} - g_{2,n}\|_0^H \leq \frac{1}{\lambda} \|g_1 - g_2\|_0^H$$

and due to (9.5.16) we can take the limit as $n \to +\infty$ and we get

$$|\varphi_1(x) - \varphi_2(x)| \leq \frac{1}{\lambda} \|g_1 - g_2\|_0^H.$$

By taking the supremum for $x \in H$, we get the dissipativity.

Thus, in order to conclude the proof we have to show that the operator $N_{\alpha,n}$ fulfills (9.5.19).

The operator $L_{\alpha,n}$ satisfies the same conditions of the operator $\mathcal{L}$ in chapter 1, thus, thanks to the Proposition 1.6.5

$$D(L_{\alpha,n}) = \left\{ \varphi \in \bigcap_{p \geq 1} W^{2,p}_{\text{loc}}(\mathbb{R}^n) \cap C_b(\mathbb{R}^n) \, ; \, \mathcal{L}_{\alpha,n}\varphi \in C_b(\mathbb{R}^n) \right\},$$

$$L_{\alpha,n}\varphi = \mathcal{L}_{\alpha,n}\varphi.$$

Now, we remark that

$$K_n(D\varphi_{1,n}(x)) - K_n(D\varphi_{2,n}(x))$$

$$= \left\langle \int_0^1 DK_n(\lambda D\varphi_{1,n}(x) + (1-\lambda)D\varphi_{2,n}(x)) \, d\lambda, D\varphi_{1,n}(x) - D\varphi_{2,n}(x) \right\rangle,$$

thus, if we set

$$U_{\alpha,n}(x) = \int_0^1 DK_n(\lambda D\varphi_{1,n}(x) + (1-\lambda)D\varphi_{2,n}(x)) \, d\lambda,$$

we have

$$\lambda(\varphi_{1,n} - \varphi_{2,n})(x) - \mathcal{L}_{\alpha,n}(\varphi_{1,n} - \varphi_{2,n})(x)$$

$$+ \langle U_{\alpha,n}(x), D(\varphi_{1,n} - \varphi_{2,n})(x) \rangle = g_{1,n}(x) - g_{2,n}(x).$$

Since the function $U_{\alpha,n}$ is uniformly continuous, as $\varphi_{1,n}$ and $\varphi_{2,n}$ belong to $C_b^1(H)$, the operator $\mathcal{N}_{\alpha,n}$ defined by

$$\mathcal{N}_{\alpha,n}\psi(x) = \mathcal{L}_{\alpha,n}\psi(x) - \langle U_{\alpha,n}(x), D\psi(x) \rangle$$

is of the same type as the operator $\mathcal{L}$ studied in the chapter 1. Therefore we can adapt the proof of the Lemma 1.6.4 to the present situation and we obtain

$$\|\varphi_{1,n} - \varphi_{2,n}\|_0^H \leq \frac{1}{\lambda} \|\lambda(\varphi_{1,n} - \varphi_{2,n}) - \mathcal{N}_{\alpha,n}(\varphi_{1,n} - \varphi_{2,n})\|_0^H = \frac{1}{\lambda}\|g_{1,n} - g_{2,n}\|_0^H.$$

$$\square$$

Proof of the Theorem 9.5.9: Let us fix $\lambda > 0$ and $\varphi_1, \varphi_2 \in D(L)$ and let us define $g_1 = \lambda\varphi_1 - N(\varphi_1)$ and $g_2 = \lambda\varphi_2 - N(\varphi_2)$. If λ_0 is the maximum between the constant introduced in the Remark 9.5.4 and the constant introduced in the Lemma 9.5.10, for any $\alpha > 0$ there exist $\varphi_{1,\alpha}, \varphi_{2,\alpha} \in D(L_\alpha)$ such that

$$(\lambda + \lambda_0) \, \varphi_{1,\alpha} - N_\alpha\varphi_{1,\alpha} = g_1 + \lambda_0 \, \varphi_1, \qquad (\lambda + \lambda_0) \, \varphi_{2,\alpha} - N_\alpha\varphi_{2,\alpha} = g_2 + \lambda_0 \, \varphi_2,$$

and according to the previous lemma

$$\|\varphi_{1,\alpha} - \varphi_{2,\alpha}\|_0^H \leq \frac{1}{\lambda + \lambda_0} \|(g_1 - g_2) + \lambda_0 (\varphi_1 - \varphi_2)\|_0^H.$$

Thus, for any $x \in H$ we have

$$|\varphi_{1,\alpha}(x) - \varphi_{2,\alpha}(x)| \leq \frac{1}{\lambda + \lambda_0} \|g_1 - g_2\|_0^H + \frac{\lambda_0}{\lambda + \lambda_0} \|\varphi_1 - \varphi_2\|_0^H.$$

Now, if $x \in E$, due to (9.5.14) we can take the limit in the left hand side as α goes to zero and we get

$$|\varphi_1(x) - \varphi_2(x)| \leq \frac{1}{\lambda + \lambda_0} \|g_1 - g_2\|_0^H + \frac{\lambda_0}{\lambda + \lambda_0} \|\varphi_1 - \varphi_2\|_0^H.$$

As φ_1 and φ_2 are continuous in H, the estimate above holds also for $x \in H$ and by taking the supremum for $x \in H$ it follows

$$\|\varphi_1 - \varphi_2\|_0^H - \frac{\lambda_0}{\lambda + \lambda_0} \|\varphi_1 - \varphi_2\|_0^H \leq \frac{1}{\lambda + \lambda_0} \|g_1 - g_2\|_0^H,$$

so that

$$\|\varphi_1 - \varphi_2\|_0^H \leq \frac{1}{\lambda} \|g_1 - g_2\|_0^H.$$

$\square$

9.5.2   Locally Lipschitz hamiltonian K

We first prove an *a priori* estimate which is crucial in order to prove the existence of solutions in the case of a locally Lipschitz hamiltonian K.

Proposition 9.5.11. *Assume that the Hypotheses 6.1, 6.2, 6.3 and 9.2 hold. Then there exists some $\mu_0 > 0$, which does not depend on K, such that for any $g \in C_b^1(H)$ and $\lambda > \mu_0$*

$$\|D\varphi(\lambda, g)\|_0^H \leq \|Dg\|_0^H. \tag{9.5.20}$$

Proof. We fix $\lambda, \mu > 0$ and $g \in C_b^1(H)$ and consider the functions $\varphi_\alpha = \varphi_\alpha(\lambda + \mu, g + \mu\,\varphi(\lambda, g))$ and $\varphi_{\alpha,n} = \varphi_{\alpha,n}(\lambda + \mu, g + \mu\,\varphi(\lambda, g))$. Since g and $\varphi(\lambda, g)$ belong to $C_b^1(H)$, then $\varphi_{\alpha,n}$ is in $C_b^2(H)$ and is a strict solution of the problem

$$(\lambda + \mu)\,\varphi - N_{\alpha,n}(\varphi) = g_n + \mu\,\varphi_n(\lambda, g),$$

where $\varphi_n(\lambda, g)(x) = \varphi(\lambda, g)(P_n x)$. The problem above can be written as

$$(\lambda + \mu)\,\varphi(x) - \frac{1}{2}\sum_{h=1}^n \lambda_h^2 D_h^2 \varphi(x) - \sum_{h,k=1}^n a_{hk} x_k D_h \varphi(x)$$

$$- \langle F_\alpha(P_n x), D\varphi(x) \rangle_H + K(P_n D\varphi(x)) = g(P_n x) + \mu\,\varphi(\lambda, g)(P_n x),$$

where $D_h\varphi(x) = \langle D\varphi(x), e_h \rangle_H$ and $a_{hk} = \langle Ae_k, e_h \rangle_H$. By differentiating with respect to x_j, setting $\psi_h = D_h\varphi$ and multiplying each side by ψ_j, we obtain

$$(\lambda + \mu)\,\psi_j^2 - \frac{1}{2}\sum_{h=1}^n \lambda_h^2 \psi_j D_h^2 \psi_j - \sum_{h,k=1}^n a_{hk} x_k \psi_j D_h \psi_j - \sum_{h=1}^n a_{h,j}\psi_h \psi_j$$

$$-\sum_{h=1}^n \langle F_{\alpha,n}, e_h \rangle_H \langle \psi_j D\psi_j, e_h \rangle_H - \sum_{h=1}^n \langle DF_{\alpha,n}\psi_j e_j, e_h \rangle_H \,\psi_h$$

$$+\sum_{h=1}^n D_h K(P_n D\varphi_{\alpha,n})\psi_j D_h \psi_j = \langle Dg_n, \psi_j e_j \rangle_H + \mu \langle D\varphi_n(\lambda, g), \psi_j e_j \rangle_H.$$

Then we sum up over j and by taking into account that

$$\psi_j D_h^2 \psi_j = \frac{1}{2}D_h^2(\psi_j^2) - (D_h\psi_j)^2,$$

if we set $z(x) = |D\varphi_{\alpha,n}(x)|_H^2$ we have

$$2(\lambda + \mu)\,z(x) - \frac{1}{2}\mathrm{Tr}\,\left[Q_n^2 D^2 z(x)\right] + \sum_{h,j=1}^n \lambda_h^2 (D_h\psi_j)^2(x) - \langle A_n x, Dz(x) \rangle$$

$$-2\langle A_n D\varphi_{\alpha,n}(x), D\varphi_{\alpha,n}(x) \rangle + \langle DK(D\varphi_{\alpha,n}(x)), Dz(x) \rangle - \langle F_\alpha(P_n x), Dz(x) \rangle$$

$$-2\langle DF_\alpha(P_n x)D\varphi_{\alpha,n}(x), D\varphi_{\alpha,n}(x) \rangle = 2\langle Dg(P_n x) + \mu D\varphi(\lambda, g)(P_n x), D\varphi_{\alpha,n}(x) \rangle.$$

Therefore, by using (4.1.6) and (6.1.28) it follows

$$2(\lambda + \mu)\,z(x) - \frac{1}{2}\mathrm{Tr}\,\left[Q_n^2 D^2 z(x)\right] - \langle A_n x, Dz(x) \rangle - \langle F_\alpha(P_n x), Dz(x) \rangle$$

$$+\langle DK(D\varphi_{\alpha,n}(x)), Dz(x) \rangle \le 2\langle Dg(x) + \mu D\varphi(\lambda, g)(P_n x), D\varphi_{\alpha,n}(P_n x) \rangle + \gamma\,z(x)$$

$$\le 2\left(\|Dg\|_0^H + \mu \|D\varphi(\lambda, g)\|_0^H\right)|D\varphi_{\alpha,n}(x)|_H + \gamma\,z(x),$$

for a suitable constant $\gamma \in \mathbb{R}$ depending only on F and A.

Now, let us consider the equation

$$du(t) = [A_n u(t) + F_{\alpha,n}(u(t)) + U_{\alpha,n}(u(t))]\,dt + Q_n\,dw(t), \qquad u(0) = P_n x, \quad (9.5.21)$$

where $U_{\alpha,n}(x) = -DK(D\varphi_{\alpha,n}(x))$, for any $x \in H$. If $g \in C_b^1(H)$, then $\varphi_{\alpha,n} \in C_b^2(H)$ and then the mapping $U_{\alpha,n} : H \to H$ is Lipschitz-continuous. This implies that there exists a unique strong solution $u_{\alpha,n}(\cdot; x) \in L^2(\Omega; C([0, +\infty); H))$ for the equation (9.5.21). If we denote by $R_t^{\alpha,n}$ the corresponding transition semigroup, it is possible to show that the solution of the problem

$$(2(\lambda+\mu)-\gamma)\,\psi(x) - \frac{1}{2}\mathrm{Tr}\,[Q_n^2 D^2\psi(x)] - \langle A_n x, D\psi(x)\rangle - \langle F_\alpha(P_n x), D\psi(x)\rangle$$

$$+\langle DK(D\varphi_{\alpha,n}(x)), D\psi(x)\rangle = 2\left(\|Dg\|_0^H + \mu\,\|D\varphi(\lambda,g)\|_0^H\right)|D\varphi_{\alpha,n}(x)|_H,$$

for any $\lambda > \gamma$ is given by

$$\psi(x) = 2\left(\|Dg\|_0^H + \mu\,\|D\varphi(\lambda,g)\|_0^H\right)\int_0^{+\infty} e^{-(2(\lambda+\mu)-\gamma)\,t}R_t^{\alpha,n}\left(|D\varphi_{\alpha,n}|_H\right)(x)\,dt$$

(see the Theorem 1.6.6 for a proof). Thus, by a comparison argument we have

$$|D\varphi_{\alpha,n}(x)|_H^2 \le \frac{1}{\lambda+\mu-\gamma/2}\left(\|Dg\|_0^H + \mu\,\|D\varphi(\lambda,g)\|_0^H\right)|D\varphi_{\alpha,n}(x)|_H,$$

and if we take $\lambda > 1 + \gamma/2 = \mu_0$, it follows

$$|D\varphi_{\alpha,n}(\lambda+\mu, g+\mu\,\varphi(\lambda,g))(x)|_H \le \frac{1}{1+\mu}\left(\|Dg\|_0^H + \mu\,\|D\varphi(\lambda,g)\|_0^H\right).$$

Due to (9.5.14) and (9.5.16), if μ is large enough we can take first the limit as n goes to infinity and then the limit as α goes to zero and for any $x \in E$ we get

$$|D\varphi(x)|_H \le \frac{1}{1+\mu}\left(\|Dg\|_0^H + \mu\,\|D\varphi(\lambda,g)\|_0^H\right).$$

As $D\varphi(\lambda,g)$ is continuous in H, the same estimate holds for $x \in H$ and then, by taking the supremum for $x \in H$ we get

$$\|D\varphi(\lambda,g)\|_0^H \le \frac{1}{1+\mu}\left(\|Dg\|_0^H + \mu\,\|D\varphi(\lambda,g)\|_0^H\right),$$

which yields (9.5.20).   □

Remark 9.5.12. It is immediate to check that the proof of the previous proposition adapts to the problem (9.5.6). Thus, there exists $\lambda_0 > 0$, which is clearly independent of $\alpha > 0$, such that for any $\lambda > \lambda_0$ and $g \in C_b^1(H)$

$$\|D\varphi_\alpha(\lambda,g)\|_0^H \le \|Dg\|_0^H. \tag{9.5.22}$$

From now on we shall assume that K fulfills the following assumption.

Hypothesis 9.3. *The hamiltonian* $K : H \to \mathbb{R}$ *is Fréchet differentiable and is locally Lipschitz-continuous, together with its derivative. Moreover,* $K(0) = 0$.

Now, for any $r > 0$ let $K_r : H \to \mathbb{R}$ be a Fréchet differentiable function such that

$$K_r(x) = \begin{cases} K(x) & \text{if } |x|_H \leq r \\[2mm] K\left(\dfrac{(r+1)x}{|x|_H}\right) & \text{if } |x|_H \geq r+1. \end{cases} \tag{9.5.23}$$

It is easy to check that K_r is Lipschitz-continuous, together with its derivative, for each $r > 0$.

Theorem 9.5.13. *Under the Hypotheses 6.1, 6.2, 6.3 and 9.3 there exists $\mu_0 > 0$ such that for any $\lambda > \mu_0$ and $g \in C_b^1(H)$ there exists a unique $\varphi(\lambda, g) \in D(L)$ which fulfills the equation (9.5.1).*

Proof. For any $r > 0$ and $g \in C_b^1(H)$ we define $\varphi_r(\lambda, g)$ as the solution of the problem $\lambda\varphi - L\varphi + K_r(D\varphi) = g$. Due to the Proposition 9.5.11 there exists $\lambda_0 > 0$ such that for any $\lambda > \lambda_0$

$$\sup_{r>0} \|D\varphi_r(\lambda, g)\|_0^H \leq \|Dg\|_0^H.$$

Thus, if we fix $r = \|g\|_1^H$, we have that $K(D\varphi_r(\lambda, g)) = K_r(D\varphi_r(\lambda, g))$ and then

$$\lambda\varphi_r(\lambda, g) - L\varphi_r(\lambda, g) + K(D\varphi_r(\lambda, g)) = g.$$

$\square$

Now we show that the operator N is closable and for any $\lambda > 0$ and $g \in C_b(H)$ there exists a unique $\varphi = \varphi(\lambda, g) \in D(\overline{N})$, where $\overline{N}$ is the closure of N, such that

$$\lambda\varphi - \overline{N}\varphi = g. \tag{9.5.24}$$

Unfortunately we are not able to give a characterization of $D(\overline{N})$ which could allow us to prove some further regularity of the solution of (9.5.24) and hence to solve the equation (9.5.1) for less regular data g.

Theorem 9.5.14. *The operator N is closable and its closure $\overline{N}$ is m-dissipative.*

Proof. The operator N is dissipative. Actually, we fix $\lambda > 0$ and $\varphi_1, \varphi_2 \in D(L)$ and we define $g_i = \lambda\varphi_i - N(\varphi_i)$, for $i = 1, 2$. If we take $r = \max\left(\|\varphi_1\|_1^H, \|\varphi_2\|_1^H\right)$, we have

$$g_i = \lambda\varphi_i - L\varphi_i + K_r(D\varphi_i), \qquad i = 1, 2.$$

Thus we can apply the Theorem 9.5.9 to the hamiltonian K_r and we get

$$\|\varphi_1 - \varphi_2\|_0^H \leq \frac{1}{\lambda}\|g_1 - g_2\|_0^H,$$

so that N is dissipative. As the domain of N is dense in $C_b(H)$, this implies that N is closable and its closure is dissipative. Therefore, since Range$(\lambda - N)$ is dense in $C_b(H)$, for λ sufficiently large, it follows that $\bar{N}$ is m-dissipative. $\square$

Chapter 10

Application to stochastic optimal control problems

In this chapter we study the minimizing problems associated with the following *cost functionals*

$$J(t, x; z) = \mathbb{E}\varphi(u(T; t, x, z)) + \mathbb{E} \int_t^T [g(u(s; t, x, z)) + k(z(s))] \, ds, \qquad (10.0.1)$$

and

$$I(x; z) = \mathbb{E} \int_0^{+\infty} e^{-\lambda s} [g(u(s; 0, x, z)) + k(z(s))] \, ds. \qquad (10.0.2)$$

Here $u(s; t, x, z)$ is the solution of the controlled system (9.1.1), φ and g are in $C_b^{0,1}(H)$, $k : H \to (-\infty, +\infty]$ is a convex lower semi-continuous function and λ is a positive constant.

The *value functions* corresponding to the cost functionals (10.0.1) and (10.0.2) are defined respectively by

$$V(t, x) = \inf \left\{ J(t, x; z); \, z \in L^2(\Omega; L^2(0, T; H)) \text{ adapted} \right\}$$

and

$$U(x) = \inf \left\{ I(x; z); \, z \in L^2(\Omega; L^2(0, T; H)) \text{ adapted} \right\}.$$

Our aim is to prove that such value functions can be identified with the solutions of the Hamilton-Jacobi-Bellman equations (9.0.1) and (9.0.2), respectively, when the hamiltonian K is the Legendre transform of k, that is

$$K(x) = \sup_{z \in H} \left\{ -\langle x, z \rangle_H - k(z) \right\}, \qquad x \in H.$$

More precisely, we show that $V(t, x) = y(T - t, x)$ and $U(x) = \varphi(x)$, for any $t \in [0, T]$ and $x \in H$.

In the first section we study the minimizing problem associated with the finite horizon cost functional (10.0.1). To this purpose, the first crucial step is proving that for any adapted control $z \in L^2(\Omega; L^2(0, T; H))$ and for any $x \in H$ and $t \in [0, T]$ the following identity holds

$$J(t, x; z) = y(T - t, x)$$

$$+ \int_t^T \mathbb{E} \left[K(Dy(T - s, u(s))) + \langle z(s), Dy(T - s, u(s)) \rangle_H + k(z(s)) \right] ds,$$

(10.0.3)

where $u(s) = u(s; t, x, z)$. In particular, due to the relation between k and K, this implies that $V(t, x) \geq y(T - t, x)$. We remark that the proof of (10.0.3) is quite delicate and we have to proceed by several approximations.

If we could prove the existence of a solution $u^*(t)$ for the *closed loop* equation

$$du(t) = [Au(t) + F(u(t)) - DK(Dy(T - t, u(t)))] \, dt + Q dw(t), \qquad u(0) = x,$$

(10.0.4)

then $z^*(t) = -DK(Dy(T - t, u^*(t)))$ would be an optimal control for the minimizing problem related to the functional (10.0.1). Actually, from well known properties of the Legendre transform, we have

$$K(x) = - \langle x, z_0 \rangle - k(z_0),$$

where $z_0 = DK(x)$. Then, if we substitute $z^*(t)$ and $u^*(t)$ in (10.0.3), we get $J(t, x; z^*) = y(T - t, x)$, due to the fact that $V(t, x) \geq y(T - t, x)$. But unfortunately as we have seen in the previous chapter we are only able to prove C^1 regularity for the solution of the Hamilton-Jacobi-Bellman equation (9.0.1), so that the function

$$H \to H, \qquad x \mapsto -DK(Dy(T - t, x)),$$

is not Lipschitz-continuous and we can not prove in general the existence of a solution for (10.0.4) which is adapted to the filtration $\mathcal{F}_t$. Actually, as the solution of the equation (9.0.1) is only C^1, the *closed loop* equation (10.0.4) admits only martingale solutions, so that we have the existence of an optimal control only in a *relaxed* sense (see [63] and [122] for the definition and main properties of relaxed controls).

In order to prove the opposite inequality $V(t, x) \leq y(t, x)$, and hence the *verification theorem* we introduce an approximating sequence of cost functionals $J_\alpha(t, x; z)$ and we prove that they satisfy a verification theorem and admit a unique optimal control, for each $\alpha > 0$. Thanks to suitable *a-priori* estimates, we show that there exists a subset $\mathcal{M}_R^2(T)$ of the space of adapted processes in $L^2(\Omega; L^2(0, T; H))$ such that for any $\alpha > 0$

$$V_\alpha(t, x) = \inf \left\{ J_\alpha(t, x; z) \; ; \; z \in \mathcal{M}_R^2(T) \right\}.$$

Moreover, we show that for any $x \in E$ the functional $J_\alpha(t, x; z)$ converges to $J(t, x; z)$ as α goes to zero, uniformly for $z \in \mathcal{M}_R$, so that

$$\lim_{\alpha \to 0} V_\alpha(t, x) \geq V(t, x).$$

Thus, by showing that

$$\lim_{\alpha \to 0} V_\alpha(t, x) = y(T - t, x)$$

we have that $y(T - t, x) \geq V(t, x)$ and the verification theorem holds for $x \in E$. The general case $x \in H$ follows by further approximation arguments.

In the second section we study the minimizing problem corresponding to the infinite horizon cost functional (10.0.2). As in the evolutionary case, in order to show that $I(x; z) \geq \varphi(x)$, for any $x \in H$ we first prove the following identity

$$I(x; z) = \varphi(x) + \mathbb{E} \int_0^{+\infty} e^{-\lambda t} \left[K(D\varphi(u(t))) + \langle z(t), D\varphi(u(t)) \rangle_H + k(z(t)) \right] dt.$$

$$(10.0.5)$$

As for (10.0.3), the proof of (10.0.5) is quite elaborate and is a consequence of several approximations. But the most delicate step is the proof of the opposite inequality. We notice that in this case, as in the finite horizon case, the existence of an optimal state would follow from the existence of a solution for the *closed loop* equation

$$du(t) = [Au(t) + F(u(t)) - DK(D\varphi(u(t)))] \, dt + Q \, dw(t), \qquad u(0) = x. \quad (10.0.6)$$

But since $D\varphi$ is only continuous, the mapping $x \mapsto -DK(D\varphi(x))$ is only continuous. Thus, as for (10.0.4), the equation (10.0.6) has only a martingale solutions $u^\star(t; x)$ and the control $z^\star(t) = -DK(D\varphi(u^\star(t; x)))$ is not adapted to the original filtration $\{\mathcal{F}_t\}$, in general, and can be interpreted as an optimal control only in a relaxed sense. Nevertheless, even if we are not able to prove the existence of an optimal control, as in the evolutionary case we show that the value function U is obtained as the limit of minima of suitable approximating cost functionals I_α, $\alpha \geq 0$, which admit unique optimal costs and unique optimal states and whose value functions coincide with the solutions of some approximating Hamilton-Jacobi-Bellman problems.

In the chapter 9 we have seen that in the elliptic case if the hamiltonian K is not Lipschitz-continuous we are able to prove the existence of a solution $\varphi(\lambda, g) \in C_b^1(H)$ for the Hamilton-Jacobi-Bellman problem, only for $\lambda > \lambda_0$ and $g \in C_b^1(H)$, with some fixed $\lambda_0 > 0$. Thus, in the application to the control problem we have to give two different formulations of the verification theorem, a stronger one which holds when K is Lipschitz-continuous and a weaker one which holds when K is only locally Lipschitz-continuous.

We conclude the chapter with the study of the finite horizon control problem in the one dimensional case, that is when $\mathcal{O}$ is a bounded interval of $\mathbb{R}$. In fact, in this case it is possible to prove the existence of a solution for the closed loop equation, so that there exist an optimal control and an optimal state for the minimizing control problem.

10.1   The finite horizon case

Let $k : H \to]-\infty, +\infty]$ be a convex lower semi-continuous function and let K be its Legendre transform, that is

$$K(x) = \sup_{z \in H} \left\{ -\langle x, z \rangle_H - k(z) \right\}, \qquad x \in H.$$

We assume that k is such that K fulfills the Hypothesis 9.1. It is possible to show that if k is strictly convex, continuously Fréchet differentiable and

$$\lim_{|z|_H \to \infty} \frac{k(z)}{|z|_H} = \infty,$$

and if $Dk : H \to H$ has a continuous inverse which is Lipschitz-continuous on bounded sets of H, then the Hypothesis 9.1 is verified. An easy example is given by

$$k(z) = \frac{1}{2} |z|_H^2, \qquad z \in H.$$

We consider here the cost functional

$$J(t, x; z) = \mathbb{E}\varphi(u(T; t, x, z)) + \mathbb{E} \int_t^T [g(u(s; t, x, z)) + k(z(s))] \, ds,$$

where $u(s; t, x, z)$ is the unique solution of the controlled system (9.1.1) at time s, starting from x at time t. We want to minimize the functional J over all adapted controls $z \in L^2(\Omega; L^2([0, T]; H))$.

The *value function* corresponding to the cost functional J is defined by

$$V(t, x) = \inf \left\{ J(t, x; z) \, : \, z \in L^2(\Omega; L^2([0, T]; H)) \text{ adapted} \right\}$$

and is related to the Hamilton-Jacobi-Bellman equation (9.4.1). Namely, we are showing that for every $t \in [0, T]$ and $x \in H$

$$V(t, x) = y(T - t, x),$$

where $y(t, x)$ is the unique mild solution of the problem (9.4.1).

For any $\alpha > 0$, we introduce the approximating cost functional

$$J_\alpha(t, x; z) = \mathbb{E}\varphi(u_\alpha(T; t, x, z)) + \mathbb{E} \int_t^T [g(u_\alpha(s; t, x, z)) + k(z(s))] \, ds, \qquad (10.1.1)$$

where $u_\alpha(s; t, x, z)$ is the unique solution of the problem (9.1.5). In what follows we will denote by $V_\alpha(t, x)$ the corresponding value function.

Lemma 10.1.1. *Assume the Hypotheses 6.1, 6.2, 6.3 and 9.1 and assume that φ and g are in $C_b^{0,1}(H)$. If $y(t,x)$ is the mild solution of the problem (9.4.1), for any $z \in L^2(\Omega; L^2([0,T]; H))$, $x \in H$ and $t \in [0,T]$, the following identity holds*

$$J(t,x;z) = y(T-t,x)$$

$$+ \int_t^T \mathbb{E}\left[K(Dy(T-s,u(s))) + \langle z(s), Dy(T-s,u(s)) \rangle_H + k(z(s)) \right] ds, \tag{10.1.2}$$

where $u(s) = u(s;t,x,z)$ is the solution of the problem (9.1.1).

Moreover, the same identity holds with $J(t,x;z)$, $y(t,x)$, $Dy(t,x)$ and $u(t)$ replaced respectively by $J_\alpha(t,x;z)$, $y_\alpha(t,x)$, $Dy_\alpha(t,x)$ and $u_\alpha(t)$.

Proof. We first assume that $\varphi, g \in C_b^1(H)$. Let $y_{\alpha,n}(t,x)$ be the solution of the problem (9.4.9) and let $u_{\alpha,n}(s) = u_{\alpha,n}(s;t,x,z)$ be the solution of the problem (9.1.9). Since $y_{\alpha,n}$ is smooth (in fact we have seen that $y_{\alpha,n} \in \mathcal{Z}_T^2$) and $u_{\alpha,n}$ is a strong solution, we can apply Itô's formula to the function $s \mapsto y_{\alpha,n}(T-s, u_{\alpha,n}(s))$, for $t \leq s \leq T$, and we get

$$d\left(y_{\alpha,n}(T-s, u_{\alpha,n}(s))\right) = \langle du_{\alpha,n}(s), Dy_{\alpha,n}(T-s, u_{\alpha,n}(s)) \rangle_H$$

$$+ \left(\frac{1}{2}\mathrm{Tr}\,[Q_n^2 D^2 y_{\alpha,n}(T-s, u_{\alpha,n}(s))] - \frac{\partial y_{\alpha,n}}{\partial t}(T-s, u_{\alpha,n}(s)) \right) ds.$$

By integrating with respect to $s \in [t,T]$ and taking the expectation, recalling that $y_{\alpha,n}(t,x)$ is the solution of (9.4.9), we get

$$\mathbb{E}\varphi(u_{\alpha,n}(T)) - y_{\alpha,n}(T-t,x) = \mathbb{E}\int_t^T [K(Dy_{\alpha,n}(T-s, u_{\alpha,n}(s)))$$

$$+ \langle z(s), Dy_{\alpha,n}(T-s, u_{\alpha,n}(s)) \rangle_H - g(u_{\alpha,n}(s))]\, ds.$$

Now, due to the Lemma 9.1.4 and to (9.4.10), we can take the limit as n goes to infinity in each side above and rearranging all terms we get

$$\mathbb{E}\varphi(u_\alpha(T)) + \mathbb{E}\int_t^T g(u_\alpha(s))\, ds$$

$$= y_\alpha(T-t,x) + \mathbb{E}\int_t^T [K(Dy_\alpha(T-s, u_\alpha(s))) + \langle z(s), Dy_\alpha(T-s, u_\alpha(s)) \rangle_H]\, ds, \tag{10.1.3}$$

which implies (10.1.2).

Now, let $\varphi, g \in C_b^{0,1}(H)$. As in the proof of the Theorem 9.4.2, we fix two sequences $\{\varphi_k\}$ and $\{g_k\}$ which are bounded in $C_b^1(H)$ and converge respectively to φ and g in $C_b(H)$. If we denote by $y_\alpha^k(t,x)$ the solutions of the problem (9.4.3) corresponding to the data φ_k and g_k, we have

$$\mathbb{E}\varphi_k(u_\alpha(T)) + \mathbb{E}\int_t^T g_k(u_\alpha(s))\,ds = y_\alpha^k(T-t,x)$$

$$+\mathbb{E}\int_t^T \left[K(Dy_\alpha^k(T-s,u_\alpha(s))) + \left\langle z(s), Dy_\alpha^k(T-s,u_\alpha(s)) \right\rangle_H \right]\,ds.$$

It is immediate to check that the sequence $\{y_\alpha^k\}$ fulfills an estimate analogous to (9.4.22) and then converges to y_α in Z_T^1, as k goes to infinity. Moreover, as estimates analogous to (9.1.3) and (9.4.19) holds for u_α and y_α^k, uniformly with respect to α, due to (6.1.12) we can apply the dominated convergence theorem and, by taking the limit for k going to infinity, we get (10.1.3), for any $\varphi, g \in C_b^{0,1}(H)$.

Now, if $x \in E$, then $u_\alpha(s) \in E$ and (9.2.4) holds. Thus, due to (9.1.6) and (9.4.4), we can take the limit as α goes to zero in each side of (10.1.3) and we easily get (10.1.2) for $x \in E$. Finally, if $x \in H$ we fix a sequence $\{x_n\} \subset E$ converging to x in H. Thanks to (9.1.4) we have that $u(s;t,x_n,z)$ converges to $u(s;t,x,z)$ in H and then, as $y(t,\cdot) \in C_b^1(H)$, we easily get (10.1.2) for any $x \in H$. $\qquad \square$

Now, we can conclude by giving the main result of this section.

Theorem 10.1.2. *Under the Hypotheses 6.1, 6.2, 6.3 and 9.1, if $\varphi, g \in C_b^{0,1}(H)$ the value function $V(t,x)$ coincides with $y(T-t,x)$, where $y(t,x)$ is the solution of the problem (9.4.2). Moreover, if $x \in E$*

$$V(t,x) = \lim_{\alpha \to 0} \min \left\{ J_\alpha(t,x;z), \ z \in L^2(\Omega; L^2(0,T;H)), \ \text{adapted} \right\},$$

where $J_\alpha(t,x;z)$ is the cost functional defined in (10.1.1).

Proof. From (10.1.2) we immediately have that $J(t,x;z) \geq y(T-t,x)$, for any $z \in L^2(\Omega; L^2(0,T;H))$, so that $V(t,x) \geq y(T-t,x)$. Now we prove the opposite inequality.

Since J_α fulfills a formula analogous to (10.1.2), we have $V_\alpha(t,x) \geq y_\alpha(T-t,x)$. In fact, we show that $V_\alpha(t,x) = y_\alpha(T-t,x)$. Actually, by a general property of the Legendre transform, for each $t \in [0,T]$ the mapping

$$H \to \mathbb{R}, \qquad z \mapsto \langle z, Dy_\alpha(T-t,u(t))\rangle_H + k(z),$$

attains its maximum for

$$z_\alpha(t) = -DK(Dy_\alpha(T-t,u(t))), \qquad t \in [0,T].$$

Thus, if we prove that the following *closed loop* equation

$$du(t) = [Au(t) + F_\alpha(u(t)) - DK(Dy_\alpha(T-t,u(t)))]\,dt + Q\,dw(t), \qquad u(0) = x$$
$$\tag{10.1.4}$$

admits a unique solution $u_\alpha^\star(t)$ and if we define

$$z_\alpha^\star(t) = -DK(Dy_\alpha(T - t, u_\alpha^\star(t))),$$

due to the analogous of (10.1.2) for J_α we have that $J_\alpha(t, x; z_\alpha^\star) = y_\alpha(T - t, x)$. As $V_\alpha(t, x) \geq y_\alpha(T - t, x)$, this implies that $V_\alpha(t, x) = y_\alpha(T - t, x)$ and $u_\alpha^\star(t)$ and $z_\alpha^\star(t)$ are respectively the unique optimal state and the unique optimal control for the minimizing problem corresponding to the functional J_α.

Assume that $\varphi, g \in C_b^1(H)$. In the previous chapter we have seen that due to (9.3.1) the solution y_α of the problem (9.4.3) belongs to Z_T^2, that is $y_\alpha(t, \cdot) \in C_b^2(H)$, for any $t \in (0, T]$ and

$$\sup_{t \in (0,T]} t^{\frac{1+\epsilon}{2}} \|D^2 y_\alpha(t, \cdot)\|_0^H < \infty.$$

Moreover

$$\|y_\alpha\|_{Z_T^2} \leq c_{\alpha,T} \left(\|\varphi\|_1^H + T \|g\|_1^H\right). \tag{10.1.5}$$

Thus, if we define for $(t, x) \in [0, T] \times H$

$$U_\alpha(t, x) = -DK(Dy_\alpha(t, x)),$$

we have that the function U_α fulfills the conditions of the Lemma 9.4.6 and then there exists a unique solution $y_\alpha^\star(t)$ for the closed loop equation (10.1.4).

Now, assume that $\varphi, g \in C_b^{0,1}(H)$. As in the proofs of the Theorem 9.4.2 and of the Lemma 10.1.1, we approximate them in $C_b(H)$ by means of two bounded sequences $\{\varphi_k\}$ and $\{g_k\}$ in $C_b^1(H)$. For each k there exists a unique solution $y_{\alpha,k}$ for the problem (9.4.3), with data φ_k and g_k. Thus, as proved above, in correspondence with each $u_{\alpha,k}$ there exists a unique solution $u_{\alpha,k}^\star(t)$ for the problem (10.1.4). Let us define $v_{h,k}^\alpha(t) = u_{\alpha,k}^\star(t) - u_{\alpha,h}^\star(t)$. We have that $v_{h,k}^\alpha(t)$ is the unique solution of the problem

$$\frac{dv}{dt}(t) = Av(t) + F_\alpha(u_{\alpha,k}^\star(t)) - F_\alpha(u_{\alpha,h}^\star(t)) - DK(Dy_{\alpha,k}(T - t, u_{\alpha,k}^\star(t)))$$

$$+ DK(Dy_{\alpha,h}(T - t, u_{\alpha,h}^\star(t))), \quad v(0) = 0.$$

Thus, by multiplying each side by $v_{h,k}^\alpha(t)$, due to the fact that F_α is Lipschitz-continuous and A is of negative type, we get

$$\frac{1}{2}\frac{d}{dt}|v_{h,k}^\alpha(t)|_H^2 \leq c_\alpha |v_{h,k}^\alpha(t)|_H^2$$
$$\tag{10.1.6}$$
$$+ |DK(Dy_{\alpha,k}(T - t, u_{\alpha,k}^\star(t))) - DK(Dy_{\alpha,h}(T - t, u_{\alpha,h}^\star(t)))|_H |v_{h,k}^\alpha(t)|_H.$$

As the sequences $\{\varphi_k\}$ and $\{g_k\}$ are bounded in $C_b^1(H)$ and DK is locally Lipschitz-continuous, due to the Proposition 9.4.8 there exists $c > 0$ independent of α, k and

h such that

$$|DK(Dy_{\alpha,k}(T-t,u^\star_{\alpha,k}(t))) - DK(Dy_{\alpha,h}(T-t,u^\star_{\alpha,h}(t)))|_H$$

$$\leq c\,|Dy_{\alpha,k}(T-t,u^\star_{\alpha,k}(t)) - Dy_{\alpha,h}(T-t,u^\star_{\alpha,h}(t))|_H, \qquad \mathbb{P}-\text{a.s.}$$

Now, for any $t > 0$ and $x, y \in H$, thanks to (10.1.5) we have

$$|Dy_{\alpha,k}(t,x) - Dy_{\alpha,k}(t,y)|_H$$

$$\leq c_{\alpha,T} t^{-\frac{1+\epsilon}{2}} \left(\|\varphi_k\|^H_1 + T\,\|g_k\|^H_1 \right) |x-y|_H \leq c_{\alpha,T} t^{-\frac{1+\epsilon}{2}} |x-y|_H,$$

for some constant independent of k. Moreover, since all arguments used in the proof of the Theorem 9.4.2 can be adapted to y_α, we have

$$\sup_{y \in H} |Dy_{\alpha,k}(t,y) - Dy_{\alpha,h}(t,y)|_H \leq c_{\alpha,T} t^{-\frac{1+\epsilon}{2}} \left(\|\varphi_k - \varphi_h\|^H_0 + T\,\|g_k - g_h\|^H_0 \right).$$

Therefore, we get

$$\left| DK(Dy_{\alpha,k}(T-t,u^\star_{\alpha,k}(t))) - DK(Dy_{\alpha,h}(T-t,u^\star_{\alpha,h}(t))) \right|_H$$

$$\leq c_{\alpha,T}(T-t)^{-\frac{1+\epsilon}{2}} \left(|v^\alpha_{h,k}(t)|_H + \|\varphi_k - \varphi_h\|^H_0 + T\,\|g_k - g_h\|^H_0 \right),$$

and from (10.1.6) we can easily conclude

$$\frac{d}{dt} |v^\alpha_{h,k}(t)|^2_H \leq c_{\alpha,T} \left(1 + (T-t)^{-\frac{1+\epsilon}{2}} \right) |v^\alpha_{h,k}(t)|^2_H$$

$$+ c_{\alpha,T}(T-t)^{-\frac{1+\epsilon}{2}} \left(\|\varphi_k - \varphi_h\|^H_0 + T\,\|g_k - g_h\|^H_0 \right)^2.$$

Due to the Gronwall lemma this yields

$$|u^\star_{\alpha,k}(t) - u^\star_{\alpha,h}(t)|^2_H \leq c_{\alpha,T} \left(\|\varphi_k - \varphi_h\|^H_0 + T\,\|g_k - g_h\|^H_0 \right)^2, \qquad \mathbb{P}-\text{a.s.}$$

so that the sequence $\{u^\star_{\alpha,k}\}$ converges to some $u^\star_\alpha$ in $C([0,T];H)$, $\mathbb{P}$-a.s. and in the mean-square, and clearly $u^\star_\alpha$ is the unique solution of the closed loop (10.1.4).

Since $z^\star_\alpha(t) = -DK(Dy_\alpha(T-t,u^\star_\alpha(t)))$, thanks to (10.1.5) there exists a constant R such that

$$\sup_{\alpha > 0} \sup_{t \in [0,T]} |z^\star_\alpha(t)|_H = R, \qquad \mathbb{P}-\text{a.s.}$$

This means that, if we define the set $\mathcal{M}_R$ as in the Lemma 9.1.3, then for any $\alpha > 0$

$$y_\alpha(T-t,x) = V_\alpha(t,x) = \inf\{ J_\alpha(t,x;z) \; ; \; z \in \mathcal{M}_R \}. \qquad (10.1.7)$$

By using (9.4.4) we have that if $\varphi, g \in C^1_b(H)$, then for any $t \in [0,T]$ and $x \in E$

$$\lim_{\alpha \to 0} V_\alpha(t,x) = \lim_{\alpha \to 0} y_\alpha(T-t,x) = y(T-t,x).$$

Moreover,

$$|J_\alpha(t,x;z) - J(t,x;z)| \leq \|\varphi\|_1^H \mathbb{E} |u_\alpha(T) - u(T)|_E + \|g\|_1^H \int_0^t \mathbb{E} |u_\alpha(s) - u(s)|_E \, ds,$$

so that, due to the Lemma 9.1.3 we have

$$\lim_{\alpha \to 0} \sup_{z \in \mathcal{M}_R} |J_\alpha(t,x;z) - J(t,x;z)| = 0.$$

Therefore, as the convergence of J_α to J is uniform for $z \in \mathcal{M}_R$, thanks to (10.1.7) we easily have

$$y(T - t, x) = \lim_{\alpha \to 0} V_\alpha(t,x) = \inf \{ J(t,x;z) \; ; \; z \in \mathcal{M}_R \} \geq V(t,x).$$

Since $y(T - t, x) \leq V(t,x)$, this implies that $y(T - t, x) = V(t,x)$, for $\varphi, g \in C_b^1(H)$ and $x \in E$.

Now, if $x \in H$ and $\{x_n\}$ is a sequence in E converging to x in H, by using (9.1.4) we can prove that

$$\lim_{n \to +\infty} \sup_{z \in \mathcal{M}_R} |J(t,x_n;z) - J(t,x;z)| = 0.$$

Therefore, since $y(t, x_n)$ converges to $y(t, x)$, the conclusion of the theorem follows for any $x \in H$. Finally, if $\varphi, g \in C_b^{0,1}(H)$ and $\{\varphi_k\}$ and $\{g_k\}$ are any two bounded sequences in $C_b^1(H)$, converging respectively to φ and g in $C_b(H)$, we have

$$\lim_{k \to +\infty} \mathbb{E} \varphi_k(u(T)) + \mathbb{E} \int_0^t (g_k(u(s)) + k(z(s))) \, ds = J(t,x;z),$$

uniformly with respect to z, and then, thanks to (9.4.22), the theorem holds for any $\varphi, g \in C_b^{0,1}(H)$. $\square$

10.2   The infinite horizon case

For any $\lambda > 0$ and $g \in C_b(H)$ we consider the following *cost functional*

$$I(x;z) = \mathbb{E} \int_0^{+\infty} e^{-\lambda t} [g(u(t;x,z)) + k(z(t))] \, dt, \qquad (10.2.1)$$

where $u(t;x,z) = u(t;0,x,z)$ is the unique solution of the system (9.1.1) at time t starting from x at time 0. The corresponding value function is defined as

$$U(x) = \inf \{ I(x;z); \; z \in L^2(\Omega; L^2(0, +\infty; H)) \text{ adapted} \}.$$

Our aim is proving that if φ is the unique solution of the Hamilton-Jacobi-Bellman equation (9.0.2), then $V(x) = \varphi(x)$, for any $x \in H$. To this purpose we first prove a preliminary result.

Lemma 10.2.1. *Assume the Hypotheses 6.1, 6.2, 6.3 and 9.3. If $\varphi = \varphi(\lambda, g)$ is a solution of the problem (9.5.1) which belongs to $C_b^1(H)$ and if $u(t) = u(t; 0, x, z)$ is the solution of the controlled system (9.1.1), then*

$$I(x; z) = \varphi(x) + \mathbb{E} \int_0^{+\infty} e^{-\lambda t} \left[K(D\varphi(u(t))) + \langle z(t), D\varphi(u(t)) \rangle_H + k(z(t)) \right] dt.$$
$$(10.2.2)$$

Proof. If $r = \|D\varphi(\lambda, g)\|_0^H$ and if K_r is defined as in (9.5.23), we have $K(D\varphi(x)) = K_r(D\varphi(x))$, for any $x \in H$, so that the problem (9.5.1) can be rewritten as

$$\lambda \varphi - L\varphi + K_r(D\varphi) = g.$$

Now, we fix a sequence $\{g_k\} \subset C_b^1(H)$ converging to g in $C_b(H)$ and for any $k, n \in \mathbb{N}$ and $\alpha > 0$ we denote by $\varphi_{\alpha,n}^k = \varphi_{\alpha,n}^k(\lambda + \mu, g_k + \mu \varphi(\lambda, g))$ the solution of the problem

$$(\lambda + \mu) \varphi - L_{\alpha,n}\varphi + K_{r,n}(D\varphi) = g_{k,n} + \mu \varphi_n(\lambda, g), \qquad (10.2.3)$$

where $K_{r,n}(x) = K_r(P_n x)$, $g_{k,n}(x) = g_k(P_n x)$ and $\varphi_n(\lambda, g)(x) = \varphi(\lambda, g)(P_n x)$, for any $x \in H$, and μ is a positive constant to be determined later on. As g_k and $\varphi(\lambda, g)$ are continuously differentiable, due to the Lemma 9.5.8 we have that $\varphi_{\alpha,n}^k$ belongs to $C_b^2(H)$. Then, since $u_{\alpha,n}(t; x, z)$ is a strong solution of the problem (9.1.9), we can apply Itô's formula to the mapping $t \mapsto e^{-\lambda t} \varphi_{\alpha,n}^k(u_{\alpha,n}(t))$ and we get

$$d\left(e^{-\lambda t} \varphi_{\alpha,n}^k(u_{\alpha,n}(t))\right) = e^{-\lambda t} \left\langle D\varphi_{\alpha,n}^k(u_{\alpha,n}(t)), Q_n \, dw(t) \right\rangle_H$$
$$(10.2.4)$$
$$+ e^{-\lambda t} \left((\mathcal{L}_{\alpha,n} - \lambda)\varphi_{\alpha,n}^k(u_{\alpha,n}(t)) + \left\langle P_n z(t), D\varphi_{\alpha,n}^k(u_{\alpha,n}(t)) \right\rangle_H \right).$$

Recalling that $\varphi_{\alpha,n}^k$ is the solution of (10.2.3) and (9.3.10) holds, we have

$$(\mathcal{L}_{\alpha,n} - \lambda)\varphi_{\alpha,n}^k = \mu \varphi_{\alpha,n}^k + K_{r,n}(D\varphi_{\alpha,n}^k) - g_{k,n} - \mu \varphi_n(\lambda, g).$$

Then, by integrating with respect to $t \in [0, T]$ each side of (10.2.4) and taking the expectation, we get

$$e^{-\lambda T} P_T^{\alpha,n} \varphi_{\alpha,n}^k - \varphi_{\alpha,n}^k = \mu \, \mathbb{E} \int_0^T e^{-\lambda t} \left(\varphi_{\alpha,n}^k - \varphi(\lambda, g) \right) (u_{\alpha,n}(t)) \, dt$$

$$+ \mathbb{E} \int_0^T e^{-\lambda t} \left(K_r(D\varphi_{\alpha,n}^k(u_{\alpha,n}(t))) - g_k(u_{\alpha,n}(t)) + \left\langle z(t), D\varphi_{\alpha,n}^k(u_{\alpha,n}(t)) \right\rangle_H \right) dt.$$

Due to the Lemma 9.1.4 and to (9.5.16), if μ is large enough we can take the limit as n goes to infinity and we get

$$e^{-\lambda T} P_T^\alpha \varphi_\alpha^k(x) - \varphi_\alpha^k(x) = \mu \, \mathbb{E} \int_0^T e^{-\lambda t} \left(\varphi_\alpha^k - \varphi(\lambda, g) \right) (u_\alpha(t)) \, dt$$

$$= \mathbb{E} \int_0^T e^{-\lambda t} \left(K_r(D\varphi_\alpha^k(u_\alpha(t))) - g_k(u_\alpha(t)) + \left\langle z(t), D\varphi_\alpha^k(u_\alpha(t)) \right\rangle_H \right) dt,$$

where $\varphi_\alpha^k = \varphi_\alpha^k(\lambda + \mu, g_k + \mu\,\varphi(\lambda, g))$ is the solution of the problem

$$(\lambda + \mu)\varphi - L_\alpha\varphi + K_r(D\varphi) = g_k + \mu\,\varphi(\lambda, g).$$

By taking the limit as T goes to infinity, this yields

$$-\varphi_\alpha^k = \mu\,\mathbb{E}\int_0^{+\infty} e^{-\lambda t}\left(\varphi_\alpha^k(u_\alpha(t)) - \varphi(\lambda, g)(u_\alpha(t))\right)\,dt$$

$$+\mathbb{E}\int_0^{+\infty} e^{-\lambda t}\left(K_r(D\varphi_\alpha^k(u_\alpha(t))) - g_k(u_\alpha(t)) + \left\langle z(t), D\varphi_\alpha^k(u_\alpha(t))\right\rangle_H\right)\,dt.$$

$$(10.2.5)$$

For any $k, h \in \mathbb{N}$ and $\alpha > 0$ we have

$$\varphi_\alpha^k - \varphi_\alpha^h = R(\lambda + \mu, L_\alpha)\left(g_k - g_h - \left(K_r(D\varphi_\alpha^k) - K_r(D\varphi_\alpha^h)\right)\right)$$

and then, due to (9.5.2)

$$\|\varphi_\alpha^k - \varphi_\alpha^h\|_1^H \le \rho(\lambda + \mu)\left(\|g_k - g_h\|_0^H + c_r\|D\varphi_\alpha^k - D\varphi_\alpha^h\|_0^H\right),$$

where c_r is the Lipschitz constant of K_r. Therefore, if μ is sufficiently large we have $\rho(\lambda + \mu)c_r < 1$, for any $\lambda > 0$, so that

$$\|\varphi_\alpha^k - \varphi_\alpha^h\|_1^H \le \frac{\rho(\lambda + \mu)}{1 - \rho(\lambda + \mu)c_r}\|g_k - g_h\|_0^H. \tag{10.2.6}$$

This means that the sequence $\{\varphi_\alpha^k\}$ converges to some φ_α in $C_b^1(H)$ and it is immediate to check that such φ_α coincides with $\varphi_\alpha(\lambda + \mu, g + \mu\,\varphi(\lambda, g))$. Then, by taking the limit as k goes to infinity in (10.2.5), due to the dominated convergence theorem we can conclude that

$$-\varphi_\alpha = \mu\,\mathbb{E}\int_0^{+\infty} e^{-\lambda t}\left(\varphi_\alpha(u_\alpha(t)) - \varphi(\lambda, g)\right)(u_\alpha(t)))\,dt$$

$$(10.2.7)$$

$$+\mathbb{E}\int_0^{+\infty} e^{-\lambda t}[K_r(D\varphi_\alpha(u_\alpha(t))) - g(u_\alpha(t)) + \langle z(t), D\varphi_\alpha(u_\alpha(t))\rangle_H]\,dt.$$

If $x \in E$ and $z \in L^p(\Omega; L^\infty(0, +\infty; H))$, with p as in the Theorem 9.1.2, we can use (9.1.6) and (9.5.14) and by taking the limit as α goes to zero we have

$$\varphi(x) + \mathbb{E}\int_0^{+\infty} e^{-\lambda t}[K(D\varphi(u(t))) - g(u(t)) + \langle z(t), D\varphi(u(t))\rangle_H]\,dt = 0.$$

Notice that here we have replaced K_r by K, as we fixed $r = \|D\varphi(\lambda, g)\|_0^H$. Since $\varphi \in C_b^1(H)$ and $u(t; x, z)$ depends continuously of $x \in H$ and $z \in L^2(\Omega; L^2(0, +\infty; H))$, the same identity holds for $x \in H$ and $z \in L^2(\Omega; L^2(0, +\infty; H))$ and then, rearranging all terms, we get (10.2.2). $\qquad\square$

Theorem 10.2.2. *Assume that the Hypotheses 6.1, 6.2, 6.3 and 9.3 hold. Then there exists μ such that for any $\lambda > \mu$ and $g \in C_b^1(H)$ the value function U corresponding to the cost functional (10.2.1) coincides with the solution $\varphi(\lambda, g)$ of the Hamilton-Jacobi-Bellman equation (9.5.1).*

Moreover, for any $x \in E$ we have

$$U(x) = \lim_{\alpha \to 0} \min \left\{ I_\alpha(x; z) ; \ z \in L^2(\Omega; L^2(0, +\infty; H)) \text{ adapted} \right\},$$

where $\{I_\alpha(x, z)\}$ is a sequence of suitable cost functionals which admit unique optimal controls and states and whose value functions U_α coincide with the solution of the problems

$$(\lambda + \mu)\varphi - L_\alpha \varphi + K_r(D\varphi) = g + \mu \varphi(\lambda, g),$$

for some $\mu > 0$ large enough and $r = \|D\varphi(\lambda, g)\|_0^H$.

Proof. In the Theorem 9.5.13 we have seen that if $\lambda > \mu$ and $g \in C_b^1(H)$ there exists a unique solution $\varphi(\lambda, g) \in C_b^1(H)$ for the equation (9.5.1). Due to (10.2.2) and to the definition of K, we have that $U(x) \geq \varphi(\lambda, g)(x)$, for any $x \in H$. Now, we try to prove the opposite inequality. To this purpose we proceed by approximation.

We fix $r = \|D\varphi(\lambda, g)\|_0^H$ and μ as in the proof of the previous proposition such that

$$\rho(\lambda + \mu)c_r = c\left((\lambda + \mu)^{\frac{\epsilon - 1}{2}} + (\lambda + \mu)^{-1}\right)c_r < 1$$

and the limit (9.5.16) holds. For any $\alpha > 0$, we define the cost functional

$$I_\alpha(x; z) = \mathbb{E} \int_0^{+\infty} e^{-\lambda t} \left[g(u_\alpha(t; x, z)) + k(z(t))\right] dt$$

$$+ \mu \mathbb{E} \int_0^{+\infty} e^{-\lambda t} \left(\varphi(\lambda, g) - \varphi_\alpha\right)(u_\alpha(t; x, z)) dt$$

$$+ \mathbb{E} \int_0^{+\infty} e^{-\lambda t} \left[K(D\varphi_\alpha(u_\alpha(t; x, z))) - K_r(D\varphi_\alpha(u_\alpha(t; x, z)))\right] dt,$$

where $\varphi_\alpha = \varphi_\alpha(\lambda + \mu, g + \mu \varphi(\lambda, g))$ is the solution of the problem

$$(\lambda + \mu)\varphi - L_\alpha \varphi + K_r(D\varphi) = g + \mu \varphi(\lambda, g).$$

We denote by $U_\alpha(x)$ the corresponding value function. Thanks to (10.2.7) we easily have that $U_\alpha(x) \geq \varphi_\alpha(x)$, for any $x \in H$. In fact, it is possible to show that $U_\alpha(x) = \varphi_\alpha(x)$. Indeed, for each $x \in H$ the function

$$H \to \mathbb{R}, \qquad z \mapsto -\langle z, D\varphi_\alpha(x)\rangle_H - k(z)$$

attains its maximum at $z = -DK(D\varphi_\alpha(x))$. Then, if we show that the *closed loop equation*

$$du(t) = [Au(t) + F_\alpha(u(t)) - DK(D\varphi_\alpha(u(t)))] \, dt + Q \, dw(t), \qquad u(0) = x \quad (10.2.8)$$

has a unique adapted solution $u_\alpha^\star(t)$, we have that in correspondence of the control

$$z_\alpha^\star(t) = -DK(D\varphi_\alpha(u_\alpha^\star(t)))$$

it holds $I_\alpha(x, z_\alpha^\star) = \varphi_\alpha(x)$, so that $U_\alpha(x) \leq \varphi_\alpha(x)$. This means that $U_\alpha(x) = \varphi_\alpha(x)$ and there exists a unique optimal control and a unique optimal state for the minimizing problem corresponding to the cost functional $I_\alpha(x; z)$.

If $g \in C_b^1(H)$, due to the Lemma 9.5.8 $\varphi_\alpha \in C_b^2(H)$, so that, as we are assuming DK locally Lipschitz-continuous, the mapping

$$U_\alpha : H \to H, \qquad x \mapsto -DK(D\varphi_\alpha(x))$$

is Lipschitz-continuous. This implies that the closed loop equation admits a unique solution and for any $\alpha > 0$ the optimal control corresponding to the functional $I_\alpha(x; z)$ is $z_\alpha^\star(t) = -DK(D\varphi_\alpha(u_\alpha^\star(t)))$. Thus, since DK is bounded on bounded sets and (9.5.22) holds, there exists $R > 0$ such that

$$\sup_{\alpha>0} \sup_{t\geq 0} |z_\alpha^\star(t)|_H = R, \qquad \mathbb{P} - \text{a.s.}$$

and then

$$U_\alpha(x) = \inf \left\{ I_\alpha(x; z) \, : \, z \in \mathcal{M}_R \right\},$$

where $\mathcal{M}_R$ is the subset of admissible controls such that

$$\sup_{t\in [0,T]} |z(t)|_H \leq R, \qquad \mathbb{P} - \text{a.s.}$$

Now, recalling the Proposition 9.5.6, for any $x \in E$ we have

$$\lim_{\alpha\to 0} U_\alpha(x) = \lim_{\alpha\to 0} \varphi_\alpha(\lambda + \mu, g + \mu\varphi(\lambda, g))(x) = \varphi(\lambda, g)(x).$$

Thus, if we show that

$$\lim_{\alpha\to 0} \sup_{z\in \mathcal{M}_R} |I_\alpha(x; z) - I(x; z)| = 0, \qquad (10.2.9)$$

it immediately follows that $U(x) = \varphi(\lambda, g)(x)$, for $x \in E$.

Due to the Lemma 9.1.3, we have that

$$\lim_{\alpha\to 0} \mathbb{E} \, |g(u_\alpha(t; x, z)) - g(u(t; x, z))| = 0,$$

uniformly for (t, x) in bounded sets of $[0, +\infty) \times E$ and $z \in \mathcal{M}_R$. Hence, if we fix $\epsilon > 0$ and $M > 0$ such that

$$\int_M^{+\infty} e^{-\lambda t} \, dt \leq \frac{\epsilon}{2 \, \|g\|_0^H},$$

we have

$$\mathbb{E} \int_0^{+\infty} e^{-\lambda t} \left[g(u_\alpha(t; x, z)) - g(u(t; x, z)) \right] dt$$

$$\leq \epsilon + \int_0^M e^{-\lambda t} \mathbb{E} \left| g(u_\alpha(t; x, z)) - g(u(t; x, z)) \right| dt,$$

and, due to the arbitrariness of $\epsilon > 0$

$$\lim_{\alpha \to 0} \sup_{z \in \mathcal{M}_R^2} \mathbb{E} \int_0^{+\infty} e^{-\lambda t} \left[g(u_\alpha(t; x, z)) - g(u(t; x, z)) \right] dt = 0.$$

Thanks to the Lemma 9.5.6, for $j = 0, 1$ we have

$$\lim_{\alpha \to 0} \sup_{|x|_E \leq R} \left| D^j \left(\varphi_\alpha - \varphi \right)(x) \right|_{\mathcal{L}^j(H)} = 0. \tag{10.2.10}$$

Moreover, since $u_\alpha(t)$ fulfills an estimate analogous to (9.1.2), uniformly with respect to $\alpha > 0$, for any $T > 0$ we have

$$\sup_{z \in \mathcal{M}_R} \sup_{t \in [0,T]} |u_\alpha(t; x, z)|_E = c_T, \qquad \mathbb{P} - \text{a.s.}$$

for some constant $c_T > 0$ independent of α. Thus, by using the same arguments as above, it follows

$$\lim_{\alpha \to 0} \sup_{z \in \mathcal{M}_R} \mathbb{E} \int_0^{+\infty} e^{-\lambda t} (\varphi_\alpha - \varphi)(u_\alpha(t; x, z)) \, dt = 0.$$

Finally, due to (9.5.22), the sequence $\{\varphi_\alpha\}$ is bounded in $C_b^1(H)$. Then, since K and K_r are bounded on bounded sets and $K(D\varphi(x)) = K_r(D\varphi(x))$, for any $x \in H$, by using (10.2.10) and by arguing as above, we have

$$\lim_{\alpha \to 0} \sup_{z \in \mathcal{M}_R} \mathbb{E} \int_0^{+\infty} e^{-\lambda t} \left| K(D\varphi_\alpha(u_\alpha(t; x, z))) - K_r(D\varphi_\alpha(u_\alpha(t; x, z))) \right| dt = 0.$$

Therefore, we can conclude that (10.2.9) holds and $U(x) = \varphi(x)$, for any $x \in E$.

Now, assume that $x \in H$. We fix a sequence $\{x_n\} \subset E$ converging to x in H. For each $n \in \mathbb{N}$ we have $U(x_n) = \varphi(x_n)$. We have

$$I(x_n; z) - I(x; z) = \mathbb{E} \int_0^{+\infty} e^{-\lambda t} \left[g(u(t; x_n, z)) - g(u(t; x, z)) \right] dt$$

and then, due to (9.1.4) we easily get

$$\lim_{n \to +\infty} \sup_{z \in \mathcal{M}_R} |I(x_n; z) - I(x; z)| = 0.$$

As $\varphi \in C_b^1(H)$, this allows to conclude that $U(x) = \varphi(x)$, for any $x \in H$. $\square$

In the section 9.5 we have seen that if we assume the hamiltonian K to be Lipschitz-continuous, then for any $\lambda > 0$ and $g \in C_b(H)$ there exists a unique solution $\varphi(\lambda, g)$ in $D(L) \subset C_b^1(H)$ for the problem (9.0.2). This allows us to reformulate the Theorem 10.2.2 in the case of Lipschitz hamiltonian.

Theorem 10.2.3. *Assume that the Hypotheses 6.1, 6.2, 6.3 and 9.2 hold. Then for any $\lambda > 0$ and $g \in C_b^{0,1}(H)$ the value function U corresponding to the cost functional (10.2.1) coincides with the solution $\varphi(\lambda, g)$ of the Hamilton-Jacobi-Bellman equation (9.0.2).*

Moreover, for any $x \in E$ we have

$$U(x) = \lim_{\alpha \to 0} \min \left\{ I_\alpha(x; z) ; \; z \in L^2(\Omega; L^2(0, +\infty; H)) \text{ adapted} \right\},$$

where $\{I_\alpha(x, z)\}$ is a sequence of suitable cost functionals which admit unique optimal costs and states and whose value functions U_α coincide with the solution of the problems

$$(\lambda + \mu)\varphi - L_\alpha \varphi + K(D\varphi) = g + \mu \, \varphi(\lambda, g),$$

for some $\mu > 0$.

Proof. The proof is identical to that of the Theorem 10.2.2. The only difference lies in the fact that here we have to show that for any $g \in C_b^{0,1}(H)$ the approximating closed loop equation

$$du(t) = [Au(t) + F_\alpha(u(t)) - DK(D\varphi_\alpha(u(t)))] \, dt + Q \, dw(t), \qquad u(0) = x,$$

admits a unique adapted solution $u_\alpha^\star(t)$.

To this purpose, if $g \in C_b^{0,1}(H)$ we can find a bounded sequence $\{g_k\} \subset C_b^1(H)$ converging to g in $C_b(H)$. For each $k \in \mathbb{N}$ there exists a unique solution $\varphi_{\alpha,k}$ for the Hamilton-Jacobi-Bellman problem

$$(\lambda + \mu) \, \varphi - L_\alpha \varphi + K(D\varphi) = g_k + \mu\varphi(\lambda, g),$$

where μ is the constant introduced in the previous theorem. Then, since $g_k + \mu\varphi(\lambda, g) \in C_b^1(H)$, as proved above the corresponding closed loop equation has a unique solution $u_{\alpha,k}^\star(t)$. If we show that for any $T > 0$ the sequence $\{u_{\alpha,k}^\star\}$ converges to some $u_\alpha^\star$ in $C([0, T]; H)$, $\mathbb{P}$-a.s. and in mean-square, then we easily have that $u_\alpha^\star$ is the solution of the closed loop equation (10.2.8).

For $k, h \in \mathbb{N}$ we define $v_\alpha^{k,h}(t) = u_{\alpha,k}^\star(t) - u_{\alpha,h}^\star(t)$. We have that $v_\alpha^{k,h}$ is solution of the problem

$$\frac{dv}{dt}(t) = \; Av(t) + F_\alpha(u_{\alpha,k}^\star(t)) - F_\alpha(u_{\alpha,h}^\star(t)) - DK(D\varphi_{\alpha,k}(u_{\alpha,k}^\star(t)))$$

$$+ DK(D\varphi_{\alpha,h}(u_{\alpha,h}^\star(t))), \qquad v(0) = 0$$

By multiplying each side by $v_\alpha^{k,h}(t)$ and recalling that A is of negative type and F_α is dissipative, we have

$$\frac{1}{2}\frac{d}{dt}|v_\alpha^{k,h}(t)|_H^2 \leq c\,|v_\alpha^{k,h}(t)|_H^2$$

$$+|DK(D\varphi_{\alpha,k}(u_{\alpha,k}^\star(t))) - DK(D\varphi_{\alpha,h}(u_{\alpha,h}^\star(t)))|_H|v_\alpha^{k,h}(t)|_H.$$

Since DK is Lipschitz-continuous, we have

$$|DK(D\varphi_{\alpha,k}(u_{\alpha,k}^\star(t))) - DK(D\varphi_{\alpha,h}(u_{\alpha,h}^\star(t)))|_H$$

$$\leq c\left|D\varphi_{\alpha,k}(u_{\alpha,k}^\star(t)) - D\varphi_{\alpha,h}(u_{\alpha,h}^\star(t))\right|_H.$$

Moreover, for each $x,y \in H$ we have

$$|D\varphi_{\alpha,k}(x) - D\varphi_{\alpha,k}(y)|_H \leq c\,\|\varphi_{\alpha,k}\|_2^H|x-y|_H$$

and then, since the sequence $\{g_k\}$ is bounded in $C_b^1(H)$, from (9.5.17) it follows

$$|D\varphi_{\alpha,k}(u_{\alpha,k}^\star(t)) - D\varphi_{\alpha,k}(u_{\alpha,h}^\star(t))|_H \leq c_\alpha\,|v_\alpha^{k,h}(t)|_H.$$

If μ is large enough, as in the proof of the Theorem 10.2.2, we have

$$\|D\varphi_{\alpha,k} - D\varphi_{\alpha,h}\|_0^H \leq c_\alpha\,\|g_k - g_h\|_0^H$$

and this yields

$$|DK(D\varphi_{\alpha,k}(u_{\alpha,k}^\star(t))) - DK(D\varphi_{\alpha,h}(u_{\alpha,h}^\star(t)))|_H \leq c_\alpha\,|v_\alpha^{k,h}(t)|_H + c_\alpha\,\|g_k - g_h\|_0^H.$$

Thus from the Young inequality we have

$$\frac{1}{2}\frac{d}{dt}|v_\alpha^{k,h}(t)|_H^2 \leq c_\alpha\,|v_\alpha^{k,h}(t)|_H^2 + c_\alpha\left(\|g_k - g_h\|_0^H\right)^2$$

and by the Gronwall lemma we get

$$\sup_{t\in[0,T]}|u_{\alpha,k}^\star(t) - u_{\alpha,h}^\star(t)|_H \leq c_{\alpha,T}\|g_k - g_h\|_0^H, \qquad \mathbb{P}-\text{a.s.},$$

for some constant $c_{\alpha,T}$. This means that $u_{\alpha,k}^\star$ converges to some $u_\alpha^\star$ in $C([0,T];H)$, $\mathbb{P}$-a.s and in mean-square, and it is not difficult to check that $u_\alpha^\star$ is the solution of the closed loop corresponding to the datum g. $\qquad\square$

10.3   Existence of the optimal control in the one dimensional case

In some particular cases the closed loop equation admits a unique solution and then there exist a unique optimal control and a unique state for the control problem. We show how it is possible to prove this fact in the case of the finite horizon control problem.

Theorem 10.3.1. *Assume the hypotheses of the Theorem 10.1.2 and take the space dimension $d = 1$.*

1. If

$$\sup_{\xi \in \overline{\mathcal{O}}} \sup_{\sigma \in \mathbb{R}^n} \frac{|D_\sigma^j f(\xi, \sigma)|}{|\sigma|^{3-j}} < \infty, \qquad j \leq 2,$$

then for any $\varphi, g \in C_b^{0,1}(H)$ and $x \in H$ there exists a unique optimal control for the minimizing problem associated with the finite horizon functional J defined in (10.0.1) . Furthermore, the optimal control $z^\star$ is related to the corresponding optimal state $y^\star$ by the feedback formula

$$z^\star(t) = -DK(D_x V(T - t, y^\star(t))), \qquad t \in [0, T].$$

2. If DK can be extended as a Lipschitz-continuous mapping from $E^\star$ into itself, then the same conclusion of 1 holds for any $x \in E$.

Proof. We first prove 1. As we have seen in the proof of the Theorem 10.1.2, the only thing we have to show is that the closed loop equation (10.0.4) admits a unique solution, for any $\varphi, g \in C_b^1(H)$. And to this purpose we have to show that the derivative with respect to x of the solution $y(t, x)$ of the problem (9.0.1) is Lipschitz-continuous and for any $x, z \in H$ and $t > 0$

$$|Dy(t, x) - Dy(t, z)|_H \leq c_T (t \wedge 1)^{-\frac{1+\varepsilon}{2}} |x - z|_H, \tag{10.3.1}$$

for some positive constant c_T. Actually, if we define for $(t, x) \in [0, T] \times H$

$$U(t, x) = -DK(Dy(T - t, x)),$$

the function U verifies the conditions of the Lemma 9.4.6 and then there exists a unique solution $y^\star(t)$ for the closed loop equation (10.0.4). As shown in the proof of the Theorem 10.1.2, the general case of $\varphi, g \in C_b^{0,1}(H)$ follows by approximation.

If we show that the function $D(P_t \varphi)$ is Lipschitz-continuous, for any $\varphi \in C_b^1(H)$ and $t > 0$, and

$$|D(P_t \varphi)(x) - D(P_t \varphi)(z)|_H \leq c (t \wedge 1)^{-\frac{1+\varepsilon}{2}} \|\varphi\|_1^H |x - z|_H, \tag{10.3.2}$$

then (10.3.1) follows, for some constant c_T depending only on g, φ and T. Actually, since $y(t, x)$ is the solution of the problem (9.4.2), we have

$$Dy(t, x) - Dy(t, z) = DP_t\varphi(x) - DP_t\varphi(z) + \int_0^t [DP_{t-s}g(x) - DP_{t-s}g(z)] \, ds$$

$$- \int_0^t [DP_{t-s}K(Dy(s, \cdot))(x) - DP_{t-s}K(Dy(s, \cdot))(z)] \, ds,$$

so that, thanks to (10.3.2)

$$|Dy(t, x) - Dy(t, z)|_H \leq c \left((t \wedge 1)^{-\frac{1+\epsilon}{2}} \|\varphi\|_{0,1}^H + \int_0^t ((t - s) \wedge 1)^{-\frac{1+\epsilon}{2}} \|g\|_{0,1}^H \, ds \right.$$

$$\left. + \int_0^t ((t - s) \wedge 1)^{-\frac{1+\epsilon}{2}} \|K(Dy(s, \cdot))\|_{0,1}^H \, ds \right) |x - z|_H.$$

Now, thanks to (9.4.18)

$$\sup_{t \in [0,T]} \|y(t, \cdot)\|_1^H \leq c_T \left(\|\varphi\|_1^H + T \|g\|_1^H \right)$$

and then, as K is locally Lipschitz-continuous

$$\sup_{s \in [0,T]} \|K(Dy(s, \cdot))\|_{0,1}^H < \infty.$$

This yields (10.3.1).

Since $D(P_t\varphi)$ is given by the formula (7.3.1), according to (7.1.4), (10.3.2) immediately follows once one proves that $\mathbb{P}$-a.s.

$$|v(t; x_1, h) - v(t; x_2, h)|_H^2 + \int_0^t |Q^{-1}(v(s; x_1, h) - v(s; x_2, h))|_H^2 \, ds \leq c_T |h|_H^2 |x_1 - x_2|_H^2,$$

$$(10.3.3)$$

We first assume that $x_1, x_2 \in E$ and define $z(t) = v(t; x_1, h) - v(t; x_2, h)$. We have that z is the unique solution of the problem

$$\begin{cases} \dfrac{d}{dt} z(t) = [Az(t) + DF(y(t; x_1))z(t)] \, dt \\[2mm] \qquad\qquad + (DF(y(t; x_1)) - DF(y(t; x_2))) v(t; x_2, h) \, dt, \qquad z(0) = 0, \end{cases}$$

Thus, we have

$$\frac{1}{2} \frac{d}{dt} |z(t)|_H^2 + |z(t)|_{D((-A)^{1/2})}^2 \leq c |z(t)|_H^2$$

$$+ |\langle (DF(y(t; x_1)) - DF(y(t; x_2))) v(t; x_2, h), z(t) \rangle_H|.$$

Due to the Sobolev embedding theorem, for any $\delta > 0$ we have

$$|\langle (DF(y(t; x_1)) - DF(y(t; x_2))) \, v(t; x_2, h), z(t) \rangle_H| \leq$$

$$|z(t)|_{D((-A)^{(1+\delta)/4})} \, |(DF(y(t; x_1)) - DF(y(t; x_2))) \, v(t; x_2, h)|_1 \, ,$$

In the chapter 6 we have proved that

$$\sup_{x \in E} |y(t; x)|_E \leq k(t)(t \wedge 1)^{-\frac{1}{2m}}, \qquad \mathbb{P} - \text{a.s.},$$

for some process $k(t)$ having has all moments finite. Thus, since

$$\sup_{x \in H} |v(t; x, h)|_H \leq c(t)|h|_H,$$

for some continuous increasing function $c(t)$, from (7.2.5) with $\theta = 1$, by interpolatingn we get

$$|\langle (DF(y(t; x_1)) - DF(y(t; x_2))) \, v(t; x_2, h), z(t) \rangle_H| \leq$$

$$k^{2m-1}(t)|z(t)|_{D((-A)^{1/2})}^{(1+\delta)/2} |z(t)|_H^{(1-\delta)/2} |x_1 - x_2|_H |h|_H \, c(t) \, (t \wedge 1)^{-(2m-1)/2m}.$$

As we can write

$$(t \wedge 1)^{-(2m-1)/2m} = (t \wedge 1)^{-(1-\delta)/2} (t \wedge 1)^{-\frac{1}{2}(1+\delta-1/m)},$$

thanks to the Young inequality we get

$$\frac{1}{2} \frac{d}{dt} |z(t)|_H^2 + |z(t)|_{D((-A)^{1/2})}^2 \leq c \, |z(t)|_H^2 + \frac{1}{2} |z(t)|_{D((-A)^{1/2})}^2$$

$$+ c \, |x_1 - x_2|_H^2 |h|_H^2 (t \wedge 1)^{-(1-\delta)} + c(t) \, (t \wedge 1)^{-2(1+\delta-1/m)/(1-\delta)} \, |z(t)|_H^2,$$

where $c(t)$ is a process having all moments finite. Now, if $m < 2$ it is possible to find some $\delta \in (0, 1)$ such that

$$2(1 + \delta - 1/m)/(1 - \delta) < 1,$$

and then, by using the Gronwall Lemma, (10.3.3) follows.

Concerning the proof of 2 we recall that in the chapter 6 it has been proved that for any $\varphi \in C_b^1(E) \supset C_b^1(H)$ and $t > 0$ it holds

$$|D(P_t\varphi)(x) - D(P_t\varphi)(y)|_{E^*} \leq c \, (t \wedge 1)^{-\frac{1+\epsilon}{2}} \|\varphi\|_1^E |x - y|_E, \qquad x, y \in E.$$

Then, as before we have that $y(t, \cdot) \in C_b^1(E)$ and

$$|Dy(t, x) - Dy(t, y)|_{E^*} \leq c \, (t \wedge 1)^{-\frac{1+\epsilon}{2}} |x - y|_E, \qquad (10.3.4)$$

where the constant c depends only on g, φ and T. This makes possible to prove that the closed loop equation admits a unique mild solution. Actually, due to the Sobolev embedding theorem, as the dimension d equals 1, for any $\epsilon > 0$ we have that $D((-A)^{1/4+\epsilon})$ is continuously embedded into E and then

$$\left| \int_0^t e^{(t-s)A} DK(u(T-s,y(s;x))) \, ds \right|_E$$

$$\leq c \int_0^t \left| e^{(t-s)A} DK(u(T-s,y(s;x))) \right|_{D((-A)^{1/4+\epsilon})} ds$$

$$\leq c \int_0^t (t-s)^{-1/2-2\epsilon} \left| DK(u(T-s,y(s;x))) \right|_{(D((-A)^{1/4+\epsilon}))^*}$$

$$\leq c \int_0^t (t-s)^{-1/2-2\epsilon} \left| DK(u(T-s,y(s;x))) \right|_{E^*}.$$

Therefore, as DK is Lipschitz-continuous on E^* and (10.3.4) holds, it is easy to show that the closed loop equation admits a unique mild solution. □

Notice that if $K(x) = |x|_H^2$, then $DK(x) = x$, so that DK can be extended as a Lipschitz-continuous mapping from E^* into itself.

Appendix A

Dissipative mappings

We recall here some basic properties of the subdifferential of the norm in a Banach space E and some general properties of dissipative mappings on E. For more details and the proofs of the quoted results we refer to [37].

A.1 Subdifferential of the norm

Let E be a Banach space and let E^* be its dual space. The mapping

$$|\cdot|_E : E \to \mathbb{R}, \quad x \mapsto |x|_E$$

is convex, then it is possible to define its *subdifferential* by setting for any $x \in E$

$$\partial |x|_E = \{ x^* \in E^* \, ; \, |x + y|_E \geq |x|_E + \langle y, x^* \rangle, \, \forall y \in E \}.$$

It is not difficult to show that

$$\partial |x|_E = \{ x^* \in E^* \, ; \, D_- |x|_E \cdot y \leq \langle y, x^* \rangle \leq D_+ |x|_E \cdot y, \, \forall y \in E \},$$

where

$$D_+ |x|_E \cdot y = \lim_{h \to 0+} \frac{|x + hy|_E - |x|_E}{h},$$

$$D_- |x|_E \cdot y = \lim_{h \to 0-} \frac{|x + hy|_E - |x|_E}{h}.$$

The set $\partial |x|_E$ is convex and closed and for any $x \neq 0$ it is given by

$$\partial |x|_E = \{ x^* \in E^* : \langle x, x^* \rangle = |x|_E, \, |x^*|_{E^*} = 1 \}.$$

Due to the Hahn-Banach theorem, $\partial |x|_E$ is a non-empty set. Moreover

$$D_+ |x|_E \cdot y = \max \{ \langle y, x^* \rangle : x^* \in \partial |x|_E \},$$

$$D_- |x|_E \cdot y = \min \{ \langle y, x^* \rangle : x^* \in \partial |x|_E \}.$$

Remark A.1.1.   1. If E is a Hilbert space and $x \neq 0$, then the set $\partial |x|_E$ consists of a unique element

$$\partial |x|_E = \left\{ \frac{x}{|x|_E} \right\}.$$

If $x = 0$ then $\partial |0|_E = \{ y \in E ; \; |y|_E = 1 \}$.

2. If $E = C(\overline{D})$, where $\mathcal{O}$ is a bounded open set, for any $x \in E$ we define $M_x = \{ \xi \in \overline{\mathcal{O}} : |x(\xi)| = |x|_E \}$. Then we have that $\mu \in \partial |x|_E$ if and only if μ is a Radon measure on $\overline{\mathcal{O}} \subset \mathbb{R}^d$ with $|\mu| = 1$, the support of μ is included in M_x and

$$\int_A \operatorname{sgn} x(\xi) \, \mu(d\xi) \geq 0, \quad A \in \mathcal{B}(\overline{\mathcal{O}}).$$

In particular, the element

$$\delta_x = \left\{ \begin{array}{ll} \delta_{\xi_0} & \text{if } x(\xi_0) = |x|_E, \\[2mm] -\delta_{\xi_0} & \text{if } x(\xi_0) = -|x|_E, \end{array} \right.$$

belongs to $\partial |x|_E$.

3. Analogously, if $E = C(\overline{\mathcal{O}}; \mathbb{R}^n)$, endowed with the norm

$$|x|_E^2 = \sum_{k=1}^{n} \sup_{\xi \in \overline{\mathcal{O}}} |x_k(\xi)|^2,$$

for any $x, y \in E$ we define

$$\langle \delta_x, y \rangle_E = \left\{ \begin{array}{ll} \dfrac{1}{|x|_E} \displaystyle\sum_{k=1}^{n} y_k(\xi_k) x_k(\xi_k) & \text{if } x \neq 0, \\[4mm] \delta_0 & \text{if } x = 0, \end{array} \right. \tag{A.1.1}$$

where ξ_k are suitable points in $\overline{\mathcal{O}}$ such that

$$|x_k(\xi_k)| = \sup_{\xi \in \overline{\mathcal{O}}} |x_k(\xi)|, \quad 0 \leq k \leq n,$$

and δ_0 is any element in the unitary ball of $E^\star$. By proceeding as before it is possible to show that $\delta_x \in \partial |x|_E$.

In what follows we will also need the following simple result.

Proposition A.1.2. *If $x, y \in E$, then the following statements are equivalent*

1. *For any $\alpha \geq 0$ it holds $|x|_E \leq |x + \alpha y|_E$*

2. *There exists $x^* \in \partial|x|_E$ such that $\langle y, x^* \rangle \geq 0$.*

Now, let us fix $u : [0, T] \to E$, for some $T > 0$. Next proposition shows how to compute the derivative (whenever it exists) of the mapping $t \in [0, T] \mapsto |u(t)|_E \in \mathbb{R}$.

Proposition A.1.3. *If $u : [0, T] \to E$ is differentiable at t_0, then the function $\gamma : [0, T] \to \mathbb{R}$, $t \mapsto |u(t)|_E$ is differentiable on the right and on the left at t_0 and we have*

$$\frac{d^+}{dt} \gamma(t_0) = D_+|u(t_0)|_E \cdot \frac{d^+}{dt} u(t_0) = \max \left\{ \langle u'(t_0), x^* \rangle_E : x^* \in \partial|u(t_0)|_E \right\}$$

$$\frac{d^-}{dt} \gamma(t_0) = D_-|u(t_0)|_E \cdot \frac{d^-}{dt} u(t_0) = \min \left\{ \langle u'(t_0), x^* \rangle_E : x^* \in \partial|u(t_0)|_E \right\}.$$

A.2   Dissipative mappings

A mapping $f : D(f) \subseteq E \to E$ is *dissipative* if for any $x, y \in D(f)$ and $\alpha \geq 0$ it holds

$$|x - y|_E \leq |x - y - \alpha(f(x) - f(y))|_E.$$

Due to the proposition A.1.2 we have that $f : D(f) \subseteq E \to E$ is dissipative if and only if for any $x, y \in D(f)$ there exists $z^* \in \partial|x - y|_E$ such that

$$\langle f(x) - f(y), z^* \rangle \leq 0.$$

Proposition A.2.1. *Let $f : E \to E$ be a continuous dissipative mapping. Then for any $\alpha > 0$ the mapping $E \to E$, $x \mapsto x - \alpha f(x)$, is a bijection.*

Thanks to the previous proposition, we can introduce the *Yosida approximations* of f, by setting for any $\alpha > 0$

$$f_\alpha(x) = f(J_\alpha(x)), \quad x \in E,$$

where the mappings J_α are defined by

$$J_\alpha(x) = (I - \alpha f)^{-1}(x), \quad x \in E.$$

Proposition A.2.2. *Let $f : E \to E$ be a continuous dissipative mapping. Then for any $\alpha > 0$ the following facts hold*

1. *for any $x, y \in E$*

$$|J_\alpha(x) - J_\alpha(y)|_E \leq |x - y|_E; \tag{A.2.1}$$

2. *f_α is dissipative and Lipschitz-continuous and for any $x, y \in E$*

$$|f_\alpha(x) - f_\alpha(y)|_E \leq \frac{2}{\alpha}|x - y|_E;$$

3. *for any* $x \in E$

$$|f_\alpha(x)|_E \leq |f(x)|_E;$$

4. *for any* $x \in E$

$$|J_\alpha(x) - x|_E \leq \alpha |f(x)|_E. \qquad (A.2.2)$$

In particular, J_α converges to the identity, as α goes to zero, uniformly on compact sets.

We conclude by discussing the regularity properties of the mapping J_α, $\alpha > 0$.

As well known, if E and F are Banach spaces and $f : E \to F$ is a bijection of class C^k, then f is a diffeomorfism from E onto F of class C^k if and only if $Df(x)$ is an isomorfism from E onto F, for any $x \in E$.

Now, let $E = \mathbb{R}^d$ and assume that f is dissipative and of class C^k, with $k \geq 1$. As stated in the proposition A.2.1, for any $\alpha > 0$ the mapping

$$h_\alpha : \mathbb{R}^d \to \mathbb{R}^d, \quad x \mapsto x - \alpha f(x),$$

is a bijection. Since f is dissipative, $Df(x) \leq 0$, for any $x \in \mathbb{R}^d$, so that

$$\langle Dh_\alpha(x)y, y \rangle = |y|^2 - \alpha \langle Df(x)y, y \rangle \geq |y|^2, \quad x, y \in \mathbb{R}^d.$$

This implies that for any $x \in \mathbb{R}^d$ and $\alpha > 0$, the mapping $Dh_\alpha(x)$ is an isomorfism from $\mathbb{R}^d$ onto $\mathbb{R}^d$ and then the function

$$J_\alpha(x) = h_\alpha^{-1}(x) = (I - \alpha f)^{-1}(x), \quad x \in \mathbb{R}^d$$

is of class C^k itself. Moreover, it is possible to compute explicitly the derivatives of J_α, for any $\alpha > 0$. Actually, if $k \in \mathbb{R}^d$ we have

$$DJ_\alpha(x)k = (I - \alpha Df(J_\alpha(x)))^{-1} k, \quad x \in \mathbb{R}^d. \qquad (A.2.3)$$

In a similar way , if $k_1, k_2, \in \mathbb{R}^d$ we have

$$D^2 J_\alpha(x)(k_1, k_2) = \alpha DJ_\alpha(x)D^2 f(J_\alpha(x))(DJ_\alpha(x)k_1, DJ_\alpha(x)k_2), \quad x \in \mathbb{R}^d, \quad (A.2.4)$$

and if $k_1, k_2, k_3 \in \mathbb{R}^d$ we have

$$D^3 J_\alpha(x)(k_1, k_2, k_3) = \alpha DJ_\alpha(x)D^3 f(J_\alpha(x))(DJ_\alpha(x)k_1, DJ_\alpha(x)k_2, DJ_\alpha(x)k_3)$$

$$+ \frac{\alpha}{2} DJ_\alpha(x) \sum_{\pi \in S_3} D^2 f(J_\alpha(x))(D^2 J_\alpha(x)(k_{\pi_1}, k_{\pi_2}), DJ_\alpha(x)k_{\pi_3}), \quad x \in \mathbb{R}^d,$$

$$(A.2.5)$$

where S_3 are the permutations on a set of three elements.

Appendix B

Weakly continuous semigroups

B.1 Definition and main properties

We describe here the notion of *weakly continuous semigroup* which has been introduced in [17]. Throughout this appendix we denote by E any separable Banach space, endowed with the norm $|\cdot|_E$.

First of all we introduce the notion of $\mathcal{K}$-convergence.

Definition B.1.1. *A sequence* $\{\varphi_n\} \subset C_b(E)$ *is* $\mathcal{K}$*-convergent to some* $\varphi \in C_b(E)$ *if*

$$\sup_{n\in\mathbf{N}} \|\varphi_n\|_0^E < \infty, \qquad and \qquad \lim_{n\to+\infty} \sup_{x\in K} |\varphi_n(x) - \varphi(x)| = 0, \tag{B.1.1}$$

for every compact $K \subset E$. *Moreover, we write*

$$\mathcal{K} - \lim_{t\to t_0} \varphi_t = \varphi$$

if for any sequence $\{t_n\}$ *converging to* t_0, *the sequence* $\{\varphi_{t_n}\}$ *fulfills (B.1.1).*

Once introduced $\mathcal{K}$-convergence, we can give the definition of weakly continuous semigroups.

Definition B.1.2. *A semigroup of bounded linear operators* P_t, $t \geq 0$, *defined on* $C_b(E)$ *is* weakly continuous *if there exists* $M > 0$ *and* $\omega \in \mathbb{R}$ *such that*

1. *the family of functions* $\{P_t\varphi ; t \in [0,T]\}$ *is equi-uniformly continuous, for any* $\varphi \in C_b(E)$ *and* $T > 0$;

2. *for every* $\varphi \in C_b(E)$

$$\mathcal{K} - \lim_{t\to 0^+} P_t\varphi = \varphi;$$

3. *for every $\varphi \in C_b(E)$ and for every sequence $\{\varphi_n\} \subset C_b(E)$ which is $\mathcal{K}$-convergent to φ, it holds*

$$\mathcal{K} - \lim_{n \to +\infty} P_t \varphi_n = P_t \varphi$$

and the limit is uniform in $t \in [0, T]$, for every $T > 0$;

4. *for any $t \geq 0$ it holds*

$$\|P_t\|_{\mathcal{L}(C_b(E))} \leq M \, e^{\omega t}.$$

Now we can introduce the notion of *infinitesimal generator* for a weakly continuous semigroup. For any $\varphi \in C_b(E)$ and $x \in E$ we define

$$F(\lambda)\varphi(x) = \int_0^{+\infty} e^{-\lambda t} P_t \varphi(x) \, dt, \qquad \lambda > \omega.$$

Proposition B.1.3. *If P_t is a weakly continuous semigroup, then the linear operator F defined above is bounded from $C_b(E)$ into itself.*

Proof. Due to the weak continuity of P_t, the function

$$[0, +\infty) \to \mathbb{R}, \quad t \mapsto e^{-\lambda t} P_t \varphi(x),$$

is continuous for any $\varphi \in C_b(E)$ and $x \in E$. Moreover, as we have

$$|P_t \varphi(x)| \leq M e^{\omega t} \|\varphi\|_0^E,$$

the integral above is convergent and the definition of $F(\lambda)$ is meaningful.
We show that $F(\lambda) \in \mathcal{L}(C_b(E))$. If $\lambda > \omega$ we have

$$|F(\lambda)\varphi(x)| \leq \int_0^{+\infty} e^{-\lambda t} |P_t \varphi(x)| \, dt \leq M \, \|\varphi\|_0^E \int_0^{+\infty} e^{-(\lambda - \omega)t} \, dt = \frac{M}{\lambda - \omega} \|\varphi\|_0^E,$$

so that $F(\lambda)\varphi$ is a bounded function and

$$\|F(\lambda)\varphi\|_0^E \leq \frac{M}{\lambda - \omega} \|\varphi\|_0^E. \tag{B.1.2}$$

Next, we fix $\epsilon > 0$. Since $\lambda - \omega > 0$, we can find $T_\epsilon > 0$ such that

$$\left(\frac{2M\|\varphi\|_0^E}{\lambda - \omega} \right) e^{-(\lambda - \omega)T_\epsilon} < \frac{\epsilon}{2}.$$

Moreover, as the family of functions $\{ P_t \varphi : t \in [0, T_\epsilon] \}$ is equi-uniformly continuous for any $\varphi \in C_b(E)$, there exists $\delta_\epsilon > 0$ such that

$$|y|_E \leq \delta_\epsilon \implies |P_t \varphi(x + y) - P_t \varphi(x)| \leq \frac{\epsilon}{2} \left(\frac{\lambda}{1 - e^{-\lambda T_\epsilon}} \right)$$

for any $t \in [0, T_\epsilon]$ and $x \in E$. Therefore, if $|y|_E \leq \delta_\epsilon$, we have

$$|F(\lambda)\varphi(x+y) - F(\lambda)\varphi(x)|$$

$$\leq \int_0^{T_\epsilon} e^{-\lambda t} |P_t \varphi(x+y) - P_t\varphi(x)|\, dt + \int_{T_\epsilon}^{+\infty} e^{-\lambda t}|P_t\varphi(x+y)P_t\varphi(x)|\, dt$$

$$\leq \frac{\epsilon}{2}\left(\frac{\lambda}{1-e^{-\lambda T_\epsilon}}\right) \int_0^{T_\epsilon} e^{-\lambda t}\, dt + 2M\|\varphi\|_0^E \int_{T_\epsilon}^{+\infty} e^{-(\lambda-\omega)t}\, dt \leq \epsilon.$$

Then, thanks to (B.1.2) we can conclude that $F(\lambda) \in \mathcal{L}(C_b(E))$, for any $\lambda > -\omega$. $\quad\square$

In the next proposition we show that $F(\lambda)$ is the resolvent of some closed operator.

Proposition B.1.4. *Assume that P_t is a weakly continuous semigroup. Then there exists a unique closed linear operator $\mathcal{A} : D(\mathcal{A}) \subseteq C_b(E) \to C_b(E)$ such that for any $\varphi \in C_b(E)$ and $\lambda > \omega$ it holds*

$$R(\lambda, \mathcal{A})\varphi = F(\lambda)\varphi.$$

Proof. By using the Fubini-Tonelli theorem, it is easy to check that for every $\lambda, \mu > \omega$

$$F(\lambda) - F(\mu) = (\mu - \lambda)F(\lambda)F(\mu), \tag{B.1.3}$$

so that F is a pseudo resolvent on $C_b(E)$. Then, by a well known result (see for instance [119, Theorem VIII 4.1]), in order to prove the proposition we have only to show that the kernel of $F(\lambda)$ is equal to $\{0\}$, for every $\lambda > \omega$. From the uniqueness of the Laplace transform, it follows that if f is any measurable and exponentially bounded function f such that

$$\int_0^{+\infty} e^{-\lambda t} f(t)\, dt = 0,$$

for a set of values λ containing a limit point, then it holds

$$f(t) = 0 \quad \text{a.s.}$$

Now, from the resolvent equation (B.1.3) we obtain

$$F(\lambda)\varphi(x) = 0 \text{ for some } \lambda > \omega \implies F(\lambda)\varphi(x) = 0 \text{ for any } \lambda > \omega$$

and then, as the function $t \mapsto P_t\varphi(x)$ is continuous for all $x \in E$, we have

$$F(\lambda)\varphi(x) = 0 \implies P_t\varphi(x) = 0, \quad t \in [0, +\infty).$$

In particular $\varphi(x) = P_0\varphi(x) = 0$.

$\square$

Definition B.1.5. *The* infinitesimal generator *of the weakly continuous semigroup* P_t *is the unique closed linear operator* $\mathcal{A} : D(\mathcal{A}) \subseteq C_b(E) \to C_b(E)$ *such that for any* $\lambda > \omega$, $\varphi \in C_b(E)$ *and* $x \in E$

$$R(\lambda, \mathcal{A})\varphi(x) = F(\lambda)\varphi(x) = \int_0^{+\infty} e^{-\lambda t} P_t \varphi(x)\, dt. \qquad (B.1.4)$$

Remark B.1.6. We could have defined the infinitesimal generator of the weakly continuous semigroup P_t as for strongly continuous semigroups by means of the incremental ratio, i.e as the linear operator $\mathcal{B} : D(\mathcal{B}) \subseteq C_b(E) \to C_b(E)$ such that

$$
\begin{cases}
D(\mathcal{B}) = \left\{ \varphi \in C_b(E) : \begin{array}{l} \text{there exists } \lim\limits_{h \to 0^+} \Delta_h \varphi(x) \quad x \in E \\ \text{and } \lim\limits_{h \to 0^+} \Delta_h \varphi(\cdot) \in C_b(E) \end{array} \right\} \\
B\varphi(x) = \lim\limits_{h \to 0^+} \Delta_h \varphi(x),
\end{cases}
$$

where Δ_h is the incremental ratio

$$\Delta_h = \frac{P_h - I}{h}, \qquad h > 0.$$

If P_t is strongly continuous, then the two definitions above coincide. Otherwise we have only the obvious inclusion

$$\mathcal{A} \subset \mathcal{B}.$$

Now we remark that from (B.1.4) it easily follows that for any $k \in \mathbb{N}$

$$\|R^k(\lambda, \mathcal{A})\|_{\mathcal{L}(C_b(E))} \le \frac{M}{(\lambda - \omega)^k}. \qquad (B.1.5)$$

Then if $D(\mathcal{A})$ were dense in $C_b(E)$, $\mathcal{A}$ would fulfill the hypotheses of the Hille-Yosida theorem and P_t would be strongly continuous. Thus, if P_t is not strongly continuous, $D(\mathcal{A})$ can not be dense in $C_b(E)$. However $D(\mathcal{A}^k)$ is dense in $C_b(E)$ with respect to the $\mathcal{K}$-convergence, for any $k \in \mathbb{N}$.

Proposition B.1.7. *Let* P_t *be a weakly continuous semigroup. Then for every* $\varphi \in C_b(E)$ *and for every* $k \in \mathbb{N}$, *there exists a sequence* $\{\varphi_n^k\}_n \subset D(\mathcal{A}^k)$ *such that*

$$\mathcal{K} - \lim_{n \to +\infty} \varphi_n^k = \varphi.$$

Proof. For every $k \in \mathbb{N}$ and for every $n \in \mathbb{N}$, we define

$$\varphi_n^k(x) = n^k R^k(n, \mathcal{A})\varphi(x), \qquad x \in E.$$

Clearly $\{\varphi_n^k\}_n \subset D(\mathcal{A}^k)$ for any $k \in \mathbb{N}$. Moreover, from (B.1.5) we have

$$\|\varphi_n^k\|_0^E \le \left(\frac{n}{n-\omega}\right)^k M \|\varphi\|_0^E \le 2M \|\varphi\|_0^E, \qquad n \in \mathbb{N}.$$

Thus we have only to show that the sequence $n^k R^k(n, \mathcal{A})\varphi$ converges to φ, uniformly on compact sets of E. As $P_t\varphi$ is $\mathcal{K}$-convergent to φ, as t goes to zero, for any fixed compact set $K \subset H$ and any fixed $\epsilon > 0$, there exists $\delta > 0$ such that

$$|t| \leq \delta \implies \sup_{x \in K} |P_t\varphi(x) - \varphi(x)| \leq \epsilon.$$

Therefore for every $n \in \mathbb{N}$ we have

$$\sup_{x \in K} |n^k R^k(n, \mathcal{A})\varphi(x) - \varphi(x)|$$

$$\leq \sup_{x \in K} \frac{n^k}{(k-1)!} \int_0^{+\infty} t^{k-1} e^{-nt} |P_t\varphi(x) - \varphi(x)|\, dt$$

$$\leq \epsilon \frac{n^k}{(k-1)!} \int_0^{\delta} t^{k-1} e^{-nt}\, dt + (1+M)\|\varphi\|_0^E \frac{n^k}{(k-1)!} \int_\delta^{+\infty} t^{k-1} e^{-(n-\omega)t}\, dt$$

$$\leq \epsilon + (1+M)\|\varphi\|_0^E n^k e^{-nt} \left(\frac{\delta^{k-1}}{n-\omega} + \ldots + \frac{\delta}{(n-\omega)^{k-1}(k-1)!} + \frac{1}{(n-\omega)^k k!} \right),$$

and our claim follows by taking the limit as $n \to +\infty$. $\square$

Remark B.1.8. If P_t is a weakly continuous semigroup on $C_b(E)$ and if $\mathcal{A} : D(\mathcal{A}) \subset C_b(E) \to C_b(E)$ is its infinitesimal generator, then $Y = \overline{D(\mathcal{A})}$ is a P_t-invariant closed subspace of $C_b(E)$ and $P_{t|_Y}$ is a strongly continuous semigroup in Y.

B.2 Differentiability of weakly continuous semigroups

If P_t is a strongly continuous semigroup and $\mathcal{A}$ is its infinitesimal generator, for any $\varphi \in D(\mathcal{A})$ we have

$$\frac{d}{dt} P_t\varphi = \mathcal{A} P_t\varphi = P_t \mathcal{A}\varphi \qquad t \geq 0.$$

For weakly continuous semigroups this is not true. Nevertheless we are showing that a similar result holds, namely for any $\varphi \in D(\mathcal{A})$ and $x \in E$ the function

$$[0, +\infty) \to \mathbb{R}, \quad t \mapsto P_t\varphi(x)$$

is differentiable and

$$\frac{d}{dt} P_t\varphi(x) = \mathcal{A} P_t\varphi(x) = P_t \mathcal{A}\varphi(x) \qquad t \geq 0.$$

Lemma B.2.1. *Let P_t be a weakly continuous semigroup and let $\mathcal{A}$ be its infinitesimal generator. Then the following statements hold*

1. for any $t \geq 0$ and $\lambda > \omega$

$$P_t R(\lambda, \mathcal{A}) = R(\lambda, \mathcal{A}) P_t; \qquad (B.2.1)$$

2. for any $\varphi \in D(\mathcal{A})$ and $t \geq 0$, $P_t \varphi \in D(\mathcal{A})$ and

$$\mathcal{A} P_t \varphi = P_t \mathcal{A} \varphi.$$

Proof. In order to prove (B.2.1) it is enough to verify that for any $\varphi \in C_b(E)$, $T > 0$ and $\lambda > \omega$

$$P_t \int_0^T e^{-\lambda s} P_s \varphi(x) \, ds = \int_0^T e^{-\lambda s} P_{t+s} \varphi(x) \, ds, \qquad x \in E. \qquad (B.2.2)$$

Indeed, assume (B.2.2). Let us fix $\varphi \in C_b(E)$ and define

$$\psi_n(x) = \int_0^n e^{-\lambda s} P_s \varphi(x) \, ds, \qquad n \in \mathbb{N}, \quad x \in E.$$

Then we have $\|\psi_n\|_0^E \leq M \|\varphi\|_0^E / (\lambda - \omega)$, for any $n \in \mathbb{N}$ and

$$\lim_{n \to +\infty} \sup_{x \in E} \left| \psi_n(x) - \int_0^{+\infty} e^{-\lambda s} P_s \varphi(x) \, ds \right| = 0.$$

From the weak continuity of P_t and from (B.2.2), this implies that for all $x \in E$, $\varphi \in C_b(E)$ and $t \geq 0$

$$P_t R(\lambda, \mathcal{A}) \varphi(x) = \lim_{n \to +\infty} P_t \psi_n(x) = \lim_{n \to +\infty} P_t \int_0^n e^{-\lambda s} P_s \varphi(x) \, ds$$

$$= \lim_{n \to +\infty} \int_0^n e^{-\lambda s} P_{t+s} \varphi(x) \, ds = \int_0^{+\infty} e^{-\lambda s} P_{t+s} \varphi(x) \, ds = R(\lambda, \mathcal{A}) P_t \varphi(x).$$

Thus, it remains to prove (B.2.2). Consider partitions $0 = s_0^h < s_1^h < \ldots < s_{j_h}^h = T$ of the interval $[0, T]$, with size $\delta_h = \max_{j \in \{1, \ldots, j_h\}} (s_j^h - s_{j-1}^h)$, for any $h \in \mathbb{N}$, and assume that $\delta_h \to 0$, as $h \to +\infty$. For every $h \in \mathbb{N}$ we define

$$I_h(\varphi)(x) = \sum_{j=1}^{j_h} (s_j^h - s_{j-1}^h) e^{-\lambda s_j^h} P_{s_j^h} \varphi(x), \qquad x \in E.$$

It is immediate to verify that $\sup_{h \in \mathbb{N}} \|I_h(\varphi)\|_0^E < +\infty$. Moreover for any compact set $K \subset H$ it holds

$$\lim_{h \to +\infty} \sup_{x \in K} \left| I_h(\varphi)(x) - \int_0^T e^{-\lambda s} P_s \varphi(x) \, ds \right| = 0. \qquad (B.2.3)$$

In fact, for every $\epsilon > 0$ we can choose $\delta_\epsilon > 0$ such that

$$|s - t| \leq \delta_\epsilon \implies \sup_{x \in K} |e^{-\lambda s} P_s \varphi(x) - e^{-\lambda t} P_t \varphi(x)| \leq \epsilon.$$

Hence, since $\lim_{h \to +\infty} \delta_h = 0$, there exists $\bar{h}$ such that $\delta_h \leq \delta_\epsilon$, for any $h \geq \bar{h}$ and then we have

$$\sup_{x \in K} |I_h(\varphi)(x) - \int_0^T e^{-\lambda s} P_s \varphi(x) \, ds|$$

$$\leq \sum_{j=1}^{j_h} \sup_{x \in K} \left| \int_{s_{j-1}^h}^{s_j^h} \left(e^{-\lambda s_j^h} P_{s_j^h} \varphi(x) - e^{-\lambda s} P_s \varphi(x) \right) \right| ds \leq \epsilon T, \quad h \geq \bar{h}.$$

This implies that for any $\varphi \in C_b(H)$

$$\mathcal{K} - \lim_{h \to 0} I_h(\varphi) = \int_0^T e^{-\lambda s} P_s \varphi(\cdot) \, ds.$$

Hence, as P_t is weakly continuous, we can conclude that

$$\int_0^T e^{-\lambda s} P_{t+s} \varphi(\cdot) \, ds = \mathcal{K} - \lim_{h \to 0} I_h(P_t \varphi)$$

$$= \mathcal{K} - \lim_{h \to 0} P_t I_h(\varphi) = P_t \int_0^T e^{-\lambda s} P_s \varphi(\cdot) \, ds.$$

Next, we prove 2). For every $\lambda > \omega$ we have

$$D(\mathcal{A}) = R(\lambda, \mathcal{A}) \left(C_b(E) \right),$$

then from (B.2.1) it follows

$$P_t D(\mathcal{A}) \subset D(\mathcal{A}).$$

Moreover if $\varphi \in D(\mathcal{A})$, there exist $\psi \in C_b(E)$ and $\lambda > \omega$ such that

$$R(\lambda, \mathcal{A})\psi = \varphi,$$

therefore from (B.2.1) we have

$$\mathcal{A} P_t \varphi = \mathcal{A} P_t R(\lambda, \mathcal{A})\psi = \mathcal{A} R(\lambda, \mathcal{A}) P_t \psi = \lambda R(\lambda, \mathcal{A}) P_t \psi - P_t \psi$$

$$= P_t \left(\lambda R(\lambda, \mathcal{A}) - I \right) \psi = P_t \mathcal{A} R(\lambda, \mathcal{A})\psi = P_t \mathcal{A}\varphi.$$

$\square$

Proposition B.2.2. *Let P_t be a weakly continuous semigroup. Then for any $\varphi \in D(A)$ and $x \in E$ we have*

$$P_t\varphi(x) = \varphi(x) + \int_0^t P_s A\varphi(x)\, ds.$$

In particular the function $t \mapsto P_t\varphi(x)$ is differentiable and

$$\frac{d}{dt}P_t\varphi(x) = P_t A\varphi(x) = A P_t\varphi(x).$$

Proof. Let us fix $\varphi \in D(A)$, $x \in E$ and $\lambda > \omega$. Then we have

$$\varphi(x) = R(\lambda, A)(\lambda I - A)\varphi(x) = \int_0^{+\infty} e^{-\lambda t} P_t(\lambda I - A)\varphi(x)\, dt$$

(B.2.4)

$$= \lambda \int_0^{+\infty} e^{-\lambda t} P_t\varphi(x)\, dt - \int_0^{+\infty} e^{-\lambda t} P_t A\varphi(x)\, dt.$$

Now we remark that if $\omega < \mu < \lambda$ we have

$$\left| e^{-\lambda t} \int_0^t P_s A\varphi(x)\, ds \right| \le e^{-\lambda t} \int_0^t |P_s A\varphi(x)|\, ds$$

$$\le e^{-(\lambda-\mu)t} \int_0^t e^{-\mu s} |P_s A\varphi(x)|\, ds \le M\, e^{-(\lambda-\mu)t} \int_0^{+\infty} e^{-(\mu-\omega)s}\, ds \|A\varphi\|_0^E,$$

and then

$$\lim_{t \to +\infty} e^{-\lambda t} \int_0^t P_s A\varphi(x)\, ds = 0.$$

Integrating by parts and taking into account of (B.2.4), this yields

$$\int_0^{+\infty} e^{-\lambda t} P_t\varphi(x)\, dt - \int_0^{+\infty} e^{-\lambda t} \int_0^t P_s A\varphi(x)\, ds\, dt = \int_0^{+\infty} e^{-\lambda t}\varphi(x)\, dt,$$

so that from the uniqueness of the Laplace transform and the continuity of the application $t \mapsto P_t\varphi(x)$ it follows

$$P_t\varphi(x) = \varphi(x) + \int_0^t P_s A\varphi(x)\, ds.$$

$\square$

Appendix C

Theorem of contractions depending on parameters

We recall here some facts concerning contractions depending on parameters. What follows is based on some results proved in [48, appendix C].

Let Λ and E be two Banach spaces and let $\mathcal{F} : \Lambda \times E \to E$ be a continuous mapping such that for a certain $\alpha \in (0,1)$

$$\|\mathcal{F}(\lambda,x) - \mathcal{F}(\lambda,y)\|_E \leq \alpha \|x - y\|_E, \quad x,y \in E, \quad \lambda \in \Lambda. \qquad (\text{C.0.1})$$

As a immediate consequence of the contraction principle, there exists a mapping $\varphi \in C(\Lambda; E)$ such that

$$\mathcal{F}(\lambda, \varphi(\lambda)) = \varphi(\lambda), \quad \lambda \in \Lambda.$$

If for any fixed $\lambda \in \Lambda$ the function $\mathcal{F}(\lambda, \cdot) : E \to E$ is Fréchet differentiable and if for any fixed $x \in E$ the mapping $\mathcal{F}(\cdot, x) : \Lambda \to E$ is also Fréchet differentiable, then the function φ is differentiable, as well. A similar result holds for higher order derivatives.

Our aim is to study the differentiability of the mapping φ, when the Fréchet derivatives of $\mathcal{F}$ do not exist. To this purpose we have to assume the following conditions.

Hypothesis C.1.   *1. For all $x \in E$, the mapping $\Lambda \to E$, $\lambda \mapsto \mathcal{F}(\lambda, x)$, is Fréchet differentiable and the derivative*

$$\frac{\partial \mathcal{F}}{\partial \lambda} : \Lambda \times E \to \mathcal{L}(\Lambda; E)$$

is continuous.

2. For all $x, y \in E$ and $\lambda \in \Lambda$, the mapping $\mathbb{R} \to E$, $h \mapsto \mathcal{F}(\lambda, x + hy)$, is differentiable and

$$\frac{d}{dh} \mathcal{F}(\lambda, x + hy)_{|h=0} = \frac{\partial \mathcal{F}}{\partial x}(\lambda, x) \cdot y,$$

where $\partial\mathcal{F}/\partial x(\lambda, x)$ is an operator in $\mathcal{L}(E)$ such that

$$\frac{\partial\mathcal{F}}{\partial x}(\cdot, \cdot)y : \Lambda \times E \to E$$

is continuous, for any fixed $y \in E$.

We have

Proposition C.0.3. *Under the hypothesis C.1, the function φ is Gâteaux differentiable at any point $\lambda \in \Lambda$ along any direction $\mu \in \Lambda$ and*

$$D\varphi(\lambda) \cdot \mu = \left[I - \frac{\partial\mathcal{F}}{\partial x}(\lambda, \varphi(\lambda)) \right]^{-1} \left(\frac{\partial\mathcal{F}}{\partial\lambda}(\lambda, \varphi(\lambda)) \cdot \mu \right). \qquad (C.0.2)$$

Moreover, for any fixed $\mu \in \Lambda$ the mapping

$$\Lambda \to E, \qquad \lambda \mapsto D\varphi(\lambda)\mu$$

is continuous.

Proof. Due to (C.0.1) and to the hypothesis C.1, for any $h \in \mathbb{R}$, $\lambda \in \Lambda$ and $x \in E$ we have

$$\left| \frac{\partial\mathcal{F}}{\partial x}(\lambda, x + hy) \right|_{\mathcal{L}(E)} \leq \alpha. \qquad (C.0.3)$$

For any $\lambda, \mu \in \Lambda$ and $h \in \mathbb{R}$ it holds

$$\varphi(\lambda + h\mu) - \varphi(\lambda) = \mathcal{F}(\lambda + h\mu, \varphi(\lambda + h\mu)) - \mathcal{F}(\lambda, \varphi(\lambda))$$

$$= \int_0^1 \frac{\partial\mathcal{F}}{\partial\lambda}(\lambda + \theta h\mu, \varphi(\lambda + h\mu))h\mu \, d\theta$$

$$+ \int_0^1 \frac{\partial\mathcal{F}}{\partial x}(\lambda, \varphi(\lambda) + \theta(\varphi(\lambda + h\mu) - \varphi(\lambda)))(\varphi(\lambda + h\mu) - \varphi(\lambda)) \, d\theta.$$

If we define

$$H_{\lambda,\mu}(h)z = \int_0^1 \frac{\partial\mathcal{F}}{\partial x}(\lambda, \varphi(\lambda) + \theta(\varphi(\lambda + h\mu) - \varphi(\lambda)))z \, d\theta, \qquad z \in E,$$

thanks to (C.0.3) we have that $\|H_{\lambda,\mu}(h)\|_{\mathcal{L}(E)} \leq \alpha < 1$, and then there exists $[I - H_{\lambda,\mu}(h)]^{-1}$ in $\mathcal{L}(E)$. This implies that

$$\varphi(\lambda + h\mu) - \varphi(\lambda) = [I - H_{\lambda,\mu}(h)]^{-1} \int_0^1 \frac{\partial\mathcal{F}}{\partial\lambda}(\lambda + \theta h\mu, \varphi(\lambda + h\mu))h\mu \, d\theta. \quad (C.0.4)$$

Due to the Hypothesis C.1-2. for any fixed $z \in E$ the mapping

$$H_{\lambda,\mu}(\cdot)z : \mathbb{R} \to E, \qquad h \mapsto H_{\lambda,\mu}(h)z$$

is continuous and then, as $\|H_{\lambda,\mu}(h)\|_{\mathcal{L}(E)} \leq \alpha < 1$, for any $h \in \mathbb{R}$, it is easy to show that the mapping

$$[I - H_{\lambda,\mu}(\cdot)]^{-1}z : \mathbb{R} \to E, \qquad h \mapsto [I - H_{\lambda,\mu}(h)]^{-1}z,$$

is continuous. Therefore, since

$$\sup_{h \in \mathbb{R}} \left\| [I - H_{\lambda,\mu}(h)]^{-1} \right\|_{\mathcal{L}(E)} \leq \frac{1}{1-\alpha}$$

and $\partial\mathcal{F}/\partial\lambda : \Lambda \times E \to \mathcal{L}(\Lambda; E)$ is continuous, if we divide by h each side of (C.0.4) and let h going to zero, we can conclude that φ is differentiable at λ along the direction μ and the directional derivative is given by (C.0.2).

The continuity of the mapping $\Lambda \to E$, $\lambda \mapsto D\varphi(\lambda)\mu$, follows from (C.0.2) and from the Hypothesis C.1-1 once one remarks that for any fixed $z \in E$ the mapping $[I - \partial\mathcal{F}/\partial x(\cdot, \varphi(\cdot))]^{-1}z : \Lambda \to E$ is continuous and

$$\sup_{\lambda \in \Lambda} \|[I - \frac{\partial\mathcal{F}}{\partial x}(\lambda, \varphi(\lambda))]^{-1}\|_{\mathcal{L}(E)} \leq \frac{1}{1-\alpha}.$$

$\square$

Remark C.0.4. In the Hypothesis C.1 it is sufficient to assume that for all $x \in E$, the mapping $\Lambda \to E$, $\lambda \mapsto \mathcal{F}(\lambda, x)$, is Gâteaux differentiable and for any fixed $\mu \in \Lambda$ the mapping

$$\frac{\partial\mathcal{F}}{\partial\lambda}(\cdot, \cdot)\mu : \Lambda \times E \to E$$

is continuous.

To get the existence of the second order directional derivatives of φ we need additional assumptions.

Hypothesis C.2.   *1. For all $x \in E$, the mapping $\Lambda \to E$, $\lambda \mapsto \mathcal{F}(\lambda, x)$, is twice differentiable and the mapping*

$$\frac{\partial^2\mathcal{F}}{\partial\lambda^2} : \Lambda \times E \to \mathcal{L}(\Lambda \times \Lambda; E)$$

is continuous.

2. For all $x, y \in E$ and $\lambda, \mu \in \Lambda$ the mapping

$$\mathbb{R} \to E, \quad h \mapsto \frac{\partial\mathcal{F}}{\partial\lambda}(\lambda, x + hy) \cdot \mu$$

is differentiable and

$$\frac{d}{dh}\left(\frac{\partial\mathcal{F}}{\partial\lambda}(\lambda, x + hy) \cdot \mu\right)_{|h=0} = \frac{\partial^2\mathcal{F}}{\partial x \partial\lambda}(\lambda, x)(\mu, y),$$

for some continuous mapping $\partial^2\mathcal{F}/\partial x\partial\lambda : \Lambda \times E \to \mathcal{L}(\Lambda \times E; E)$.

3. *For any $x, y \in E$ the mapping*

$$\Lambda \to E, \qquad \lambda \mapsto \frac{\partial \mathcal{F}}{\partial x}(\lambda, x)y,$$

is differentiable and for any $\lambda, \mu \in \Lambda$

$$\left[\frac{\partial}{\partial \lambda} \left(\frac{\partial \mathcal{F}}{\partial x}(\cdot, x)y \right)(\lambda) \right] \mu = \frac{\partial^2 \mathcal{F}}{\partial \lambda \partial x}(\lambda, x)(\mu, y),$$

for some continuous mapping $\partial^2 \mathcal{F} / \partial \lambda \partial x : \Lambda \times E \to \mathcal{L}(\Lambda \times E; E)$.

4. *There exists a subspace $G \subset E$ such that for any $x \in E$, $y, z \in G$ and $\lambda \in \Lambda$ the mapping*

$$\mathbb{R} \to E, \quad h \mapsto \frac{\partial \mathcal{F}}{\partial x}(\lambda, x + hy) \cdot z$$

is differentiable and

$$\frac{d}{dh} \left(\frac{\partial \mathcal{F}}{\partial x} F(\lambda, x + hy) \cdot z \right)_{|_{h=0}} = \frac{\partial^2 \mathcal{F}}{\partial x^2}(\lambda, x)(y, z),$$

where $\partial^2 \mathcal{F} / \partial x^2(\lambda, x) \in \mathcal{L}(G \times G; E)$ and $\partial^2 \mathcal{F} / \partial x^2(\cdot, \cdot)(y, z) : \Lambda \times E \to E$ is continuous. Moreover, for any $\lambda \in \Lambda$ the mapping $\partial^2 \mathcal{F} / \partial x^2(\lambda, \cdot) : E \to \mathcal{L}(G \times G; E)$ is bounded on bounded sets of E.

5. *For all $\lambda, \mu \in \Lambda$ and $h \in \mathbb{R}$, we have that $\varphi(\lambda + h\mu) - \varphi(\lambda) \in G$ and*

$$D\varphi(\lambda) \cdot \mu = \lim_{h \to 0} \frac{\varphi(\lambda + h\mu) - \varphi(\lambda)}{h}, \qquad in \ G. \qquad (C.0.5)$$

Proposition C.0.5. *Under the hypotheses C.1 and C.2, the function φ is twice differentiable at any point $\lambda \in \Lambda$ along any directions $\mu, \nu \in \Lambda$ and it holds*

$$D^2\varphi(\lambda)(\mu, \nu) = \left[I - \frac{\partial \mathcal{F}}{\partial x}(\lambda, \varphi(\lambda)) \right]^{-1} \left(\frac{\partial^2 \mathcal{F}}{\partial x^2}(\lambda, \varphi(\lambda))(D\varphi(\lambda)\mu, D\varphi(\lambda)\nu) \right.$$

$$\left. + \frac{\partial^2 \mathcal{F}}{\partial \lambda^2}(\lambda, \varphi(\lambda))(\mu, \nu) + \frac{\partial^2 \mathcal{F}}{\partial \lambda \partial x}(\lambda, \varphi(\lambda))(D\varphi(\lambda)\mu, \nu) + \frac{\partial^2 \mathcal{F}}{\partial x \partial \lambda}(\lambda, \varphi(\lambda))(\mu, D\varphi(\lambda)\nu) \right).$$
$$(C.0.6)$$

Moreover, the mapping

$$\Lambda \to E, \qquad \lambda \mapsto D^2\varphi(\lambda)(\mu, \nu)$$

is continuous, for any fixed $\mu, \nu \in \Lambda$. In particular, if $D^2\varphi : \Lambda \to \mathcal{L}(\Lambda \times \Lambda; E)$ is bounded on bounded subsets of Λ, we have that $D\varphi : \Lambda \to \mathcal{L}(E)$ is continuous, so that $\varphi \in C^1(\Lambda; E)$.

Proof. It is not difficult to show that for any $\lambda, \mu, \nu \in \Lambda$ and $h \in \mathbb{R}$ it holds

$$D\varphi(\lambda + h\nu)\mu - D\varphi(\lambda)\mu = J_1(h) + J_2(h)$$

$$+ \frac{\partial \mathcal{F}}{\partial x}(\lambda, \varphi(\lambda)) \left(D\varphi(\lambda + h\nu)\mu - D\varphi(\lambda)\mu\right),$$

where

$$J_1(h) = \frac{\partial \mathcal{F}}{\partial \lambda}(\lambda + h\nu, \varphi(\lambda + h\nu))\mu - \frac{\partial \mathcal{F}}{\partial \lambda}(\lambda, \varphi(\lambda))\mu$$

and

$$J_2(h) = \left(\frac{\partial \mathcal{F}}{\partial x}(\lambda + h\nu, \varphi(\lambda + h\nu)) - \frac{\partial \mathcal{F}}{\partial x}(\lambda, \varphi(\lambda)) \right) D\varphi(\lambda + h\nu)\mu.$$

Therefore, as in the proof of the previous proposition, due to (C.0.3) we have

$$D\varphi(\lambda + h\nu)\mu - D\varphi(\lambda)\mu = \left[I - \frac{\partial \mathcal{F}}{\partial x}(\lambda, \varphi(\lambda)) \right]^{-1} (J_1(h) + J_2(h)).$$

Thanks to the conditions 1 and 2 in the Hypotheses C.1 we have

$$\lim_{h \to 0} \frac{J_1(h)}{h} = \frac{\partial^2 \mathcal{F}}{\partial \lambda^2}(\lambda, \varphi(\lambda))(\nu, \mu) + \frac{\partial \mathcal{F}}{\partial x \partial \lambda}(\lambda, \varphi(\lambda))(\mu, D\varphi(\lambda)\nu).$$

As far as $J_2(h)$ is concerned, we remark that thanks to the Hypothesis C.2-5, $\varphi(\lambda + h\mu) - \varphi(\lambda) \in G$ and $D\varphi(\lambda)\mu \in G$. Then we have

$$\frac{J_2(h)}{h} = \int_0^1 \frac{\partial^2 \mathcal{F}}{\partial \lambda \partial x}(\lambda + \theta h\nu, \varphi(\lambda + h\nu)) (\nu, D\varphi(\lambda + h\nu)\mu) \, d\theta$$

$$+ \int_0^1 \frac{\partial^2 \mathcal{F}}{\partial x^2}(\lambda, \varphi(\lambda) + \theta (\varphi(\lambda + h\nu) - \varphi(\lambda))) \left(\frac{\varphi(\lambda + h\nu) - \varphi(\lambda)}{h}, D\varphi(\lambda + h\nu)\mu \right) d\theta$$

$$= I_1(h) + I_2(h).$$

According to the Hypothesis C.2-3 and to the continuity of the mapping $D\varphi(\cdot)\mu : \Lambda \to E$, we have

$$\lim_{h \to 0} I_1(h) = \frac{\partial^2 \mathcal{F}}{\partial \lambda \partial x}(\lambda, \varphi(\lambda))(\nu, D\varphi(\lambda)\mu).$$

Moreover, as

$$\sup_{\theta, h \in [0,1]} |\varphi(\lambda) + \theta (\varphi(\lambda + h\nu) - \varphi(\lambda))|_E < \infty,$$

we can use the Hypothesis C.2-4 and we have

$$\lim_{h \to 0} I_2(h) = \frac{\partial^2 \mathcal{F}}{\partial x^2}(\lambda, \varphi(\lambda)) (D\varphi(\lambda)\nu, D\varphi(\lambda)\mu).$$

Now, the continuity of the mapping $\Lambda \to E$, $\lambda \mapsto D^2\varphi(\lambda)(\mu, \nu)$, for $\mu, \nu \in \Lambda$ fixed, follows from arguments analogous to those used for $D\varphi(\lambda)\mu$ in the previous proposition.

Finally, concerning the continuity of the mapping $D\varphi : \Lambda \to \mathcal{L}(\Lambda; E)$, we have

$$\sup_{|\mu|_\Lambda \leq 1} |D\varphi(\lambda)\mu - D\varphi(\lambda_0)\mu|_E$$

$$\leq \sup_{|\mu|_\Lambda \leq 1} \int_0^1 |D^2\varphi(\theta\lambda + (1-\theta)\lambda_0)(\lambda - \lambda_0, \mu)|_E \, d\theta \leq c|\lambda - \lambda_0|_\Lambda,$$

so that

$$\lim_{\lambda \to \lambda_0} \|D\varphi(\lambda) - D\varphi(\lambda_0)\|_{\mathcal{L}(\Lambda)} = 0.$$

$\square$

For higher order differentiability we proceed with similar arguments.

Bibliography

[1] S. Agmon, LECTURES ON ELLIPTIC BOUNDARY VALUE PROBLEMS, Van Nostrand Mathematical Studies, D. Van Nostrand Company, Princeton (1965).

[2] A. Ambrosetti, G. Prodi, ANALISI NON LINEARE, Scuola Normale Superiore, Classe di Scienze, Pisa (1973).

[3] D.G. Aronson, P. Besala, *Parabolic equations with unbounded coefficients*, Journal of Differential Equations 3 (1967), pp. 1-14.

[4] M.S. Baouendi, C. Goulaouic, *Regularité analitique et itérés d'operateurs elliptiques dégénérés, applications*, Journal of Functional Analysis 9 (1972), pp. 208-248.

[5] V. Barbu, G. Da Prato, HAMILTON-JACOBI EQUATIONS IN HILBERT SPACES, Reasearch Notes in Maths, Pitman, Boston (1982).

[6] P. Besala, *On the existence of a foundamental solution for a parabolic equation with unbounded coefficients*, Ann. Polon. Math. 29 (1975) pp. 403-409.

[7] J.M. Bismut, *Martingales, the Malliavin calculus and hypoellipticity under general Hörmander's conditions*, Z. Wahrscheinlichkeitstheorie verw. Gebiete 56 (1981), pp. 469-505.

[8] H. Brezis, W. Rosenkrantz, B. Singer, *On a degenerate elliptic-parabolic equation occurring in the theory of probability*, Communications in Pure and Applied Mathematics 24 (1971), pp. 395-416.

[9] M. Campiti, G. Metafune, *Ventcel's boundary conditions and analytic semigroups*, Arch. Math. 70 (1998), pp. 377-390.

[10] M. Campiti, G. Metafune, D. Pallara, *Degenerate self-adjoint evolution equations on the unit interval*, Semigroup Forum 57 (1998), pp. 1-36.

[11] P. Cannarsa, G. Da Prato, *Second-order Hamilton-Jacobi equations in infinite dimensions*, SIAM Journal on Control and Optimization 29 (1991), pp. 474-492.

[12] P. Cannarsa , G. Da Prato, *A functional analysis approach to parabolic equations in infinite dimensions.* Journal of Functional Analysis 118 (1993), pp. 22-42.

[13] P. Cannarsa, G. Da Prato, *Direct solutions of a second-order Hamilton-Jacobi equation in Hilbert spaces,* in Stochastic Partial Differential Equations and Applications, G. Da Prato and L. Tubaro editors, Pitman Research Notes in Maths Series 208, pp. 72-85.

[14] P. Cannarsa, G. Da Prato, *Infinite-dimensional elliptic equations with Hölder continuous coefficients,* Advances in Differential Equations 1 (1996), pp. 425-452.

[15] P. Cannarsa, V. Vespri, *Generation of analytic semigroups by elliptic operators with unbounded coefficients,* SIAM Journal of Mathematical Analysis 18 (1987), pp. 857-872.

[16] P. Cannarsa, V. Vespri, *Generation of analytic semigroups in the L^p topology by elliptic operators in $\mathbb{R}^n$,* Israel Journal of Mathematics 61 (1988), pp. 235-255.

[17] S. Cerrai, *A Hille Yosida theorem for weakly continuous semigroups,* Semigroup Forum 49 (1994), pp. 349-367.

[18] S. Cerrai, F. Gozzi, *Strong solutions of Cauchy problems associated to weakly continuous semigroups,* Differential and Integral Equations 8 (1994), pp. 465-486.

[19] S. Cerrai, *Elliptic and parabolic equations in $\mathbb{R}^n$ with coefficients having polynomial growth,* Communications in Partial Differential Equations 21 (1996), pp. 281-317.

[20] S. Cerrai, *Invariant measures for a class of SDE's with drift term having polynomial growth,* Dynamic Systems and Applications 5 (1996), pp. 353-370.

[21] S. Cerrai, *Some results for second order elliptic operators having unbounded coefficients,* Differential and Integral Equations, 11 No. 4 (1998), pp. 561-588.

[22] S. Cerrai, *Differentiability with respect to initial datum for solutions of SPDE's with no Fréchet differentiable drift term,* Communications in Applied Analysis 2 (1998), pp. 249-270.

[23] S. Cerrai, *Kolmogorov equations in Hilbert spaces with non smooth coefficients,* Communications in Applied Analysis 2 (1998), pp. 271-297.

[24] S. Cerrai, *Smoothing properties of transition semigroups relative to SDE's with values in Banach spaces,* Probabability Theory and Related Fields 113 (1999), pp. 85-114.

[25] S. Cerrai, *Differentiability of Markov semigroups for stochastic reaction-diffusion equations and applications to control*, Stochastic Processes and their Applications 83 (1999), pp. 15-37.

[26] S. Cerrai, *Ergodicity of stochastic reaction-diffusion systems with polynomial coefficients*, Stochastics and Stochastics Reports 67 (1999), pp. 17-51.

[27] S. Cerrai, *Analytic semigroups and degenerate elliptic operators with unbounded coefficients: a probabilistic approach*, Journal of Differential Equations 166 (2000), pp. 151-174.

[28] S. Cerrai, *A generalization of the Bismut-Elworthy formula*, in Evolution Equations and their Applications in Physical and Life Sciences, G. Lumer and L. Weiss editors, Lecture Notes in Pure and Applied Mathematics 215, Marcel Dekker (2000), pp. 473-487.

[29] S. Cerrai, *Optimal control problems for stochastic reaction-diffusion systems with non Lipschitz coefficients*, to appear in SIAM Journal on Control and Optimization.

[30] S. Cerrai, *Stationary Hamilton-Jacobi equations in Hilbert spaces and applications to a stochastic optimal control problem*, to appear in SIAM Journal on Control and Optimization.

[31] S. Cerrai, *Classical solutions for Kolmogorov equations in Hilbert spaces*, to appear in the Proceedings of the Ascona Conference 1999 on Stochastic Analysis, Random Fields and their Applications.

[32] Ph. Clément, C.A. Timmermans, *On C_0-semigroups generated by differential operators satisfying Ventcel's boundary conditions*, Indag. Math. 89 (1986), pp. 379-387.

[33] P.L. Chow, J.L. Menaldi, *Infinite dimensional Hamilton-Jacobi-Bellmann equations in Gauss-Sobolev spaces*, Nonlinear Analysis 29 (1997), pp. 415-426.

[34] M.G. Crandall, H. Ishii, P.L. Lions, *User's guide to viscosity solutions of second order partial differential equations*, Bulletin of the AMS (N.S.) 27 (1992), pp. 1-67.

[35] Y.L. Daleckij, *Differential equations with functional derivatives and stochastic equations for generalized random processes*, Dokl. Akad. Nauk. SSSR, 166 (1965), pp. 1035-1038.

[36] Y.L. Daleckij, *Infinite-dimensional elliptic operators and related parabolic equations*, Usp. Mat. Nauk., 22 (1967), pp. 3-54.

[37] G. Da Prato, APPLICATIONS CROISSANTES ET ÉQUATIONS D'ÉVOLUTIONS DANS LES ESPACES DE BANACH, Istituto Nazionale di Alta Matematica, Institutiones Mathematicae, volume II, Academic Press, London (1976).

[38] G. Da Prato, A. Debussche, *Control of the stochastic Burgers model of turbolence*, SIAM Journal on Control and Optimization 37 (1999), pp. 1123-1149.

[39] G. Da Prato, A. Debussche, *Dynamic programming for the stochastic Burgers equation*, Annali di Matematica pura e Applicata (IV) 178 (2000), pp. 143-174.

[40] G. Da Prato, K.D. Elworthy, J. Zabczyk, *Strong Feller property for stochastic semilinear equations*, Stochastic Analysis and Applications 13 (1995), pp. 35-45.

[41] G. Da Prato, D. Gatarek, J. Zabczyk, *Invariant measure for semilinear stochastic equations*, Stochastic Analysis and Applications 10 (1992), pp. 387-408.

[42] G. Da Prato, B. Goldys, *Elliptic operators on $\mathbb{R}^d$ with unbounded coefficients*, to appear in Journal of Differential Equations.

[43] G. Da Prato, S. Kwapien, J. Zabczyk, *Regularity of solutions for linear stochastic equations in Hilbert spaces*, Stochastics and Stochastics Reports 23 (1987), pp. 1-23.

[44] G. Da Prato, A. Lunardi, *On the Ornstein-Uhlenbeck operator in spaces of continuous functions*, Journal of Functional Analysis 131 (1995), pp. 94-114.

[45] G. Da Prato, E. Pardoux, *Invariant measures for white noise driven stochastic partial differential equations*, Stochastic Analysis and Applications 13 (1995), pp. 295-305.

[46] G. Da Prato, J. Zabczyk, *Regular densities of invariant measures in Hilbert spaces*, Journal of Functional Analysys 130 (1995), pp. 427-449.

[47] G. Da Prato, J. Zabczyk, STOCHASTIC EQUATIONS IN INFINITE DIMENSIONS, Cambridge University Press, Cambridge (1992).

[48] G. Da Prato, J. Zabczyk, ERGODICITY FOR INFINITE DIMENSIONAL SYSTEMS, London Mathematical Society, Lecture Notes Series 229, Cambridge University Press, Cambridge (1996).

[49] G. Da Prato, J. Zabczyk, *Convergence to equilibrium for classical and quantum spin systems*, Probability Theory and Related Fields 103 (1995), pp. 529-552.

[50] E.B. Davies, HEAT KERNELS AND SPECTRAL THEORY, Cambridge University Press, Cambridge (1989).

[51] E. B. Davies, B. Simon, *Ultracontractivity and the heat kernel of Schrödinger operators and dirichlet laplacians*, Journal of Functional Analysis 59 (1984), pp. 335-395.

[52] A. Devinatz, *Self-adjointness of second order degenerate elliptic operators*, Indiana University Mathematics Journal 27 (1978), pp. 255-266.

[53] E.B. Dynkin, MARKOV PROCESSES, Vol.1, Academic Press and Springer-Verlag, New York, 1968.

[54] J.L. Doob, *Asymptotic properties of Markov transition probability*, Transactions of the AMS 63 (1948), pp.393-421.

[55] N. Dunford, J.T. Schwartz, LINEAR OPERATORS. PART II: SPECTRAL THEORY, Interscience Publishers, John Wiley and Sons, New York (1964).

[56] A. Eberle, UNIQUENESS AND NON UNIQUENESS OF SINGULAR DIFFUSION OPERATORS, Lecture Notes in Mathematics 1718, Springer Verlag (1999).

[57] K.D. Elworthy, X.M. Li, *Formulae for the derivatives of heat semigroups*, Journal of Functional Analysis 125 (1994), pp. 252-286.

[58] S.N.Ethier, Th.G. Kurtz, MARKOV PROCESSES, CHARACTERIZATION AND CONVERGENCE, Wiley Series in Probability and Mathematical Statistics, John Wiley and Sons, New York (1986).

[59] A. Favini, J.A. Goldstein, S. Romanelli, *An analytic semigroup associated to a degenerate evolution equation*, in STOCHASTIC PROCESS AND FUNCTIONAL ANALYSIS, J.A. Goldstein, N.E. Gretsky and J. Uhl editors, Lecture Notes in Pure and Appl. Math. 186 (1996), pp. 85-100.

[60] W. Feller, *Two singular diffusion problems*, Annals of Mathematics 54 (1951), pp. 173-182.

[61] W. Feller, *The parabolic differential equations and the associated semigroups of transformations*, Annals of Mathematics 55 (1952), pp. 468-519.

[62] F. Flandoli, B. Maslowski, *Ergodicity for the 2-D Navier-Stokes equation under random perturbations*, Communications in Mathematical Physics 171 (1995), pp. 119-141.

[63] W.H. Fleming, Nisio, *On stochastic relaxed control for partially observed diffusions*, Nagoya Mathematical Journal 93 (1984), pp. 71-108.

[64] W.H. Fleming, H.M. Soner, CONTROLLED MARKOV PROCESSES AND VISCOSITY SOLUTIONS, Springer Verlag, New York (1993).

[65] M. Freidlin, A.D. Wentzell, RANDOM PERTURBATIONS OF DYNAMICAL SYSTEMS, Springer Verlag, Berlin (1983).

[66] M. Freidlin, FUNCTIONAL INTEGRATION AND PARTIAL DIFFERENTIAL EQUATIONS, Annals of Mathematics Studies, Princeton University Press, Princeton, New Jersey (1985).

[67] M. Freidlin, MARKOV PROCESSES AND DIFFERENTIAL EQUATIONS, ASYMPTOTIC PROBLEMS, Lectures in Mathematics EHT Zürich, Birkhäuser, Basel (1996).

[68] A. Friedman, STOCHASTIC DIFFERENTIAL EQUATIONS AND APPLICATIONS, Academic Press, New York, 1975.

[69] M. Fuhrman, *On a class of stochastic equations in Hilbert spaces, solvability and smoothing properties*, Stochastic Analysis and Applications 17 (1999), pp. 43-69.

[70] I. Gyöngy, E. Pardoux, *On the regularization effect of space-time white noise on quasilinear parabolic partial differential equations*, Probability Theory Related Fields 97 (1993), pp. 211-229.

[71] F. Gozzi, *Regularity of solutions of a second order Hamilton-Jacobi equation and application to a control problem*, Communications in Partial Differential Equations 20 (1995), pp. 775-826.

[72] F. Gozzi, *Global regular solutions of second order Hamilton-Jacobi equations in Hilbert spaces with locally Lipschitz nonlinearities*, Journal of Mathematical Analysis and Applications 198 (1996), pp. 399-443.

[73] F. Gozzi, R. Monte, V. Vespri, *Generation of analytic semigroups for degenerate elliptic operators arising in financial mathematics*, to appear in Advances in Differential Equations.

[74] F. Gozzi, E. Rouy, *Regular solutions of second-order stationary Hamilton-Jacobi equations*, Journal of Differential Equations 130 (1996), pp. 201-234.

[75] L. Gross, *Potential theory in Hilbert spaces*, Journal of Functional Analysis 1 (1967), pp. 123-189.

[76] T. Haverneanu, *Existence for the dynamic programming equation of control diffusion processes in Hilbert spaces*, Nonlinear Analysis Theory, Methods and Applications 9 (1985), pp. 619-629.

[77] S. Ito, *Fundamental solutions of parabolic differential equations and boundary value problem*, Japan Journal of Mathematics 27 (1957), pp. 5-102.

[78] B. Jefferies. *Weakly integrable semigroups on locally convex spaces*, Journal of Functional Analysis 66 (1986), pp. 347-364.

[79] B. Jefferies. *The generation of weakly integrable semigroups*, Journal of Functional Analysis 73 (1987), pp. 195-215.

[80] R.Z. Khas'minskii, *Ergodic properties of recurrent diffusion processes and stabilization of the solutions of the Cauchy problem for parabolic equations*, Theory of Probabability and its Applications 5 (1960), pp.179-196.

[81] R.Z. Khas'minskii, STOCHASTIC STABILITY OF DIFFERENTIAL EQUATIONS, Sijthoff and Noordhoff (1980).

[82] A.N. Kolmogorov, *Uber die analitischen methoden in der wahrscheinlichkeitsrechnung*, Math. Ann. 104 (1931), pp. 415-458.

[83] N.V. Krylov, INTRODUCTION TO THE THEORY OF DIFFUSION PROCESSES, American Mathematical Society, Providence (1995).

[84] N.V. Krylov, M. Roeckner, J. Zabczyck, STOCHASTIC PDE'S AND KOLMOGOROV EQUATIONS IN INFINITE DIMENSIONS, G. Da Prato editor, Lecture Notes in Mathematics 1715, Springer Verlag (1999).

[85] J. Kurtzweil, *On Approximation in real Banach spaces*, Studia Mathematica 14 (1954), pp. 213-231

[86] S. Kusuoka, D. Stroock, *Some boundeedness properties of certain stationary diffusion semigroups*, Journal of Functional Analysis 60 (1985), pp. 243-264.

[87] J.M. Lasry, P.L. Lions, *A remark on regularization in Hilbert spaces*, Israel Journal of Mathematics 55 (1986), pp. 257-266.

[88] J.L. Lions, QUELQUES MÉTHODS DE RÉSOLUTION DES PROBLÈMS AUX LIMITES NON LINÉAIRES, Dunod, Gauthier-Villars, Paris (1969).

[89] P.L. Lions, *Viscosity solutions of fully nonlinear second order equations and optimal control in infinite dimensions*.

Part I: *The case of bounded stochastic evolutions*, Acta Mathematica 161 (1998), pp. 243-278.

Part II: *Optimal control of Zakai's equation*, Lecture Notes in Mathematics 1390, G. Da Prato and L. Tubaro editors, Springer Verlag (1989), pp. 147-170.

Part III: *Uniqueness of viscosity solutions for general second order equations*, Journal of Functional Analysis 86 (1989), pp. 1-18.

[90] V.A. Liskevich, M.A. Perelmuter *Analyticity of sub-markovian semigroups*, Proceedings of the AMS 123 (1995), pp. 1097-1104.

[91] A. Lunardi, ANALYTIC SEMIGROUPS AND OPTIMAL REGULARITY IN PARABOLIC PROBLEMS, Birkäuser Verlag, Basel (1995).

[92] A. Lunardi, *On the Ornstein-Uhlenbeck operator in L^2 spaces with respect to invariant measures*, Transactions of the AMS 349 (1997), pp. 155-169.

[93] A. Lunardi, *An interpolation method to characterize domains of generators of semigroups*, Semigroup Forum 53 (1996), pp. 1-29.

[94] A. Lunardi, *Schauder estimates for a class of degenerate elliptic and parabolic operators with unbounded coefficients in $\mathbb{R}^n$*, Annali della Scuola Normale Superiore di Pisa, Serie IV, Vol. 24 (1997), pp. 133-164.

[95] A. Lunardi, *Schauder theorems for linear elliptic and parabolic problems with unbounded coefficients in $\mathbb{R}^n$*, Studia Mathematica 128 (1998), pp. 171-198.

[96] A. Lunardi, V. Vespri, *Optimal L^∞ and Schauder estimates for elliptic and parabolic operators with unbounded coefficients*, Reaction-diffusion Systems (Trieste 1995), pp. 217-239, Lecture Notes in Pure and Applied Mathematics 194, Marcel Dekker editor, New York (1998).

[97] Z.M. Ma, M. Roeckner, INTRODUCTION TO THE THEORY OF (NON SYMMETRIC) DIRICHLET FORMS, Springer-Verlag (1992).

[98] R. Manthey, *Existence and uniqueness of a solution of a reaction-diffusion equation with polynomial nonlinearity and white noise disturbance*, Math. Nachr. 125 (1986), pp. 121-133.

[99] B. Maslowski *Strong Feller property for semilinear stochastic evolution equations and applications*, Proceedings of IFIP Conference on Stochastic Systems and Optimization, Warsaw 1989, Lecture Notes in Control and Information Science N.136, 210-235, J. Zabczyk editor. Springer Verlag.

[100] B. Maslowski, *On probability distributions of solutions of a semilinear stochastic PDE*, Ceskolovenska Akademie ved Mathematicky Ustar (1992).

[101] A.S. Nemirovski, S.M. Semenov, *The polynomial approximation of functions in Hilbert spaces*, Mat. Sb. (N.S.) 92 (1973), pp. 257-281.

[102] A. Pazy, SEMIGROUPS OF LINEAR OPERATORS AND APPLICATIONS TO PARTIAL DIFFERENTIAL EQUATIONS, Springer-Verlag, Berlin, New-York (1983).

[103] S. Peszat, *Existence and uniqueness of the solutions for stochastic equations on Banach spaces*, Stochastics and Stochastics Reports, 55 (1995), pp. 167-193.

[104] S. Peszat, J. Zabczyk, *Strong Feller property and irriducibility for diffusion processes on Hilbert spaces*, Annals of Probability 23 (1995), pp. 157-172.

[105] A. Piech, *A fundamental solution of the parabolic equation on Hilbert spaces*, Journal of Functional Analysis 3 (1969), pp. 85-114.

[106] E. Priola *On a class of Markov type semigroups in uniformly continuous and bounded function spaces*, Studia Mathematica 136 (1999), pp. 271-295.

[107] J. Seidler, *Ergodic behaviour of stochastic parabolic equations*, Czechoslovak Mathematical Journal 47/122 (1997), pp. 277-316.

[108] E. Sinestrari, *Accretive differential operators*, Bollettino dell'Unione Matematica Italiana 13 (1976), pp. 19-31.

[109] E.M. Stein TOPICS IN HARMONIC ANALYSIS (RELATED TO LITTLEWOOD-PALEY THEORY), Annals of Mathematics Studies, Princeton University Press, Princeton New York (1970).

[110] L. Stettner, *Remarks on ergodic conditions for Markov processes on Polish spaces*, Bullettin of the Polish Academy of Science 42 (1194), pp. 103-114.

[111] B. Stewart, *Generation of analytic semigroups by strongly elliptic operators*, Transactions of the AMS 199 (1974), pp. 141-162.

[112] B. Stewart, *Generation of analytic semigroups by strongly elliptic operators under general boundary conditions*, Transactions of the AMS 259 (1980), pp. 299-310.

[113] D.W. Stroock, *The Malliavin calculus, a functional analytic approach*, Journal of Functional Analysis 40 (1981), pp. 212-257.

[114] D.W. Stroock, S.R.S. Varadhan, MULTIDIMENSIONAL DIFFUSION PROCESSES, Springer Verlag, Berlin (1979).

[115] A. Swiech *Unbounded second order partial differential equations in infinite dimensional Hilbert spaces*, Communications in Partial Differential Equations 19 (1994), pp. 1999-2036.

[116] R. Temam, INFINITE DIMENSIONAL DYNAMICAL SYSTEMS IN MECHANICS AND PHYSICS, Springer-Verlag, New York (1988).

[117] H. Triebel, INTERPOLATION THEORY, FUNCTION SPACES, DIFFERENTIAL OPERATORS, North-Holland Amsterdam (1986).

[118] V. Vespri, *Analytic semigroups, degenerate elliptic operators and applications to non-linear Cauchy problems*, Annali di Matematica Pura e Applicata 155 (1989), pp. 353-388.

[119] K. Yosida, FUNCTIONAL ANALYSIS, Springer-Verlag, Berlin, New-York (1980).

[120] J.B. Walsh, *An introduction to stochastic partial differential equations*, Ecole d'Eté de Probabilité de Saint-Flour XIV (1984), P.L. Hennequin editor, Lectures Notes in Mathematics 1180, pp. 265-439.

[121] L. Zambotti, *Infinite dimensional elliptic and stochastic equations with Hölder continuous coefficients*, Stochastic Analysis and Applications 17 (1999), pp. 487-508.

[122] X.Y. Zhou, *On the existence of optimal relaxed controls of stochastic PDE's*, SIAM Journal on Control and Optimization 30 (1992), pp. 247-261.

Index

4. Lecture Notes are printed by photo-offset from the master-copy delivered in camera-ready form by the authors. Springer-Verlag provides technical instructions for the preparation of manuscripts. Macro packages in T_EX, L^AT_EX2e, $L^AT_EX2.09$ are available from Springer's web-pages at

http://www.springer.de/math/authors/b-tex.html.

Careful preparation of the manuscripts will help keep production time short and ensure satisfactory appearance of the finished book.

The actual production of a Lecture Notes volume takes approximately 12 weeks.

5. Authors receive a total of 50 free copies of their volume, but no royalties. They are entitled to a discount of 33.3 % on the price of Springer books purchase for their personal use, if ordering directly from Springer-Verlag.

Commitment to publish is made by letter of intent rather than by signing a formal contract. Springer-Verlag secures the copyright for each volume. Authors are free to reuse material contained in their LNM volumes in later publications: A brief written (or e-mail) request for formal permission is sufficient.

Addresses:

Professor J.-M. Morel
CMLA, Ecole Normale Supérieure de Cachan
61 Avenue du Président Wilson
94235 Cachan Cedex France
E-mail: Jean-Michel.Morel@cmla.ens-cachan.fr

Professor B. Teissier
Université Paris 7
UFR de Mathématiques
Equipe Géométrie et Dynamique
Case 7012
2 place Jussieu
75251 Paris Cedex 05
E-mail: Teissier@ens.fr

Professor F. Takens, Mathematisch Instituut,
Rijksuniversiteit Groningen, Postbus 800,
9700 AV Groningen, The Netherlands
E-mail: F.Takens@math.rug.nl

Springer-Verlag, Mathematics Editorial, Tiergartenstr. 17
D-69121 Heidelberg, Germany
Tel.: *49 (6221) 487-701
Fax: *49 (6221) 487-355
E-mail: lnm@Springer.de